The Minerals, Metals & Materials Series

Guillaume Lambotte · Jonghyun Lee ·
Antoine Allanore · Samuel Wagstaff
Editors

Materials Processing Fundamentals 2019

TMS

Springer

Editors
Guillaume Lambotte
Boston Metal
Woburn, MA, USA

Jonghyun Lee
Iowa State University
Ames, IA, USA

Antoine Allanore
Massachusetts Institute of Technology
Cambridge, MA, USA

Samuel Wagstaff
Novelis
Kennesaw, GA, USA

ISSN 2367-1181 ISSN 2367-1696 (electronic)
The Minerals, Metals & Materials Series
ISBN 978-3-030-05727-5 ISBN 978-3-030-05728-2 (eBook)
https://doi.org/10.1007/978-3-030-05728-2

Library of Congress Control Number: 2018964000

Cover illustration:
Bottom right: From Chapter "Effect of Casting Temperature on the Surface Finish of Grey Iron Castings", Izudin Dugic, page 93, Figure 6: 3-D temperature gradient results from the simulation when 100% of the mould is filled. https://doi.org/10.1007/978-3-030-05728-2_8
Bottom left: From Chapter "Optimization of the Flow Behavior of Molten Steel in Ultrahigh-Speed Billet Continuous Casting Mold", Pei Xu et al., page 66, Figure 6: Liquid surface velocity field of four SEN inner diameters. https://doi.org/10.1007/978-3-030-05728-2_6
Top right: From Chapter "Modeling of Steel–Slag–Air Three-Phase Flow in Continuous Casting Strand", Xubin Zhang, Wei Chen, Lifeng Zhang, Piotr Roman Scheller, page 19, Figure 3: Streamlines of liquid steel in the mold. https://doi.org/10.1007/978-3-030-05728-2_2

This Springer imprint is published by the registered company Springer Nature Switzerland AG
The registered company address is: Gewerbestrasse 11, 6330 Cham, Switzerland

Preface

The symposium Materials Processing Fundamentals is hosted at the Annual Meeting of The Minerals, Metals & Materials Society (TMS) as the flagship symposium of the Process Technology and Modeling Committee. It is a unique opportunity for interdisciplinary presentations and discussions about, among others, processing, sensing, modeling, multi-physics, computational fluid dynamics, and thermodynamics.

The materials covered include ferrous and non-ferrous elements, and the processes range from mining unit operations to joining and surface finishing of materials. Acknowledging that modern processes involve multi-physics, the symposium and its proceedings allow the reader to learn the methods and outcome of other fields' modeling practices, often enabling the development of practical solutions to common problems. Modeling of basic thermodynamic and physical properties plays a key role, along with computational fluid dynamics and multiphase transport and interface modeling.

Contributions to the proceedings include applications such as steel processing, modeling of steel and non-ferrous alloys treatments for properties control, multi-physics and computational fluid dynamics modeling for molten metal processes and properties measurement. Extractive, recovery, and recycling process modeling is also presented, completing a broad view of the field and practices of modeling in materials processing.

The engagement of TMS and committee members to chair sessions and review manuscripts makes this symposium and its proceedings possible. The editor and coeditors acknowledge the invaluable support and contribution of these volunteers as well as TMS staff members, in particular, Patricia Warren, Trudi Dunlap, and Matt Baker.

Guillaume Lambotte
Jonghyun Lee
Antoine Allanore
Samuel Wagstaff

Contents

Part I Modeling of Minerals and Metals Processing

Dynamic Current and Power Distributions in a Submerged Arc Furnace 3
Y. A. Tesfahunegn, T. Magnusson, M. Tangstad and G. Saevarsdottir

Modeling of Steel–Slag–Air Three-Phase Flow in Continuous Casting Strand 15
Xubin Zhang, Wei Chen, Lifeng Zhang and Piotr Roman Scheller

Dynamic Modeling of Unsteady Bulging in Continuous Casting of Steel 23
Zhelin Chen, Hamed Olia, Bryan Petrus, Madeline Rembold, Joseph Bentsman and Brian G. Thomas

Modeling on the Two-Phase Flow in a Slab Continuous Casting Strand Using Euler–Euler Approach 37
Haichen Zhou and Lifeng Zhang

Flow Control in the Model of a Continuous Caster by Using Contactless Inductive Flow Tomography 49
I. Glavinić, S. Abouelazayem, M. Ratajczak, D. Schurmann, S. Eckert, F. Stefani, J. Hlava and T. Wondrak

Optimization of the Flow Behavior of Molten Steel in Ultrahigh-Speed Billet Continuous Casting Mold 59
Pei Xu, Dengfu Chen, Shixin Wu, Hengsong Yu, MuJun Long, Sheng Yu and Huamei Duan

Part II Steel—Microstructure and Properties

A New Alloy System Having Autogenous Grain Pinning at High Temperature 73
Tihe Zhou, Hatem S. Zurob and Ronald J. O'Malley

Effect of Casting Temperature on the Surface Finish of Grey Iron Castings 87
Izudin Dugic

Carbide Precipitation of TBM Cutter Ring Steel During Tempering 97
Shaoying Li, Hanjie Guo, Mingtao Mao and Xiao Shi

Analysis of Large Inclusions in Crankshaft Steel by Ingot Casting 107
Qinghai Zhou, Jiongming Zhang and Yanbin Yin

Research on the L2 Control Model Technology of Double Cold Reduction During Continuous Annealing Process 117
Wei Guo, Hui Wang, Yanglong Li, Jie Wen, Meng Yu and Fengqin Wang

Research on Level 2 Rolling Model of Tin Plate Double Cold Reduction Process 131
Hui Wang, Wei Guo, Yanglong Li, Fei Chen, Jie Wen, Meng Yu and Fengqin Wang

Part III Alloys Processing and Properties Modeling

Numerical Modelling and Influence of Cu Addition on the Microstructure and Mechanical Properties of Additive Manufactured Ti–Al–Cu/Ti–6Al–4V Composite 143
E. T. Akinlabi, O. S. Fatoba and S. A. Akinlabi

High-Cycle Fatigue Behaviour of Ultrafine Grained 5052 Al Alloy Processed Through Cryo-Forging 153
K. K. Yogesha, Amit Joshi, Raviraj, A. Raja and R. Jayaganthan

Effect of Heat Treatment on Microstructure of Continuous Unidirectional Solidified Cu–Ni–Sn Alloy 163
Ji Hui Luo, Qin Li, Yan Hui Chen, Shu Liu, Qiu Yue Wen and Hui Min Ding

Part IV Multiphysics—Process and Properties Modeling

Modeling of Fluid Flow Effects on Experiments Using Electromagnetic Levitation in Reduced Gravity 171
Gwendolyn Bracker, Xiao Xiao, Jonghyun Lee, Marcus Reinartz, Stefan Burggraf, Dieter Herlach, Markus Rettenmayr, Douglas Matson and Robert Hyers

Optimal Stator Design for Oxide Films Shearing Found by Physical Modelling 181
Agnieszka Dybalska, Dmitry G. Eskin and Jayesh B. Patel

An Investigation on Electrodeposition of Titanium in Molten LiCl-KCl 193
Chenyao Li, Jianxun Song, Shaolong Li, Xuepeng Li, Yongchun Shu and Jilin He

Part V Extractive Process and Thermodynamic Modeling

Effect of Ultrasound on the Extraction of Silicon and Aluminum from the Metallurgical Slag of Laterite Nickel Ore 205
Pengju Zhang, Jilai Xue, Xuan Liu and Donggen Fang

Thermal Stability and Thermodynamics of the Ag_2ZnGeS_4 Compound 215
Mykola Moroz, Fiseha Tesfaye, Pavlo Demchenko, Myroslava Prokhorenko, Daniel Lindberg, Oleksandr Reshetnyak and Leena Hupa

Thermochemical Data of Selected Phases in the FeO_x–$FeSO_4$–$Fe_2(SO_4)_3$ System 227
Fiseha Tesfaye, In-Ho Jung, Min-Kyu Paek, Mykola Moroz, Daniel Lindberg and Leena Hupa

The Effect of Heat Treatment to FePt/Fe_2O_3 and FePt/Cu Magnetic Performance 241
Naidu Seetala, Deidre Henderson, Jumel Jno-Baptiste, Hao Wen and Shengmin Guo

Part VI Poster Session

High-Temperature Study of Perovskite Evaporation 253
Sergey Shornikov

Power Consumption Model for Electrolytic Preparation of Copper Powders Using Response Surface Methodology 265
Hongdan Wang, Wentang Xia and Bingzhi Ren

Tensile Properties and Microstructure of Squeeze Cast Magnesium Matrix Composite Reinforced with 35 Vol. % of AL_2O_3 Fibers 279
Luyang Ren, Xuezhi Zhang and Henry Hu

Author Index 289

Subject Index 291

About the Editors

Guillaume Lambotte is a Senior R&D Scientist at Boston Metal, a Massachusetts Institute of Technology (MIT), spin-off startup focusing on the development of an environmentally friendly and energetically efficient primary metal extraction process. He primarily focuses on computational thermodynamic modeling, electrochemistry, and high-temperature equilibrium. Prior to joining Boston Metal, he conducted research as a postdoctoral associate at the University of Massachusetts (UMass) Amherst and MIT. Before his graduate studies, he worked as a production assistant manager at Alcan Extruded Products (Crailsheim, Germany).

He obtained his bachelor degree from the European Engineer School for Materials Science (Nancy, France). He received an M.Sc. and a Ph.D. in Metallurgical Engineering from Ecole Polytechnique of Montreal (Montreal, Canada).

He is currently serving as the Chair of the TMS Process Technology and Modeling Committee and was the recipient of the 2015 TMS EPD Young Leaders Professional Development Award. In 2015, he was one of the TMS representatives at the Emerging Leaders Alliance Conference.

Jonghyun Lee is an Assistant Professor in the Department of Mechanical Engineering at Iowa State University. He has been conducting multiple industry- and government-funded projects in the field of materials processing as PI and Co-PI.

He is the recipient of the Young Leaders Professional Development Award in 2013 from The Minerals, Metals & Materials Society where he has been serving as a Co-organizer and Co-editor of the Materials Processing Fundamentals Symposium since 2014 and as Vice-Chair of the Process Modeling and Technology Committee since 2017.

Prior to joining his current institution, he was a Research Assistant Professor at the University of Massachusetts Amherst. He also had nearly 5 years of industry experience and worked as a Postdoctoral Associate for Tufts University, Medford, Massachusetts. He earned his M.S. and Ph.D. in Mechanical Engineering from the University of Massachusetts Amherst and his B.S. in the same discipline from Inha University in Incheon, South Korea.

Antoine Allanore is an Associate Professor of Metallurgy in the Department of Materials Science & Engineering at MIT. He received his higher education in Nancy (France) where he earned a chemical process engineer diploma from Ecole Nationale Supérieure des Industries Chimiques and an M.Sc. and Ph.D. from Lorraine University.

He joined MIT in 2012 as a faculty member, leading a research group that develops sustainable materials extraction and manufacturing processes. He has developed numerous alternative approaches for metals and minerals extraction and processing. With an emphasis on electrochemical methods for both analytical and processing purposes, his group combines experimental and modeling approaches to promptly investigate the ultimate state of condensed matter, the molten state. He teaches thermodynamics and sustainable chemical metallurgy at both the undergraduate and graduate levels.

He received the Vittorio de Nora Award from TMS in 2012, and the TMS Early Career Faculty Fellow Award in 2015.

Samuel Wagstaff began working in the aluminum industry at age 14 with Novelis in Spokane, Washington and now works for the same company in Kennesaw, Georgia as a Process Scientist. In 2013, he received his Bachelor of Science from Cornell University in Mechanical and Aerospace Engineering. He continued his education at the Massachusetts Institute of Technology in the Department of Materials Science and Engineering.

His Ph.D. on the minimization of macrosegregation through jet erosion of a continuously cast ingot uses a turbulent jet to reduce the uneven distribution in aluminum alloy ingots by over 70%. He finished his masters and doctorate at MIT in September 2016 after just 3 years. He has published over a dozen articles on DC casting and macrosegregation, and holds 12 patents.

Part I
Modeling of Minerals and Metals Processing

Dynamic Current and Power Distributions in a Submerged Arc Furnace

Y. A. Tesfahunegn, T. Magnusson, M. Tangstad and G. Saevarsdottir

Abstract Most submerged arc furnaces used for the production of ferroalloys run on three-phase alternating current. This affects the electrical operation of the furnace and thus it is of interest to study alternating current distributions in the system. This work presents computations of alternating electric current distributions inside an industrial submerged arc furnace for silicon production. A 3D model has been developed in ANSYS Maxwell using the eddy current solver. In each phase, electrode, central arc, crater, crater wall and side arcs that connect electrode and crater wall are taken into account. In this paper, the dynamic current distributions in different parts of the furnace, as well as skin and proximity effects in and between electrodes are presented. Moreover, active and reactive power distributions in various components of the furnace are quantified.

Keywords Current distribution · Current paths · Power distributions
Submerged arc furnace

Y. A. Tesfahunegn (✉) · G. Saevarsdottir
School of Science and Engineering, Reykjavik University, Menntavegur 1, 101, Reykjavik, Iceland
e-mail: yonatant@ru.is

G. Saevarsdottir
e-mail: gudrunsa@ru.is

T. Magnusson
United Silicon, Stakksbraut 9, 230, Reykjanesbæ, Iceland
e-mail: tm@silicon.is

M. Tangstad
Department of Materials Science and Engineering, Norwegian University of Science and Technology, 7491 Trondheim, Norway
e-mail: merete.tangstad@ntnu.no

G. Lambotte et al. (eds.), *Materials Processing Fundamentals 2019*, The Minerals, Metals & Materials Series, https://doi.org/10.1007/978-3-030-05728-2_1

Introduction

The current distribution in the submerged arc furnace is critical to good operation in the silicon metal production process. Phase current or resistance are among the most important control parameters, but for modern silicon metal or ferrosilicon furnaces, there is no mechanism to measure the actual current distribution. Metallurgists operate the furnaces based on the analysis of limited data at hand. Recent dig-outs of industrial furnaces have expanded available information on location-dependent charge properties, this enables more realistic modelling of electrical conditions in the furnace than previously possible. Having proper data makes the developed numerical models reliable in predicting the furnace behavior. This will enhance the understanding of critical process parameters and allow more accurate furnace control.

The current distribution is not well known for silicon furnaces and cannot be directly measured. Sævarsdottir et al. [1] calculated that the arc could be a maximum of 10–15 cm in length, based on magnetohydrodynamics (MHD) arc modelling. Although there have been publications on this subject [2], results from an accurate model where the current distribution can be calculated have not been published to date.

The geometry of the zones in a silicon furnace is dependent on the operation history, and hence a number of different geometries, sizes and compositions are possible in the various parts of the furnace. A report from recent excavations of industrial furnaces published by Tranell et al. [3] described the various zones in a FeSi furnace. Myrhaug [4] reported similar features from a pilot scale excavation operating around 150 kW. Tangstad et al. [5] published results from the excavation of industrial furnaces, where the interior of the furnace is divided into zones depending on the materials and their degree of conversion. Mapping the material distribution gives a basis for quantifying the location-dependent physical properties of the charge materials such as the electrical conductivity.

Complete numerical modeling of submerged arc furnace (SAF) requires electrical, chemical, thermal and fluid flow considerations. In this paper, we only consider the electrical aspect, which needs electrical conductivity of the different parts of the furnace. Some works have been done to address this issue. Krokstad [6] outlined an experimental method and published data on the electrical conductivity of silicon carbide and Vangskåsen [7] looked in detail at the metal producing mechanisms. Mølnås [8] and Nell and Joubert [9] have also published data on dig-out samples and material analysis that are relevant. These are some of the essential inputs necessary to set up a reasonably realistic modeling domain with correct physical properties to model the current and power distributions within a furnace, and this opens a unique opportunity to create a model which enables understanding of the current and power distributions in the furnace. These results can be used in the development of furnace control strategies that can enable improved silicon recovery and current efficiency.

The recent developments of electrical numerical modeling include several features of the furnace. Tesfahunegn et al. [10, 11] developed a 3D numerical furnace model that contains electrodes, main arcs, side arcs, crater wall, crater, and other parts using ANSYS Fluent electric potential solver. The authors showed results for current distribution with or without taking into account the main arc. As a continuation of their work, they have implemented a vector potential method using a user-defined function in ANSYS Fluent environment to calculate dynamic current distributions [12, 13]. Their model is only able to consider electrodes and capable of predicting skin and proximity effects. Other researchers have developed different numerical models for SAF based on Computational Fluid Dynamics (CFD) and Finite Element Method (FEM). Herland et al. [14] studied proximity effects in large FeSi and FeMn furnaces using FEM. In their model, they have included different parts of the furnaces. Dhainaut [15] presented computations of electric field in SAF using CFD. The author showed the effect of contact resistance by studying the contact between two coke particles before dealing with a full-scale furnace. The furnace is partitioned in layers to consider different materials and no assumption has been made on the current path. Bezuidenhout et al. [16] applied CFD on a three-phase electric smelting furnace to investigate the electrical aspects, thermal and flow behavior. They showed relationships between electrode positions, current distribution and slag electrical resistivity. Darmana et al. [17] developed a modeling concept applicable for SAFs using CFD that considers various physical phenomena such as thermodynamics, electricity, hydrodynamics, heat radiation and chemical reactions. Wang et al. [18] investigated the thermal behavior inside three different electric furnaces for MgO production.

This paper presents computations of alternating current and power distributions inside an industrial submerged arc furnace for silicon metal production. A 3D model has been developed in ANSYS Maxwell [19] using the eddy current solver. Electrode, main arc, crater, crater wall, and side arc that connects the electrode and crater wall are taken into account for each phase. Other furnace parts such as carbon block, steel shell, and aluminum block are also incorporated.

The Process

In the silicon production process, quartz and carbon materials, that are called charge, are fed into a submerged arc furnace. Three electrodes penetrate the charge from above. Electric heating from the current provides the energy to charge through the electrodes, each of which carries one of the three phases of 50 Hz AC current, canceling out at a star point in the charge.

The overall reaction for producing Silicon metal is

$$SiO_2 + 2C = Si + 2CO(g) \tag{1}$$

This reaction, however, takes a series of sub-reactions, changing the properties of the charge along the way as intermediary reaction products are formed. The current passes from the electrodes through the raw material charge and an electric arc burning at the tip of the electrode. The arc, which consists of thermal plasma in the range of 10,000–20,000 K [20], provides heat for energy-consuming silicon-producing reaction (4), while the SiC-forming reaction and SiO(g) condensation reactions (2) and (3) take place at a lower temperature higher up in the furnace, see Schei et al. [21].

$$\mathrm{SiO(g) + 2C = SiC + CO(g)} \tag{2}$$

$$\mathrm{2SiO(g) = Si + SiO_2} \tag{3}$$

$$\mathrm{SiO_2 + SiC = SiO(g) + CO(g) + Si(l)} \tag{4}$$

It is essential for the silicon recovery in this process that there is a balance between the high-temperature reactions (4) and the low-temperature reactions (2) and (3). Therefore, it is necessary that sufficient heat is released in the arc to drive reaction (4), while a certain part should be released in the raw material charge to drive reaction (2). The stoichiometry of reaction (4) is affected by temperature, and the ratio is decreased at higher temperature, which above 1900 °C enables a high silicon recovery. In the silicon process, it is the electric arc that creates sufficiently high temperature; therefore, sufficient arcing is important for good silicon recovery.

Computational Model

In this section, we describe the mathematical modeling, the furnace geometry, material properties, mesh generation and boundary conditions.

Mathematical Modeling

In this paper, we will focus only on the electrical aspects of SAF. The 3D electrical model is developed in ANSYS Maxwell [19] using eddy current solver, which is suitable for low-frequency devices and phenomena. It solves sinusoidally varying magnetic fields in the frequency domain. The frequency domain solution assumes frequency to be the same throughout the domain. Induced fields such as skin and current proximity effects are also considered. It is a quasi-static solver. To solve for the magnetic field $\boldsymbol{H}$, the solver computes the values as follows [19]:

$$\nabla \times \left(\frac{1}{\sigma + j\omega\varepsilon} \nabla \times \boldsymbol{H} \right) = -j\omega\mu\boldsymbol{H} \tag{5}$$

where σ, ω, μ and ε are electrical conductivity, circular frequency, magnetic permeability and electrical permittivity. The magnetic permeability is typically given by $\mu = \mu_r\mu_0$, where $\mu_0 = 4\pi \times 10^{-7}$[H/m] is the constant magnetic permeability of vacuum and μ_r [–] is the relative magnetic permeability. Once Eq. (5) is solved, the electric field (***E***) and the electric current density (***J***) are solved using Faraday's and Ampere's laws. Also ***J*** and ***E*** are related by Ohm's law. The equation is solved using the finite element method.

Furnace Geometry and Material Properties

The computational domain is based on the actual design of a 32 MW industrial furnace with AC frequency of 50 Hz. A simplified schematic drawing of the furnace is shown in Fig. 1. The furnace is partitioned into different zones based on the material properties. Included in the modeling are the furnace lining, three electrodes, charge, molten material, three arcs below electrodes, side arcs, and three craters with crater walls made of carbides. The geometry of each electrode is considered as a truncated right conical shape. The upper surface of the electrode is the base of the cone with radius equals to the radius of the electrode. The radius of the bottom surface of the electrodes changes as the slope of the slant height changes. We assume that several concentrated side arcs are distributed around the circumference of the electrode near the tip electrodes, and the circular distances between each side arc are held constant. With this configuration, the number of side arcs increases linearly with the circumference of the electrode.

For brevity, a section of the furnace and one electrode are depicted in Fig. 1. For each phase, two types of arcs are introduced. The main-arc, burning below the electrode, with an arc length of 10 cm and diameter of 5 cm [2], and some shorter side arcs connecting the crater wall to the side of the electrode. The curvature of the three crater walls is assumed to be a circular section with a diameter of 100 cm [22]. Each

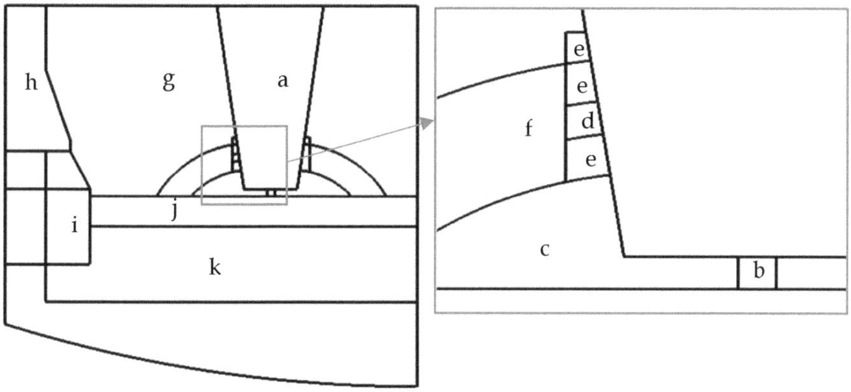

Fig. 1 Schematic of the industrial Silicon SAF with different zones (**a**) electrode, (**b**) arc, (**c**) crater, (**d**) side arc, (**e**) gap, (**f**) carbide, (**g**) charge, (**h**) alumina brick, (**i**) carbon block and carbide, (**j**) molten material, and (**k**) carbon block

Table 1 Electrical conductivity of different zones

Zones	Electrical conductivity [S/m]
Electrode [6]	225,000
Arc [22]	7000
Crater	$1e^{-14}$
Carbide [6]	400
Charge	0.15, 15
Molten material [23]	1,388,900
Carbon block [6]	225,000
Alumina brick	$1e^{-14}$
Steel shell [14]	$6.3e^{+10}$

of the zones is assumed to have constant electrical conductivity. The conductivity of each zone is taken from various literature sources and summarized in Table 1.

Mesh Generation and Boundary Conditions

Mesh generation is a crucial part of any computational method. It has a significant influence on the runtime and memory use of simulation, as well as the accuracy and stability of the solution. Since the eddy current solver utilizes an adaptive mesh refinement algorithm, the material volumes described in Section "Furnace Geometry and Material Properties" were meshed according to the method. This type of meshing technique provides automated mesh refinement capability based on reported energy error in simulation.

The model boundary conditions were imposed based on the positions of the surfaces in the model. Two types of boundary conditions are required, i.e., the natural and Neumann. The natural boundary condition is used for interface between objects. It describes the natural variation from one material to the next one, as defined by material property. The Neumann boundary condition is applied for exterior boundary of solution domain and the $\boldsymbol{H}$ field is tangential to the boundary and flux cannot cross it. To impose appropriate boundary conditions on the $\boldsymbol{H}$ field, a large far-field around the furnace which is filled with air is created. The top surface of the three electrodes is excited by current with equivalent value of $I_{rms} = 99\,kA$. The phase shift between electrodes is 120°.

Numerical Cases

In this section, we determine the current and power distributions inside the furnace described in Section "Furnace Geometry and Material Properties" as well as other parameters, such as resistance, power factor, and voltage of the system. We consider

Table 2 Two simulation groups

Category	Number of side arcs		Charge conductivity	
	8	14	0.15 S/m	15 S/m
Main arcs	✓	✓	✓	✓
No main arcs	✓	✓	✓	✓

three factors. The first factor is the number of side arcs with two levels (8 and 14), the second aspect is the charge conductivity with two levels (0.15 and 15 S/m) and the third element is the consideration of the main arcs with two levels (with main arcs and without main arcs). Hence, a total of 8 simulation cases have been performed. For discussion purposes, we group them into two categories based on the third factor. We only vary the other two factors, i.e., number of side arcs and charge conductivity. The two categories are summarized in Table 2.

For all cases, the phase current has the same value. This means that with changing domain configuration the total resistance changes, and thus the voltage for the system. Some of the cases represent realistic phase resistance in the system while others do not, and the goal with this effort is to gain a qualitative understanding of the governing mechanisms for the current and power distributions in the system. For all cases, the simulations were performed by adaptive meshing algorithm using energy error as a convergence criterion. The energy error was set to 2%. For all cases, the initial mesh size is ~$0.7e^{+06}$ elements and the simulation is converged the mesh size is ~$1.5e^{+06}$ elements. The simulation time per a case on average is around 3 h.

Since the results that are required for this study are not directly obtained from the simulation, we need to perform postprocessing. The current is calculated from current density by integrating on the surface of interest. The active power density, $p\,[W/m^3]$, given by $p = |\boldsymbol{J}|^2/2\sigma$, and the reactive power density $q\,[W/m^3]$, given by $q = (\pi f/\mu)|\boldsymbol{B}|^2$. By integrating the respective power densities over different material domain and the entire furnace, we obtain active power, $P[M\,W]$ and reactive power, $Q[M\,W]$. Once the active and reactive powers of the furnace are calculated, others results such as power factor (PF) and resistance (R) of the system can be calculated.

Figure 2 shows the resulting nonuniform current density on the three electrodes due of skin and proximity effects. Figure 3 shows the total current through electrode and the main arc at different height of the furnace. The vertical axis is a normalized current, which is the fraction of the phase current in the electrode and arc. The horizontal axis is dimensionless furnace height, which is the ratio between a given height and the total height of the furnace. In this paper, we define the total height of the furnace from the bottom of the furnace to the top of the electrodes. In Fig. 3a, main arc is considered whereas in Fig. 3b is not included. In both figures, the charge conductivity and the number of side arcs are varying as shown in Table 2. Irrespective of the magnitude of reduction, the current is decreasing from the top of the electrode to the bottom as the charge conductivity increases. Moreover, the current passed to the main-arc (Fig. 3a) is also decreased as the number of side arcs is increased.

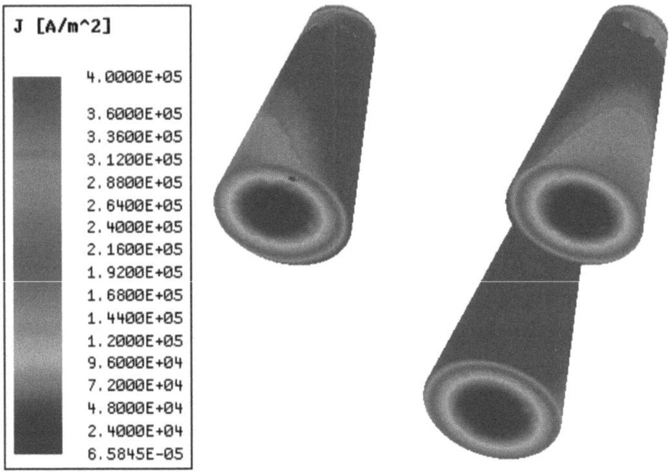

Fig. 2 Current density in the electrodes

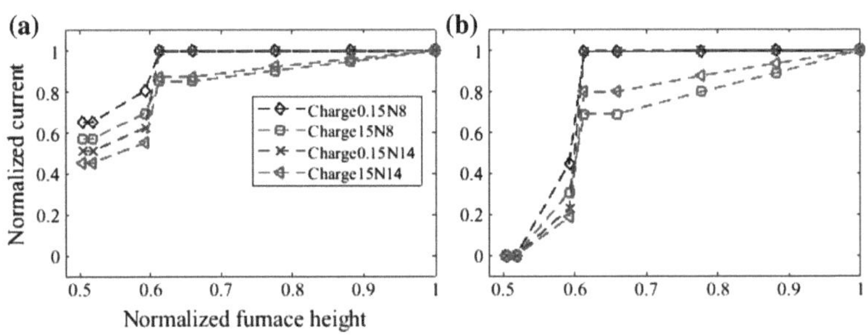

Fig. 3 Normalized current passing through electrode and main arc as a function of normalized height from the furnace bottom to the top of electrodes: (**a**) with main arc, and (**b**) without main arc

Table 3 shows the active and reactive power distributions in different zones for eight side arcs with and without main arcs consideration. Besides, the charge conductivity is varying. When the main arcs are included and charge conductivity is low, most of the power is accumulated in the main arcs and crater wall, while some power is deposited in the remaining zones. However, when the charge conductivity is changed by the order of two magnitudes, the active power in the charge is increased by the same order of magnitude while decreasing in the main arcs and crater wall. Without the main arcs, we can see the same trend except for no power in the main arcs. The main contributors to the reactive power are the far-field, charge, and electrodes. The other materials have some contributions. Since we have not included the electric components outside the furnace, such as bus bars and flexibles, the total value of the reactive power could be higher than the reported values. The simulation results for the 14 side arcs are not reported as in Table 3 since we saw the same trend. Instead, the results are summarized for all simulations results as shown in Table 4. Having

Table 3 Active and reactive power distributions in different zones for eight side arcs setup

	With main arcs				No main arcs			
Zones	Charge cond. 0.15		Charge cond. 15		Charge cond. 0.15		Charge cond. 15	
	P [MW]	Q [MW]	P [MW]	Q [MW]	P [MW]	Q [MW]	P [MW]	Q [MW]
Electrode	2.36	1.88	2.15	1.75	2.07	1.73	1.80	1.55
Main arcs	23.07	0.02	17.25	0.02	0.00	0.00	0.00	0.00
Side arcs	0.55	0.00	0.38	0.00	4.64	0.00	2.24	0.00
Crater wall	10.30	0.42	7.42	0.32	83.94	0.25	40.23	0.13
Crater	0.00	0.40	0.00	0.30	0.00	0.11	0.00	0.06
Charge	0.06	6.79	4.64	5.84	0.45	6.78	20.30	4.88
Molten Si	0.15	0.13	0.12	0.10	0.06	0.07	0.04	0.04
Carbon block	0.03	0.03	0.03	0.02	0.03	0.03	0.03	0.02
Farfield	0.00	13.00	0.00	13.00	0.00	13.00	0.00	12.99
Steel shell	0.01	0.14	0.01	0.14	0.01	0.15	0.01	0.13
Alumina brick	0.00	0.18	0.00	0.17	0.00	0.18	0.00	0.16
Total	36.53	23.00	32.00	21.66	91.18	22.29	64.63	19.97

main arcs show that resistance of the system is sensitive to the change of charge conductivity and the number of side arcs. Without the main arcs, the resistance in the furnace is increased by 100–150%, compared with corresponding simulation cases (Table 4). Most furnaces are operated to strive towards constant resistances. The variations in conductivity conditions in the furnace are met by moving the electrodes up and down. From these simulations, we see how the phase resistance can change with either the conductivity of the charge is changed and (or) exist main arcs and(or) side arcs. One of the assumptions that we made in the simulations is that for each case the charge conductivity is uniform. In a real furnace, however, the charge conductivity is increasing as it moves from the top of the furnace to the bottom. Overall the trend that can be observed is that increasing the system conductivity will result in a reduction of the system resistance.

Conclusions

This paper presents computations of dynamic current and power distributions inside an industrial submerged arc furnace for silicon production. A 3D model has been developed in ANSYS Maxwell using eddy current solver. Electrodes, main arcs, crater, crater wall, and side arcs that connect electrode and crater wall are considered

Table 4 Summary of power distributions and other results of all cases

Cases	With main arcs				No main arcs			
	P [MW]	Q [MW]	PF [/]	R [mΩ]	P [MW]	Q [MW]	PF [/]	R [mΩ]
Charge0.15_Nside8	36.53	23.00	0.85	1.24	91.18	22.29	0.97	3.10
Charge15_Nside8	32.00	21.66	0.83	1.09	64.63	19.74	0.96	2.20
Charge0.15_Nside14	29.00	21.23	0.81	0.99	51.39	20.73	0.93	1.75
Charge15_Nside14	25.77	20.37	0.78	0.78	40.19	19.50	0.9	1.37

for each phase. In this paper, the current distributions in the electrodes and main arcs and the power distributions in different parts of the furnace are presented by varying the charge conductivity, the number of side arcs and with and without main arcs. The presented model is able to capture skin and proximity effects. It was observed that the resistance of the furnace is sensitive to changes in charge conductivity, number of side arcs and existence of main arcs. When main arcs are present, most of the power is accumulated in the main arcs and crater wall for both high and low charge conductivities. It is the conductivity in the crater wall that determines the resistance in the volume at the side-arc attachment and limits the side-arc current. Thus, without main arcs, a significant portion of the power is placed in the crater and charge depending on the charge conductivity value, but the overall resistance in the system is unrealistically high. It is seen that most of the reactive power in the furnace resides in the charge and far-field and depends on the overall current in the system. It is observed that a more narrow current path tends to increase the reactive power in the furnace and thus reduce the power factor. However as the phase resistance and thus real-power dissipation is much more sensitive to the current path, the power factor is much higher for the cases without the main arcs.

Acknowledgements The Icelandic Technology development fund is greatly acknowledged for their funding of this work.

References

1. Sævarsdottir GA, Bakken JA, Sevastyanenko VG, Liping Gu (2011) High power ac arcs in metallurgical furnaces. High Temp Mater Processes 15(3)
2. Saevarsdottir GA, Bakken JA (2010) Current distribution in submerged arc furnaces for silicon metal/ferrosilicon production. In: Proceedings INFACON12
3. Tranell G, Andersson M, Ringdalen E, Ostrovski O, Stenmo JJ (2010) Reaction zones in a FeSi75 furnace—results from an industrial excavation. In: INFACON XII, pp 709–715
4. Myrhaug EH (2003) Non-fossil reduction materials in the silicon process -properties and behavior. Ph.D. thesis, NTNU
5. Tangstad M, Ksiazek M, Andersen JE (2014) Zones and materials in the Si furnace. In: Proceedings: Silicon for the chemical and solar industry XII, Trondheim, Norway, June, pp 24–27
6. Krokstad M (2014) Electrical resistivity of industrial SiC crusts. M.Sc. thesis, NTNU
7. Vangskåsen J (2012) Metal-producing mechanisms in the carbothermic silicon process. M.Sc. thesis, NTNU
8. Mølnås H (2010) Investigation of SiO condensate formation in the silicon process, Project report in TMT 4500. NTNU, Norway
9. Nell J, Joubert C (2013) Phase chemistry of digout samples from a ferrosilicon furnace. In: Infacon prceedings Kazakhstan
10. Tesfahunegn YA, Magnusson T, Tangstad M, Saevarsdottir G (2018) Effect of electrode shape on the current distribution in submerged arc furnaces for silicon production—A modelling approach. J South Afr Inst Min Metall 118(6):595–600
11. Tesfahunegn, YA, Magnusson, T, Tangstad, M, Saevarsdottir, G (2018) Effect of carbide configuration on the current distribution in submerged arc furnaces for silicon production—A modelling approach. In: Nastac L, Pericleous K, Sabau A, Zhang L, Thomas B (eds) CFD modeling and simulation in materials processing 2018. TMS 2018. The Minerals, Metals & Materials Series. Springer, Cham, pp 175–185

12. Tesfahunegn, YA, Magnusson, T, Tangstad, M, Saevarsdottir, G (2018) Dynamic current distribution in the electrodes of submerged arc furnace using scalar and vector potentials. In: Shi Y et al (eds) Computational science—ICCS 2018. Lecture Notes in Computer Science, vol 10861. Springer, Cham, pp 518–527
13. Tesfahunegn, YA, Magnusson, T, Tangstad, M, Saevarsdottir, G (2018) The effect of frequency on current distributions inside submerged arc furnace. In: Paper presented at the IEEE MTT-S international conference on numerical and electromagnetic and multiphysics modeling and optimization. Reykjavik, Iceland, 08–11 Aug 2018
14. Herland EV, Sparta M, Halvorsen SA (2018) 3D models of proximity effects in large FeSi and FeMn furnaces. J South Afr Inst Min Metall 118(6):607–618
15. Dhainaut M (2004) Simulation of the electric field in a submerged arc furnace. In: INFACON X, pp 605–613
16. Bezuidenhout JJ, Eksteen JJ, Bardshaw SM (2009) Computational fluid dynamic modelling of an electric furnace used in the smelting of PGM containing concentrates. Miner Eng 22:995–1006. https://doi.org/10.1016/j.mineng.2009.03.009
17. Darmana D, Olsen JE, Tang K, Ringldalen E (2012) Modelling concept for submerged arc furnaces. In: Paper presented at the ninth international conference on CFD in the minerals and process industries CSIRO. Melbourne, Australia, 10–12 Dec
18. Wang Z, Fu Y, Wang N, Feng L (2014) 3D numerical simulation of electrical arc furnaces for the MgO production. J Mater Process Technol 214:2284–2291. https://doi.org/10.1016/j.jmatprotec.2014.04.033
19. Maxwell, ver. 18.0 (2018) ANSYS Inc., Southpointe, 275 technology drive, Canonsburg, PA 15317
20. Saevarsdottir G, Bakken J, Sevastyanenko V, Liping G (2011) High power ac arcs in metallurgical furnaces. High Temp Mater Processes 15(3)
21. Schei A, Tuset JK, Tveit H (1998) Production of high silicon alloys. Tapir Forlag, Trondheim
22. Sævarsdottir GA (2002) High current ac arcs in silicon and ferrosilicon furnaces. Ph.D. thesis, NTNU
23. Sasaki H, Ikari A, Terashima K, Kimura S (1995) Temperature dependence of the electrical resistivity of molten silicon. Jpn J Appl Phys. https://doi.org/10.1143/JJAP.34.3426

Modeling of Steel–Slag–Air Three-Phase Flow in Continuous Casting Strand

Xubin Zhang, Wei Chen, Lifeng Zhang and Piotr Roman Scheller

Abstract In the current study, a three-dimensional mold model was established by Fluent software to investigate the fluid flow of three phases (steel–slag–air) in the mold. A quarter of the mold was simulated through the k-ε model, volume of fluid (VOF) model, solidification model and continuum surface force (CFS) method. The interfacial tension between liquid steel and liquid slag and the oscillation of the mold were added into the model to show the 3D steel–slag interface. The liquid steel exiting from the submerged entry nozzle (SEN) existed as the upper backflow and lower backflow, and flowed towards the wide face and the SEN. The largest speed on the steel–slag interface was located at approximately 0.25 m from the narrow face, which was approximately 0.15 m/s. Under the influence of the upper backflow and the movement of the shell, the slag on the steel–slag interface moved from the narrow face to the SEN, and infiltrated into the gap, which affected the lubrication in the gap.

Keywords Three-phase flow · Steel–slag interface · Simulation
Continuous casting

Introduction

In the continuous casting process [1], the liquid steel in the tundish moved through the SEN into the mold [2] and then existed as double-roll flow or single-roll flow patterns [3] with different casting parameters. The powder was added into the mold successively and existed as liquid slag, solid slag and powder from the bottom up

X. Zhang · W. Chen · L. Zhang (✉) · P. R. Scheller
School of Metallurgical and Ecological Engineering, University of Science and Technology Beijing (USTB), Beijing 100083, China
e-mail: zhanglifeng@ustb.edu.cn

X. Zhang · W. Chen · L. Zhang · P. R. Scheller
Beijing Key Laboratory of Green Recycling and Extraction of Metal, University of Science and Technology Beijing (USTB), Beijing 100083, China

G. Lambotte et al. (eds.), *Materials Processing Fundamentals 2019*, The Minerals, Metals & Materials Series, https://doi.org/10.1007/978-3-030-05728-2_2

[4, 5]. The slag covering the liquid steel could protect the steel from oxidation, slow the heat transfer from the steel to the air, and absorb the inclusion from the steel. Meanwhile, the slag could infiltrate into the gap between the steel shell and the copper plate, and exist as lubricant [6], and the phenomena near the meniscus from 2D mold model [7] were shown in Fig. 1. However, the flow of the liquid steel and slag could be affected by the interaction. The slag entrapment [8] was easy to occur with the high speed of the upper backflows.

It was reported that the curved meniscus formed with the interfacial tension between the steel and slag [9]. The meniscus solidification led to the formation of hooks, which could entrap bubbles, inclusions and slags, deteriorating the surface quality of the slab [10]. Hence, the understanding of the slag movement in the mold was extremely important. However, most studies about the fluid flow in the mold only involved the flow of the steel.

In the current study, a three-dimensional mold model was established to investigate the fluid flow and heat transfer of three phases (steel–slag–air) in the mold. The double-roll flow pattern in the mold was revealed, and the velocity of the slag on the steel–slag interface and slag–air interface was obtained. The interfacial tension between liquid steel and liquid slag and the oscillation of the mold were added into the model to show the 3D steel–slag interface.

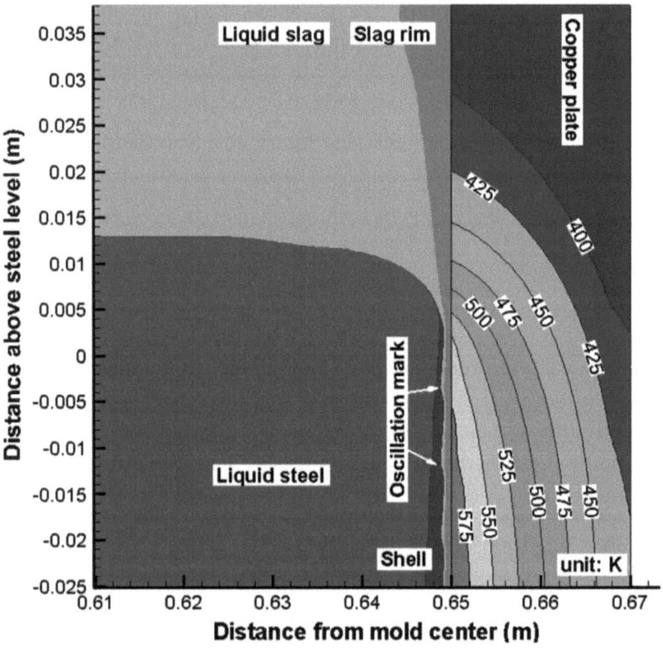

Fig. 1 Phenomena near meniscus from 2D mold model [7]

Mathematical Modeling

In the current study, a three-dimensional mathematical model was established to investigate the three-phase (Steel–slag–air) fluid flow in the mold, and the velocity on the steel–slag interface was focused on. The k-ε model, VOF (volume of fluid) model, solidification model and CFS method were applied in the model, and the fluid flow, heat transfer and the solidification of the steel were calculated through solving the continuity equation, Navier–Strokes equations, and energy equation [11]. The interfacial tension between the liquid steel and slag was considered to reveal the shape of steel–slag interface. The oscillation of the mold was added to obtain the infiltration of the slag into the gap between the steel shell and the copper plate.

The mesh and simulation conditions of the mold model are shown in Fig. 2. The computation domain included a quarter of the submerged entry nozzle (SEN), a quarter of the mold with the length of 0.9 and 0.5 m below the exit of the mold to reduce the amount of computation. In order to investigate the flow velocity near the meniscus, the finer mesh of 50 μm was applied near the initial steel–slag interface and the wall of the mold, as shown in Fig. 2(a). The total mesh number was 1462044. In Fig. 2(b), the initial thickness of the air and slag above the steel were both 50 mm. The velocity inlet (1.512 m/s) was applied at the inlet of the SEN, and the velocity was calculated on the basis of mass conservation. The temperature of the steel at the inlet was 1830 K. The pressure outlet was applied at the bottom of the model, and the turbulent kinetic energy and dissipation rate were 0.0001 m^2/s^2 and 0.0001 m^2/s^3, respectively. The free surface was applied on the top of the mold, the symmetry conditions were applied on the central section, and other boundaries were applied as non-slip wall. Simulation parameters were shown in Table 1, and other details could be found elsewhere [12]. The simulated shape of steel–slag interface was compared with the measured shape of hook lines to validate the accuracy of the model [13].

Fluid Flow in the Mold

The streamline of liquid steel in the mold is exhibited in Fig. 3. Under the current simulation condition, the liquid steel existed as double-roll flow pattern in the mold. When the liquid steel exiting from the SEN rushed to the narrow face, the liquid steel flowed towards the wide face, and then the upper and lower backflows formed. The speed of the upper and lower backflows was below 0.4 m/s. In Fig. 4, the vector of several sections in the mold is demonstrated to show the velocity of the steel and slag. Under the influence of the upper backflow, the lower part of the slag moved from the narrow face to the SEN. On the contrary, the upper part of the slag flowed from the SEN to the narrow face, when neglecting the powder slag. The speed near the steel–slag interface is larger than that near the slag–air interface. Hence, a circulation might also be found in the slag zone.

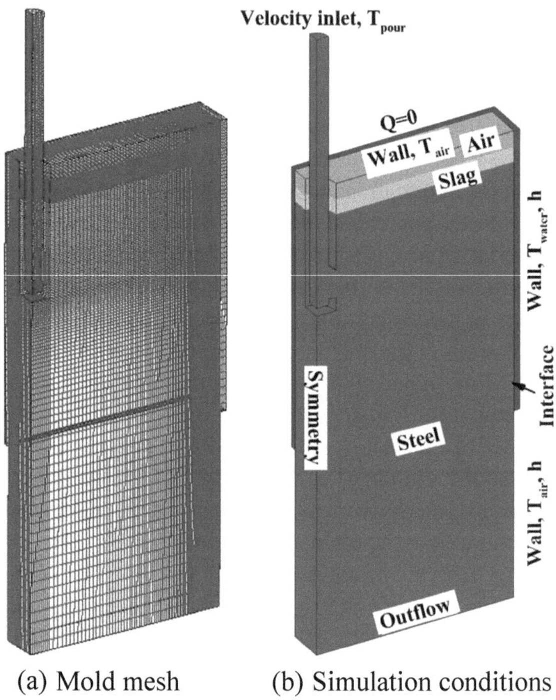

Fig. 2 Mesh and simulation conditions of the mold model

Table 1 Simulation parameters

Parameters	Value	Parameters	Value
Mold section	1300 mm × 247 mm	Density of liquid steel	7020 kg/m^3
Simulation length	1400 mm	Viscosity of liquid steel	0.0063 kg/(m s)
Submerged depth of SEN	180 mm	Density of liquid slag	2500 kg/m^3
Inner diameter of port	85 mm	Viscosity of liquid slag	0.262 kg/(m s)
Air thickness	50 mm	Specific heat of steel	750 J/(kg K)
Slag thickness	50 mm	Latent heat of steel	270,000 J/kg
Interfacial tension of slag–steel	1.3 N/m	Mold oscillation mode	Sinusoidal
Casting speed	1.45 m/min	Oscillation frequency	3 Hz
Contact angle between steel and slag	46°	Oscillation stroke	6 mm

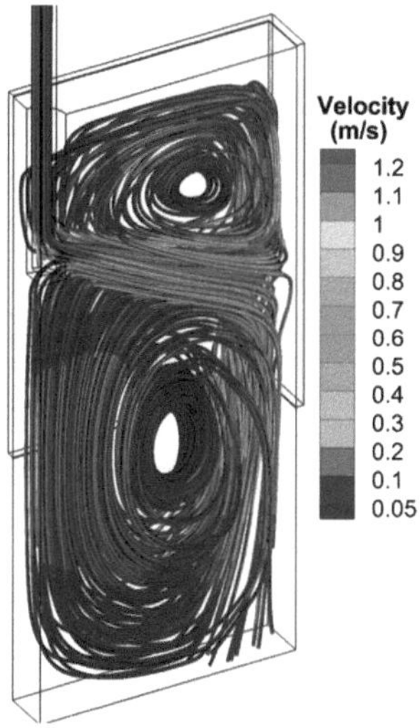

Fig. 3 Streamlines of liquid steel in the mold

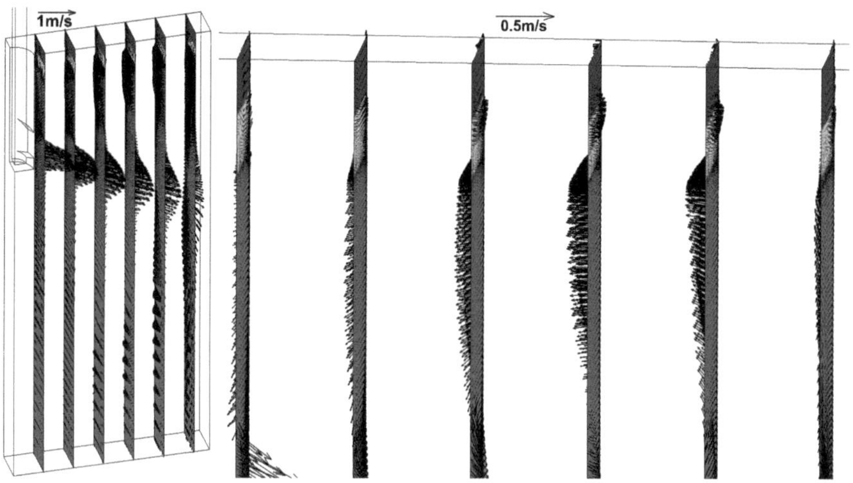

Fig. 4 Vector of several sections in the mold

Velocity and Profile of the Steel–Slag Interface

In order to investigate the movement of the slag above the steel, the flow velocity on the steel–slag interface and slag–air interface is shown in Fig. 5. Under the influence of the upper backflow, the slag on the steel–slag interface flowed from the narrow face to the SEN, while the slag on the slag–air interface flowed from the SEN to the narrow face. In Fig. 5(a), the slag flowed from the point near the narrow face of the

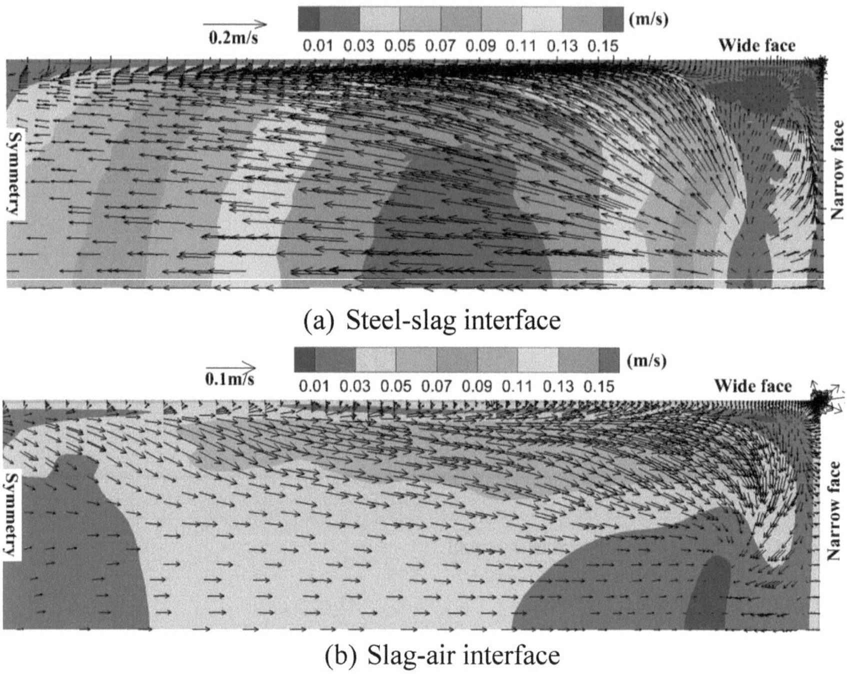

Fig. 5 Flow velocity on the steel–slag and slag–air interface

mold to the direction of the wide face and the SEN, which might be related to the movement of the slag into the gap between the shell and the mold. The largest speed on the steel–slag interface was located at approximately 0.25 m from the narrow face, where slag entrapment was apt to occur. The speed was approximately 0.15 m/s. In Fig. 5(b), the slag on the slag–air interface flowed from the wide face and the SEN to the narrow face, and the speed near the wide face was larger than that near the central section. The speed was within 0.09–0.11 m/s.

The shape of the 3-D steel–slag interface near the corner is exhibited in Fig. 6. With the oscillation of the mold and the interfacial tension between the steel and slag, the steel–slag interface existed as the shell surface on side face and meniscus on the top face. Under the influence of the upper backflow and the movement of the shell, the slag moved from the narrow face to the SEN, and infiltrated into the gap, which affected the lubrication in the gap.

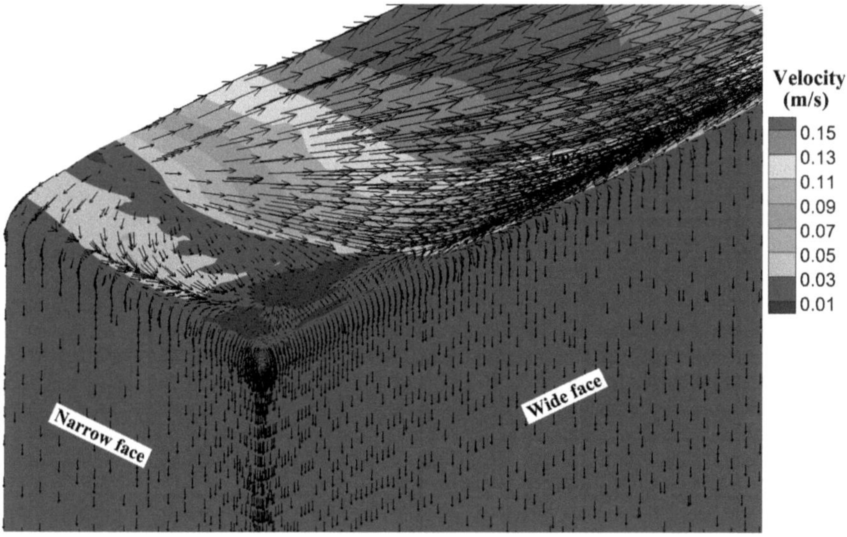

Fig. 6 Profile of steel–slag interface

Conclusions

In the current study, a three-dimensional mold model was established to investigate the three-phase fluid flow and the velocity of the slag in the mold. The conclusions were reached as follows:

(1) Under the influence of the upper backflow, the lower part of the slag moved from the narrow face to the wide face and the SEN. On the contrary, the upper part of the slag flowed from the wide face and the SEN to the narrow face. Hence, a circulation might also be found in the slag zone.
(2) The largest speed on the steel–slag interface was located at approximately 0.25 m from the narrow face. The speed was approximately 0.15 m/s, which might be related to the slag entrapment.
(3) The slag on the steel–slag interface moved from the narrow face to the SEN, and infiltrated into the gap between the shell and copper plate, which affected the lubrication in the gap.

Acknowledgements The authors are grateful for support from the National Science Foundation China (Grant No. U1860206), the Fundamental Research Funds for the Central Universities (Grant No. FRF-TP-15-001C2), Beijing Key Laboratory of Green Recycling and Extraction of Metals (GREM) and the High Quality Steel Consortium (HQSC) at the School of Metallurgical and Ecological Engineering at University of Science and Technology Beijing (USTB), China.

References

1. Mizoguchi S, Ohashi T, Saeki T (1981) Continuous casting of steel. Annu Rev Mater Sci 11(1):151–169
2. Mills KC, Ramirezlopez P, Lee PD, Santillana B, Thomas BG, Morales R (2014) Looking into continuous casting mould. Ironmaking Steelmaking 41(4):242–249
3. Liu Z, Sun Z, Li B (2017) Modeling of quasi-four-phase flow in continuous casting mold using hybrid Eulerian and Lagrangian approach. Metall Mater Trans B 48(2):1248–1267
4. Mills KC, Fox AB (2003) The role of mould fluxes in continuous casting-so simple yet so complex. ISIJ Int 43(10):1479–1486
5. Mills KC, Fox AB, Li Z, Thackray RP (2005) Performance and properties of mould fluxes. Ironmaking Steelmaking 32(1):26–34
6. Meng YA, Thomas Brian G (2003) Modeling transient slag-layer phenomena in the shell/mold gap in continuous casting of steel. Metall Mater Trans B 34(5):707–725
7. Zhang X, Chen W, Scheller PR, Ren Y, Zhang L (2018) Mathematical modeling of initial solidification and slag infiltration at the meniscus of slab continuous casting mold. JOM
8. Hibbeler LC, Thomas BG (2013) Mold slag entrainment mechanisms in continuous casting molds. Iron Steel Technol 10(10):121–136
9. Ramirez-Lopez PE, Lee PD, Mills KC, Santillana B (2010) A new approach for modelling slag infiltration and Solidification in a continuous casting mould. Isij Int 50(50):1797–1804
10. Sengupta J, Shin H-J, Thomas BG, Kim S-H (2006) Micrograph evidence of meniscus solidification and sub-surface microstructure evolution in continuous-cast ultralow-carbon steels. Acta Mater 54(4):1165–1173
11. Wang Y, Zhang L (2011) Fluid flow-related transport phenomena in steel slab continuous casting strands under electromagnetic brake. Metall Mater Trans B 42(6):1319–1351
12. Zhang X, Chen W, Zhang L (2017) A coupled model on fluid flow, heat transfer and solidification in continuous casting mold. China Foundry 14(5):416–420
13. Zhang X, Chen W, Yang W, Zhang L (2018) Study of oscillation marks and hooks at the corner in continuous casting steel slabs. In: 7th International congress on science and technology of steelmaking

Dynamic Modeling of Unsteady Bulging in Continuous Casting of Steel

Zhelin Chen, Hamed Olia, Bryan Petrus, Madeline Rembold, Joseph Bentsman and Brian G. Thomas

Abstract Mold level fluctuations caused by unsteady bulging of the solidifying shell affect the quality of the steel and stable operation of the continuous steel casting process. A dynamic bulging model, which captures the behavior of the 2-D longitudinal domain through interpolation of multiple 1-D moving slices, is used to calculate the transient bulging profile, volume changes caused by unsteady bulging, and the accompanying level fluctuations in the mold. The liquid steel flow rate through the SEN into the tundish is calculated with a stopper-position-based model. These two models are combined to investigate mold level fluctuations in a thin-slab caster under real casting conditions. The model is verified by comparing the simulation results with transient measurements in a commercial thin-slab caster.

Keywords Continuous casting · Unsteady bulging · Mold level fluctuation
Stopper rod flow model · Dynamic bulging model

Introduction

In the continuous casting of steel, bulging is an important phenomenon where the internal ferrostatic pressure, partially restrained by the support rolls, causes the partially solidified shell to bulge outward between each pair of rolls. Bulging is directly responsible for internal cracks, centerline segregation, and permanent slab-width variations [1–3]. It also increases roll forces and roll wear. In addition, time variations of the bulged shape may cause volume changes of the molten steel contained within the solidifying shell in the strand, leading to mold level fluctuations. Such fluc-

Z. Chen · J. Bentsman
University of Illinois at Urbana-Champaign, 1206 W Green St, Urbana, IL 61801, USA

H. Olia · B. G. Thomas (✉)
Colorado School of Mines, 1610 Illinois St, Golden, CO 80401, USA
e-mail: bgthomas@mines.edu

B. Petrus · M. Rembold
Nucor Steel Decatur, 4301 Iverson Blvd, Trinity, AL 35673, USA

G. Lambotte et al. (eds.), *Materials Processing Fundamentals 2019*, The Minerals, Metals & Materials Series, https://doi.org/10.1007/978-3-030-05728-2_3

tuations may correspond to the time period between certain roll pitches, or certain roll diameters in different zones of the caster [3, 4].

It is well known that excessive mold level fluctuations lead to strand surface cracks and even breakouts. Mold level control has been studied for many years [5, 6] to minimize these problems. However, the mold level control system has difficulty recognizing and responding to dynamic bulging and cannot prevent it. Other methods, such as increasing the spray cooling in the secondary cooling region to decrease the surface temperature and thicken the shell, and adoption of non-uniform roll pitch have been proposed to reduce dynamic bulging problems [3, 7].

Previous work to study bulging has focused mainly on steady bulging, using Finite Element Analysis (FEM) [1, 8, 9]. A few recent studies have measured unsteady bulging in the casting machine using position detectors between rolls [3, 4, 7]. Such detectors are useful for model validation but are very difficult for everyday online use, especially for thin-slab casters, and have never been used to help mold level control systems.

To the authors' best knowledge, this work is the first attempt to develop a dynamic bulging model that is calculated fast enough for implementation into real-time online control systems and validated with plant measurements. A dynamic model, ConOffline [10–13], which captures the behavior of the 2-D longitudinal domain through interpolation of multiple 1-D moving slices, is used to calculate the bulging amplitudes. A new dynamic volume model then calculates the volume change induced by dynamic bulging and stopper rod movement. This model is verified with plant measurements and applied to gain new insight into the dynamic bulging phenomenon.

Model Description

Heat Transfer and Steady Bulging Model

First, shell thickness and temperature distribution in the continuous cast strand were predicted using ConOffline, an off-line version of the ConOnline model [10], which has been validated and used in many previous studies [11–14]. ConOffline solves the transient heat conduction equation within many transverse slices through the center of the strand using an explicit finite-difference method in a Lagrangian reference frame, which moves with the steel in the z-direction at the casting speed v_c:

$$\rho c_p^* \frac{\partial T}{\partial t} = \frac{\partial}{\partial x}\left(k \frac{\partial T}{\partial x}\right) \tag{1}$$

where x is the thickness direction, T is temperature, and temperature-dependent properties are density ρ, thermal conductivity k, and effective specific heat c_p^* which includes the latent heat, L_f and f_s is the solid fraction:

$$c_p^* = c_p + L_f \frac{df_s}{dT} \tag{2}$$

The caster and casting conditions simulated in this work were based on the thin-slab (90 mm) caster at Nucor Steel, Decatur. Average heat flux in the mold was based on an empirical correlation from Duvvuri [15]:

$$Q_m\left[MW/m^2\right] = 1.197(v_c)^{0.544} \tag{3}$$

Heat flux from the spray water in the secondary cooling zones was based on Nozaki's empirical correlation [16]:

$$h_{spray} = 0.3925 \times Q_{water}^{0.55} \times \left(1 - 0.0075 \times T_{spray}\right) \tag{4}$$

where $Q_{water}\left(L/m^2\right)$ is water flux in the spray zone and T_{spray} is the water temperature. Heat transfer in secondary cooling is a subject of ongoing research, and other relations are available and used at different casters.

ConOffline simulates $N = 200$ slices simultaneously; each slice starts at the meniscus at a different time to achieve a fixed z-direction spacing between slices. Then, the maximum bulging amplitude, δ_{max}, within each roll pitch is found from the following equation based on fitting many FEM simulations, proposed by Yu [8], knowing the local strand surface temperature, shell thickness, roll pitch, and ferrostatic pressure calculated using $P = \rho gh$.

$$\delta_{\text{max}}(mm) = 7.1496 \times 10^{-34} \frac{L_{(mm)}^{6.5} P_{(MPa)}^{1.993} T_{surf\,(^oC)}^{8.766}}{d_{(mm)}^{5.333}} \tag{5}$$

where P: ferrostatic pressure, L: roll pitch, T_{surf}: strand surface temperature, and d: solidified shell thickness.

Dynamic Bulging Model for Unsteady Bulging

Bulging of the strand involves both static and dynamic components according to time variations of the bulging which accompany the movement of the strand. Figure 1 illustrates steady and unsteady bulging. With steady bulging, the surface profile of strand is constant with time, and each portion of the steel shell follows the bulged profile, wiggling like a snake as it moves down with the casting speed. Although the strand bulges outwards between rolls and is pushed back beneath rolls, the mold level stays constant because the volume of molten steel obtained inside the solidifying shell does not change with time. In contrast, with unsteady bulging, the contorted shape of the strand surface becomes partially frozen, so the bulged profile moves down the caster. When the steady bulged shape of the solidified shell, called the "concave profile", moves between the rolls, the strand must be squeezed inwards. The total volume of molten steel contained within the strand decreases which causes the mold

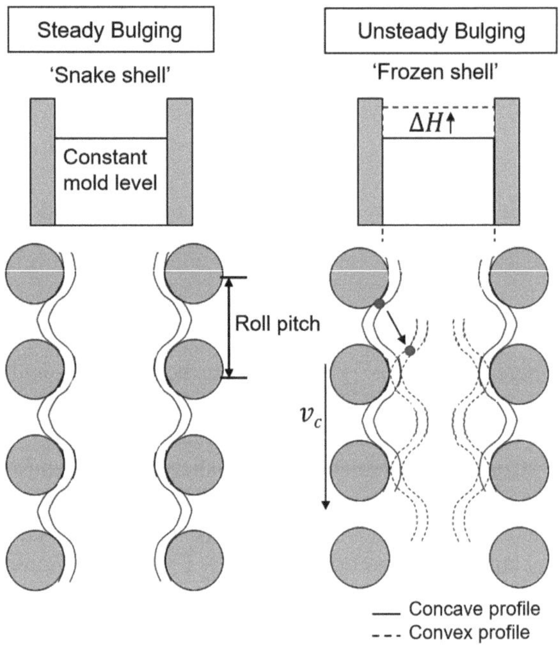

Fig. 1 Schematic illustration of steady, unsteady dynamic bulging

level to rise. Continuing down the caster, the strand experiences repeated transverse expansions and contractions, resulting in repeated vertical mold level fluctuations.

To model this behavior, molten steel flow is divided into three components: inflow into the mold through the stopper rod, outflow due to downward movement of the solidifying shell at the casting speed, and liquid steel flow due to transverse movement of the solidified shell. The latter component is found by tracking changes in the volume of the molten steel inside the solidified shell according to the shape of the bulged strand, the extent to which it is partially "frozen", and its downward movement. This volume change, ΔV_b, will induce mold level fluctuations, which the mold level control system will attempt to compensate. The resulting movement of the stopper rod, which changes the inlet steel flow from the tundish induces transient volume changes, ΔV_s. Any time variation in casting speed will introduce further volume changes, ΔV_o. As shown in Fig. 2, ignoring other effects, the following volume conservation equation gives the total time-varying volume change of the strand, ΔV_m:

$$\Delta V_m = -\Delta V_b + \Delta V_s - \Delta V_o \tag{6}$$

This total volume change of molten steel also can be found from the measured mold level fluctuations, ΔH:

$$\Delta V_m(t) = \Delta H(t) W D \tag{7}$$

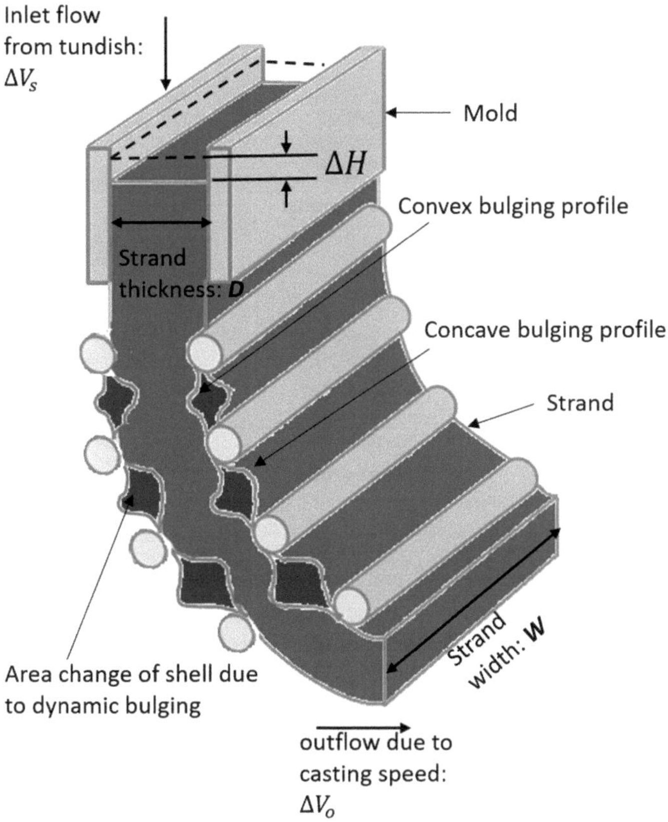

Fig. 2 Schematic diagram of evaluation of volume change by unsteady bulging

where W is strand width, D is strand thickness, and average $\Delta H(t)$ is estimated by the mold level sensor. In the steady-state, or "snaking shell" case, the volume change induced by bulging is zero, i.e. $\Delta V_b = 0$. In the partially frozen shell case, it depends on both the inter-roll bulging profile along the entire strand, and how that shape changes as it moves down the caster. In a slab caster with relatively constant width, this simplifies to

$$\Delta V_b = \sum A_i W \tag{8}$$

where A_i is area change in ith roll pitch induced by repeated transverse expansions and contractions of the strand.

The dynamic bulging model in this section is developed for the unsteady bulging case. For simplicity, the maximum bulging location is assumed to be at the center of the roll pitch, midway between rolls, and the bulging profile was very roughly approximated as triangular. Figure 3 shows the two extreme cases in unsteady bulging

Fig. 3 Concave and convex profile of triangular bulging profile

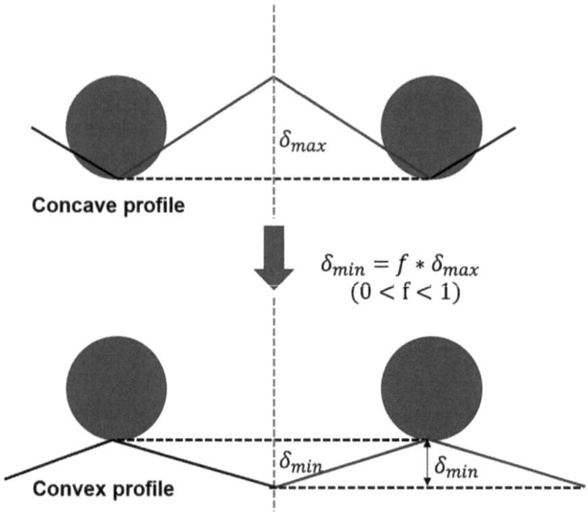

over one roll pitch: concave and convex profiles. Due to creep deformation, the bulging amplitude is assumed to decrease after the bulged 'peak' moves from midway between rolls to lie directly under the next roll. The following assumptions were made for simplicity: (A1) this decrease in height happens linearly from concave profile to convex profile, and vice versa; (A2) $\delta_{min} = f\delta_{max}$, f is set to be 0.5 in this study; (A3) The bulging at each roll pitch is independent.

To calculate the time evolution of the x–z area change of the bulging profile with strand movement, the concave profile of steady bulging was taken as the initial state. The passing line of steel strand was taken as the zero reference. Denoting z as the strand movement along the casting direction starting from time t = 0, the area change in a single roll pitch L can be calculated by

$$A_i(z) = \begin{cases} 0.5\left(\left(L - 2z'\right)\left(1 - \frac{2z'}{L}\right) - \frac{4z'^2}{L}\right)\left(1 - (1-f)\frac{2z'}{L}\right)\delta_{\max}, \\ z \in [nL, \frac{L}{2} + nL) \\ 0.5\left(-2\left(L - z'\right) * \left(1 - \frac{2z' - L}{L}\right) + \frac{\left(2z' - L\right)^2}{L}\right)\left(f + (1-f)\frac{2z' - L}{L}\right)\delta_{\max}, \\ z \in [\frac{L}{2} + nL, L + nL) \end{cases} \tag{9}$$

where $z' = z \bmod L$, $n = 1, 2, \ldots\ldots$. The distance that the strand moved, $z(t)$, can be calculated from casting speed $v_c(t)$:

$$z(t) = \int_0^t v_c(\tau)d\tau \tag{10}$$

Based (A3), the total volume change due to bulging can be calculated by

$$\Delta V_b(t) = \sum A_i(z(t))W \tag{11}$$

Stopper Rod Flow Model

A pressure drop—flow rate model for stopper rod systems (PFSR) is used to estimate the time-dependent inlet steel flow rate, based on the input stopper position history recorded at the plant. PFSR is a stand-alone MATLAB-based program that solves a system of Bernoulli equations for flow rate and pressure distribution down the entire system from tundish top surface, through SEN, to the mold top surface. Details of the model can be found in [17].

During continuous casting, the stopper rod is susceptible to erosion. To minimizing the effect of erosion, data were collected from temporal regions in the plant measurements where there was constant casting speed followed by a sudden change (increase or decrease of casting speed), and finally followed by another time period of constant casting speed. Throughput differences within this short period should be explained by the stopper rod movement.

The average erosion rate was estimated for the entire casting sequence, according to flow rate changes during times of constant speed. After accounting for the erosion, the stopper rod data were extrapolated to zero flow rate at zero stopper position. After further calibration and validation of PFSR with 18 sets of measurements, the following linear interpolation of the stopper rod inlet flow rate by PFSR was obtained.

$$Q_s(t) = 0.5924 * h(t) + 0.764 \tag{12}$$

where Q_s(ton/min) is the inlet flow through stopper rod opening and $h(t)$(mm) is a stopper rod opening. This linear interpolation is believed only valid for a narrow range of stopper rod (3.2–4.7 mm) and flow rate [2.66–3.53 (ton/min)], which happens to be relevant for typical casting conditions at the plant.

Computational Procedure

The calculation procedure of these models is as follows:

- ConOffline calculates the maximum bulging amplitude at each roll pitch based on recorded or specified casting conditions including slab geometry, casting speed, spray flow rates, steel grade, etc., for each time step, Δt(1 s in this study), based on the shell thickness and surface temperature profiles down the strand from Eqs. (1–5).

- Next, the dynamic bulging model calculates the mold level fluctuation caused by volume changed due to bulging, $\Delta V_b(t)$ from Eqs. (6–12) knowing the local strand surface temperature, shell thickness, roll pitch, ferrostatic pressure, and casting speed.
- Inflow volume induced by stopper rod movements are calculated from the inlet steel flow based on Eq. (13), knowing the stopper rod position measurement $h(t)$ as follows:

$$\Delta V_s(t) = Q_s(t)\Delta t \tag{13}$$

- Outflow volume due to downward movement at the casting speed, $v_c(t)$, are found from:

$$\Delta V_o(t) = v_c(t) W D \rho \Delta t \tag{14}$$

  Since the casting speed is relatively constant during this simulation, the variation of $\Delta V_o(t)$ is almost negligible.
- The total volume change history in the mold can be calculated from Eq. (7) knowing the mold level position measurement.

Simulation Results and Discussion

Casting Conditions

The casting scenario simulated is shown in Fig. 4. The first plot shows the casting speed history, which is relatively constant around 3.1 m/min. The spray flow rate in secondary cooling zones 5 and 6 were increased at around 1000 s because large bulging in these 2 zones was suspected. The last two plots are measured mold level position and stopper rod position, which show evidence of significant dynamic bulging followed by a time interval where only small bulging is suspected. Steel composition was changed at around 2700 s, which appears to be at least partly responsible.

Stopper Rod Flow Model Results

First, the new volume model was applied to the measured scenario, including only the effect of stopper rod movements, and neglecting any dynamic bulging. A comparison

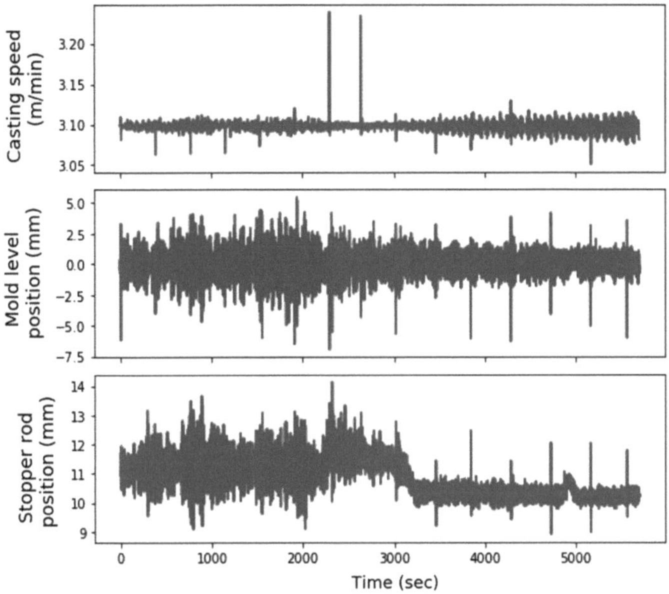

Fig. 4 Simulated casting scenario

of the calculated and measured mold level position histories are shown in Fig. 5 for two time intervals: 500–510 s (dynamic bulging suspected) and 3400–3410 s (no bulging problems suspected). Differences between the calculated mold level from the stopper rod flow model and the measured mold level should be explained by dynamic bulging. When dynamic bulging is suspected to be large (left figures), the model greatly over predicts the level fluctuations. This indicates that in the real caster, the level control system was compensating for those changes. This is also indicated by the difference between measured throughput, which was constant, and the model predicted inlet flow, which appears to have been almost exactly compensating for the significant dynamic bulging. When dynamic bulging was small, (right figures), the predicted mold level is similar to the measurements indicating that the mild level fluctuations are caused by stopper rod movement and surface waves.

To better illustrate the importance of dynamic bulging, the model prediction of the mold level (considering stopper rod movement only) was subtracted from the measured mold level signal, i.e.$\Delta V_m - \Delta V_s + \Delta V_o$, and converted into $\Delta H(t)$ using Eq. (7). This new prediction, "model prediction with no level control", represents the expected mold level history if the stopper opening had been held constant (i.e., with no level control system). This prediction is shown in Fig. 6 for the 500–520 s time interval, where it reveals much larger level fluctuations than the actual measured mold level. This confirms the suspicion that dynamic bulging must have been very severe during this time interval. Furthermore, it shows that the stopper rod movement due to the level control system was helping to decrease the mold level fluctuations. Note,

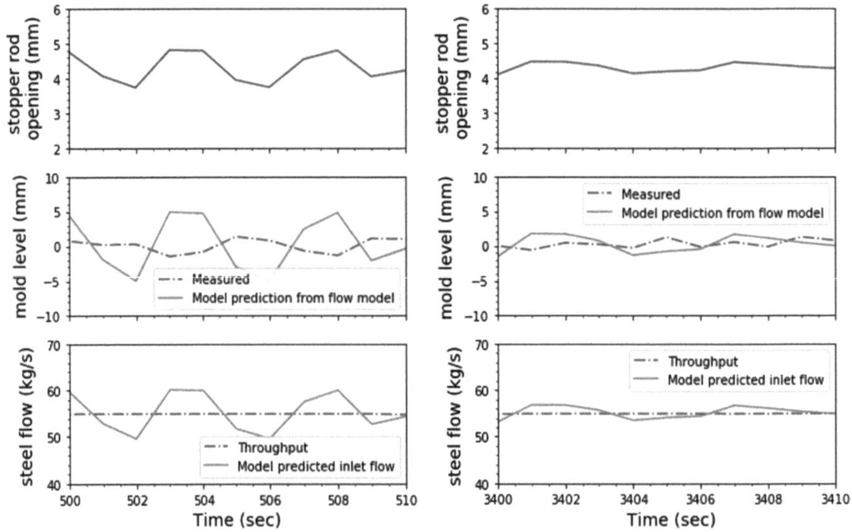

Fig. 5 Measured mold level position and calculated mold level position due to the variation of inlet steel flow

Fig. 6 Comparison of measured and calculated mold level histories, including dynamic bulging

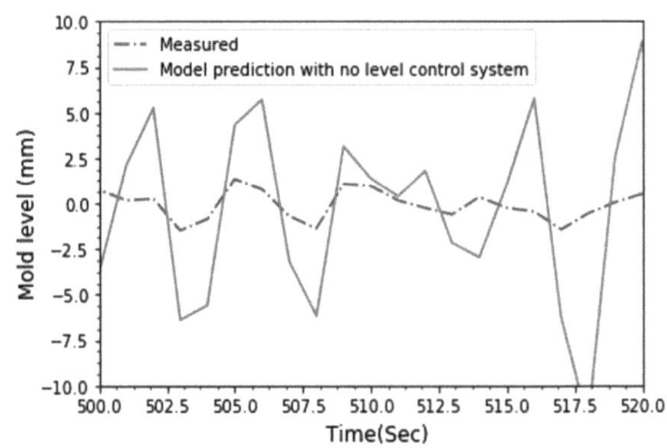

however, that there is still much room for improvement, if the unsteady bulging could be decreased or eliminated, or the mold level control system could be improved.

Dynamic Bulging Model Results

Next, to predict dynamic bulging, the bulging amplitude calculated from Eq. (5) was simulated for each roll pitch, and used to predict the mold level fluctuation induced by dynamic bugling. The "model prediction with no level control" explained in the Section "Stopper Rod Flow Model Results", which is the differences between

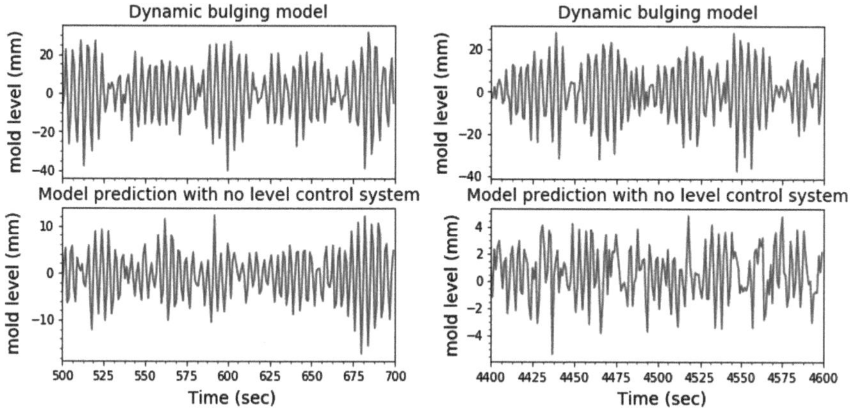

Fig. 7 Mold level prediction with dynamic bugling model and mold level prediction with no control system for 500–700 s and 4400–4600 s

the calculated mold level from the stopper rod flow model and the measured mold level, was used as a comparison to the dynamic bulging model prediction. Figure 7 shows this two predictions for time intervals of 500–700 s and 4400–4600 s. In the case of 500–700 s, significant dynamic bulging was suspected, the magnitude of the prediction from the dynamic bulging model is two times bigger than the model prediction with no level control system. We argue that this difference is caused by assuming dynamic bugling in all zones in the dynamic bulging model, while in the real caster, it may be steady bulging in certain regions and dynamic unsteady bulging in other regions. The trend of the two signals shows a qualitative match with similar clustering of the peaks. However, for 4400–4600 s, the two signals do not match. This suggests that after the steel grade change, the new shell becomes more susceptible to creep, so the bulging profile reverts from dynamic to more steady bulging, with smaller f-value, and thus reduces the volume changes and mold level fluctuations. Indeed, with minimal dynamic bulging, the measured signal is more likely to be caused by random turbulent flow and surface waves, which cannot be controlled by stopper rod movement.

Power spectrum density (PSD) analysis of the signals in Fig. 7 is shown in Fig. 8. The period used for Fast Fourier Transform is 200 s. When bad dynamic bulging is suspected (500–700 s), the PSD of the measured signal shows that the main frequency is the frequency corresponding to the roll pitch in zone 5, while the PSD of the estimation shows matching frequency at zone 5, and zone 6. When small dynamic bulging is suspected (4400–4600 s), the PSD of the measurement shows no matching frequency of roll pitch in spray zones, and it appears to be random noise. This also supports the hypothesis that there is steady bulging for 4400–4600 s. The PSD of the estimation still shows the main frequency of the roll pitch of zone 5, and 6 but

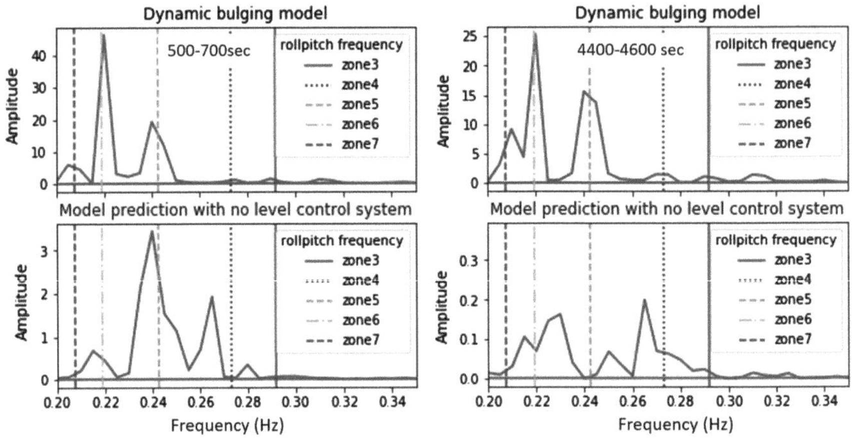

Fig. 8 Power spectrum analysis of the measured and estimated mold level position-time signals shown in Fig. 7

with smaller magnitude. This is because the dynamic bulging model was simulating a case of unsteady bulging throughout the caster. Therefore, it should not match the steady bulging results.

Conclusions

A dynamic volume model has been developed to relate the volume changes from dynamic unsteady bulging and stopper rod movements to level fluctuations in the continuous slab-casting process. The model incorporates separate submodels to predict steady-state bulging and the relation between flow rate and pressure drops according to stopper rod position. Model predictions of the history of mold level variations caused by unsteady bulging matches reasonably with measurements of mold level history in a thin-slab caster. Power Spectrum Density analysis shows that for the studied case, dynamic bulging in zone 5 appears to be responsible for the level fluctuations, as the dynamic volume model is able to catch the zone 5 frequency peaks. This work confirms that dynamic bulging is responsible for significant mold level fluctuations in the plant under some circumstances. Furthermore, the model shows that the mold level control system only partly compensates for the dynamic bulging, making mold level fluctuation only slightly less severe than would have occurred with a constant stopper position. This leaves a lot of room for future improvement of mold level control systems. This work is only a preliminary first step to model unsteady dynamic bulging. To fully understand dynamic bulging, much further work is needed.

Acknowledgements This work was supported by NSF Grant #1300907, NSF INTERN DCL #1747876, and the Continuous Casting Center at the Colorado School of Mines. Special thanks are extended to Nucor Steel Decatur for providing caster data and casting conditions.

References

1. Dalin J, Chenot J (1988) Finite element computation of bulging in continuously cast steel with a viscoplastic model. Int J Numer Methods Eng 25(1):147–163
2. Barber B, Lewis BA, Leckenby BM (1985) Finite-element analysis of strand deformation and strain distribution in solidifying shell during continuous slab casting. Ironmak Steelmak 12(4):171–175
3. Ohno H, Miki Y, Nishizawa Y (2016) Generation mechanism of unsteady bulging in continuous casting-1-Development of method for measurement of unsteady bulging in continuous casting. ISIJ Int 56(10):1758–1763
4. Yoon U-S, Bang I-W, Rhee JH, Kim S-Y, Lee J-D, Oh KH (2002) Analysis of mold level hunching by unsteady bulging during thin slab casting. ISIJ Int 42(10):1103–1111
5. Manayathara TJ, Tsao T-C, Bentsman J (1996) Rejection of unknown periodic load disturbances in continuous steel casting process using learning repetitive control approach. IEEE Trans Control Syst Technol 4(3):259–265
6. Dussud M, Galichet S, Foulloy LP (1998) Application of fuzzy logic control for continuous casting mold level control. IEEE Trans Control Syst Technol 6(2):246–256
7. Lee JD, Yim CH (2000) The mechanism of unsteady bulging and its analysis with the finite element method for continuously cast steel. ISIJ Int 40(8):765–770
8. Yu L (2000) FEM analysis of bulging between rolls in continuous casting. M.S. thesis, University of Illinois Urbana-Champaign
9. Miyazawa K, Schwerdtfeger K (1979) Computation of bulging of continous of bulging in continuously cast slabs wiht simple bending theory. Ironmak Steelmak 6(2):68–74
10. Petrus B, Zheng K, Zhou X, Thomas BG, Bentsman J (2011) Real-time, model-based spray-cooling control system for steel continuous casting. Metall Mater Trans B 42(1):87–103
11. Chen Z, Bentsman J, Thomas BG, Matsui A (2017) Study of spray cooling control to maintain metallurgical length during speed drop in steel continuous casting. Iron Steel Technol 14(10):92–103
12. Petrus B, Chen Z, Bentsman J, Thomas BG (2018) Online recalibration of the state estimators for a system with moving boundaries using sparse discrete-in-time temperature measurements. IEEE Trans Automat Contr 63(4):1090–1096. https://doi.org/10.1109/TAC.2017.2736950
13. Chen Z, Bentsman J, Thomas BG (2018) Bang-Bang free boundary control of a Stefan problem for metallurgical length maintenance. Paper presented at the Am Control Conf., Milwaukee, 27–29 June 2018. https://doi.org/10.23919/acc.2018.8431904
14. Petrus B, Hammon D, Miller M, Williams B, Zewe A, Chen Z, Bentsman J, Thomas BG (2015) New method to measure metallurgical length and application to improve computational models. Iron Steel Technol 12(12):58–66
15. Duvvuri P, Petrus B, Thomas BG (2014) Correlation for mold heat flux measured in a thin slab casting mold. Paper presented at the AISTech—Iron and steel technology conference, Indianapolis, 5–8 May 2014
16. Nozaki T, Matsuno J, Murata K, Ooi H, Kodama M (1978) A secondary cooling pattern for preventing surface cracks of continuous casting slab. Trans. Iron Steel Inst Jpn 18(6):330–338
17. Olia H, Thomas BG (2018) Flow rate—stopper position model of NUCOR caster using Pressure Drop Flow Rate Model for Stopper Rod Flow Control Systems (PFSR). Report presented at CCC annual meeting, Golden

Modeling on the Two-Phase Flow in a Slab Continuous Casting Strand Using Euler–Euler Approach

Haichen Zhou and Lifeng Zhang

Abstract The stability of flow field in the mold has a great effect on the quality of the final continuous casting product. In the current study, the multiphase flow in a slab continuous casting strand was systematically studied using a full scale numerical simulation via Euler–Euler approach, and the effect of the operational parameters including casting speed, gas flow rate, and the submergence depth of the submerged entry nozzle (SEN) was investigated. The study showed that the lower casting speed tended to generate more single roll flow in the strand. With the decrease in the gas flow rate, the flow pattern was evolved from a single roll to a complex flow and then to a double roll. For a fixed gas flow rate and a fixed casting speed, the deeper submergence depth generated more double roll pattern in the strand.

Keywords Continuous casting · Flow pattern · Multiphase flow
Euler–Euler approach · Nail board experiment

Introduction

The final quality product of the continuous cast steel slab is directly related to the flow pattern of liquid steel in the mold. An undesired flow pattern may cause a costly "break-out" happens [1], quality problems at the meniscus by crack formation [2], and line defects such as surface slivers, blisters, pencil pipes by the entrainment of bubbles and mold slag into the solidified shell or internal defects in the rolled product [3–5].

H. Zhou · L. Zhang (✉)
School of Metallurgical and Ecological Engineering, University of Science and Technology Beijing (USTB), Beijing 100083, China
e-mail: zhanglifeng@ustb.edu.cn

H. Zhou · L. Zhang
Beijing Key Laboratory of Green Recycling and Extraction of Metal, University of Science and Technology Beijing (USTB), Beijing 100083, China

G. Lambotte et al. (eds.), *Materials Processing Fundamentals 2019*, The Minerals, Metals & Materials Series, https://doi.org/10.1007/978-3-030-05728-2_4

The flow pattern in the mold is associated with many factors, such as casting speed, gas fraction and submergence depth of the SEN, etc. Thomas et al. [6] studied the effect of the volume fraction and velocities of the gas on the liquid flow, and they found that increasing gas injection rate can intensely change the flow pattern. Kubo et al. [7] concluded that molten steel flow patterns can be controlled by balancing the molten steel throughput and the argon gas flow rate. Yuan et al. [8] studied the behavior of flow transients and asymmetries during nominally steady-state flow conditions with the large eddy simulation (LES) computational approach. Liu et al. [9] simulated multiphase flow of steel and gas bubbles in the SEN and mold utilizing Eulerian–Eulerian approach.

In this paper, the effect factors on flow pattern in the continuous casting strand were discussed in terms of casting speed, gas flow rate and the submergence depth of SEN. The recommend practices related to the operational parameters can improve the flow pattern in the continuous casting strand.

Mathematical Model and Simulation Methods

Governing Equations

The continuity and momentum equations are as follows:

$$\frac{\partial}{\partial t}\left(\alpha_q \rho_q\right) + \nabla \cdot \left(\alpha_q \rho_q \mathbf{u_q}\right) = 0 \tag{1}$$

$$\frac{\partial\left(\alpha_q \rho_q \mathbf{u_q}\right)}{\partial t} + \nabla \cdot \left(\alpha_q \rho_q \mathbf{u_q} \mathbf{u_q}\right) = -\nabla \cdot \left(\alpha_q \tau_q\right) - \alpha_q \nabla P + \alpha_q \rho_q \mathbf{g} + \mathbf{F}_{pq} \tag{2}$$

where α, ρ, t, $\mathbf{u}$, τ, P and $\mathbf{g}$ are the volume fraction, density, time, velocity, shear stress, pressure, and gravity acceleration, respectively. The subscripts q and p denote the liquid and gas phase, respectively.

The realizable $k - \varepsilon$ turbulence model is employed to study the two-phase flow inside the mold. Then, the turbulent kinetic energy k and the dissipation rate ε of kinetic energy for the primary phase are

$$\frac{\partial}{\partial t}\left(\rho_q \alpha_q k\right) + \nabla \cdot \left(\rho_q \alpha_q \mathbf{u_q} k\right) = -\nabla \cdot \left(\alpha_q \frac{\mu_{T,I}}{\sigma_k} \nabla k\right) + \alpha_q (G_k - \rho \varepsilon) \tag{3}$$

$$\frac{\partial}{\partial t}\left(\rho_q \alpha_q \varepsilon\right) + \nabla \cdot \left(\rho_q \alpha_q \mathbf{u_q} \varepsilon\right) = -\nabla \cdot \left(\alpha_q \frac{\mu_{T,I}}{\sigma_k} \nabla \varepsilon\right) + \alpha_q \rho_q \left(C_1 S \varepsilon - C_2 \frac{\varepsilon^2}{k + \sqrt{\nu \varepsilon}}\right) \tag{4}$$

where G_k represents the generation of turbulence kinetic energy, and S is the modulus of the mean rate of strain tensor. The model constants are $C_{1\varepsilon} = 1.44$, $C_2 = 1.9$, $\sigma_k = 1.0$, and $\sigma_\varepsilon = 1.2$.

In the right-hand side of Eq. (2), the last term represents the interfacial forces between two phases as follows:

$$\mathbf{F}_{q,p} = -\mathbf{F}_{p,q} = \mathbf{F}_D + \mathbf{F}_L + \mathbf{F}_{TD} + \mathbf{F}_{VM} \tag{5}$$

where $\mathbf{F}_D$ is the drag force, $\mathbf{F}_L$ is the lift force, $\mathbf{F}_{TD}$ is the turbulent dispersion force and $\mathbf{F}_{VM}$ is the virtual mass force. These forces are described in detail in the work of Ávila-Ortiz et al. [10].

Boundary Conditions and Numerical Simulation Strategies

The computational domain and mesh are shown in Fig. 1. The fluid properties and operating conditions used in numerical simulation are given in Table 1. A velocity was prescribed at the inlet boundary condition and the gas volume fraction was set at the inlet. The hot argon flow rate was used in the simulation. The top surface of the mold was degassing boundary conditions that allow gas to leave, but the liquid steel is not. The outlet ports in the bottom wall were defined as a constant pressure condition. The wall boundaries were defined as no-slip wall. The convergence criteria were set to 1×10^{-4} for the flow variables. Time step was small at the beginning and typical set as 5×10^{-4} s.

Results and Discussion

Numerical Simulation Verification

To validate the accuracy of the numerical flow model, the simulations were run to match the flow conditions in a commercial steel caster and validated by nailboard dipping method [9]. Parameters used in the experiment and simulation are as shown in Table 1. Figure 2 quantitatively compares the calculated surface velocities with the plant nail dipping tests. It can be found that the surface velocities of the simulation and experiment have similar trend with experiments. Therefore, these verification results indicate that the numerical flow model is valid and reliable in this work.

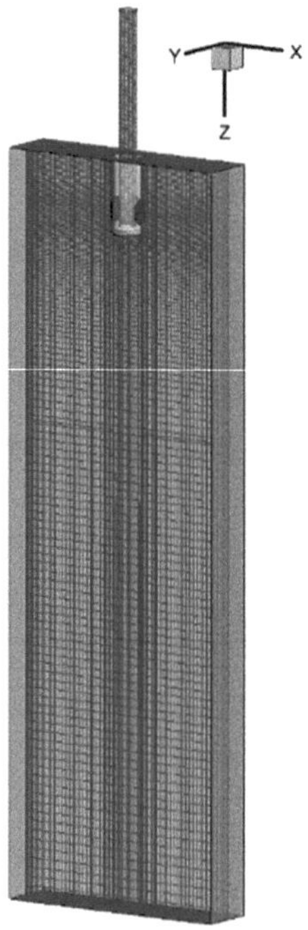

Fig. 1 Schematics and mesh structures of the simulation domain

Table 1 Parameters [11] used in the simulation of the actual continuous casting mold

Parameter	Value	Parameter	Value
Width of mold (mm)	1000	Casting speed (m/min)	1.7
Thickness of mold (mm)	237	Gas density @1803 K (kg/m^3)	0.26
Length of the computational domain (mm)	3000	Gas viscosity @1803 K (kg/m s)	8.1×10^{-5}
Diameter of SEN (mm)	78	Gas flow rate (SLPM)	12.5
Length of SEN (mm)	900	Liquid density @1803 K (kg/m^3)	7100
Exit angle of nozzle (°)	20	Liquid viscosity @1803 K (kg/m s)	0.0064
Height of SEN port (mm)	90	Gravity constant (m/s^2)	9.8
Width of SEN port (mm)	70	Interfacial tension (N/m)	1.6
Submergence depth of SEN (mm)	170		

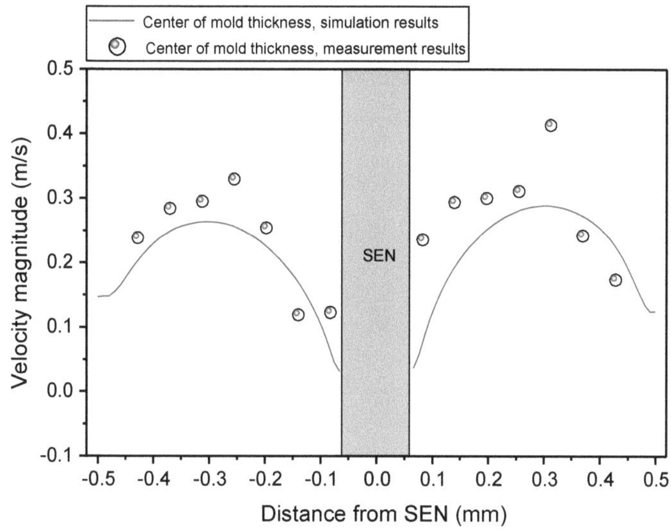

Fig. 2 The experimental and simulation results of surface velocities

Influence of Casting Speed on the Flow Pattern

The flow patterns were first reported by Andrzejewski et al. [12]. The single roll flow pattern means that an intensive surface stream is directed from the nozzle to the narrow side of the mould. The double roll flow pattern involves the inclined inlet stream splits into an ascending and a descending branch when impacting the narrow mould side. The lower roll is rotating clockwise and the upper roll counterclockwise. There is no surface stream flow from the nozzle to the narrow side of the mould. The complex roll flow pattern fell in between the single roll and the double roll flow pattern.

Figure 3 shows a typical instantaneous velocity field along the center plane of the vertical part of the calculation domain. For a constant mold width, submergence depth and a fixed gas flow rate, the flow pattern in the mold tends towards from single roll to double roll flow pattern with increasing casting speed. In Fig. 3. (a), the low casting speed makes liquid velocity small when the jet from the SEN port exit. The jet bent up more and has more tendency to form a single roll flow pattern. Increasing speed to 1.3 m/min causes the recirculation region near both the SEN and narrow wall to appear, and a complex flow pattern, as shown in Fig. 3. (b). In Fig. 3. (c) shows that the high casting speed generates a strong jet, which travels to the liquid pool and splits into two loops after impinging on the narrow wall. Therefore, higher casting speeds favour double roll flow pattern.

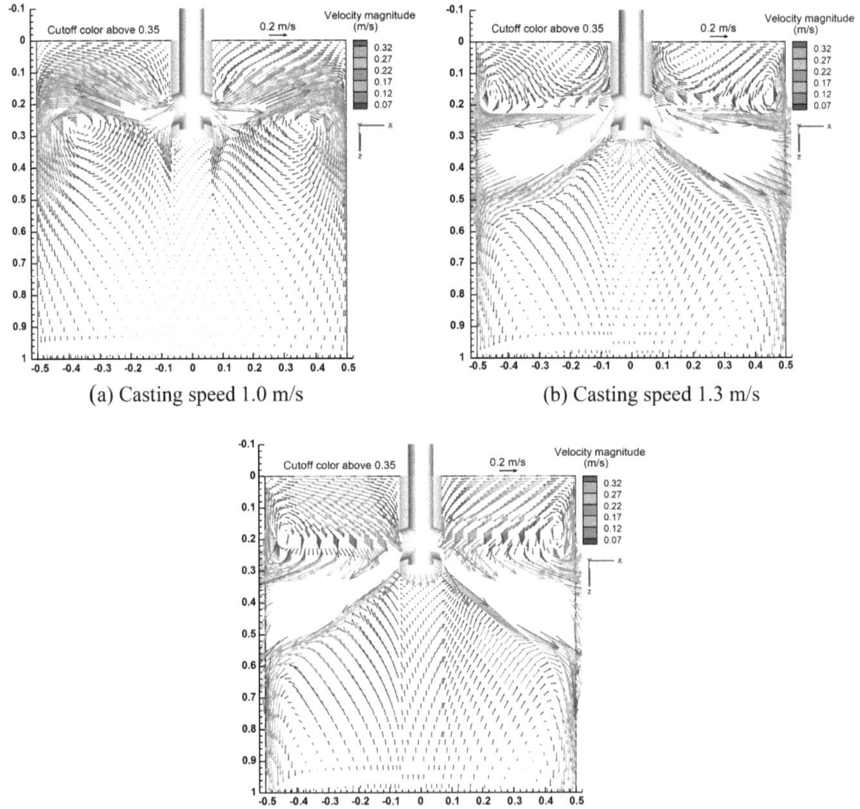

(a) Casting speed 1.0 m/s

(b) Casting speed 1.3 m/s

(c) Casting speed 1.7 m/s

Fig. 3 Velocity fields at center plane of the mold with different casting speeds

Influence of Gas Volume Fraction on the Flow Pattern

Simulation results inside the mold with different gas flow rates in Fig. 4 present that the flow pattern will evolve through single roll, complex and double roll with the increasing gas volume fraction. In Fig. 4 (a), the jet can reach the narrow face at the low gas flow rate of 10 standard liter per minute (SLPM). When the gas flow rate becomes higher, the exiting jet towards the top surface owing to large buoyancy force and more gas bubbles rising near the SEN which oppose the flow towards the SEN from the narrow wall. This reverse flow alters the double roll flow pattern towards a complex or even single-roll flow pattern, as shown in Fig. 4(b) and (c), respectively. With the gas flow rate increasing, the exiting jet will push the mold flux towards the narrow wall under the effect of the large buoyancy force and entrap slag droplets which cause slivers and blisters in the solidified steel shell.

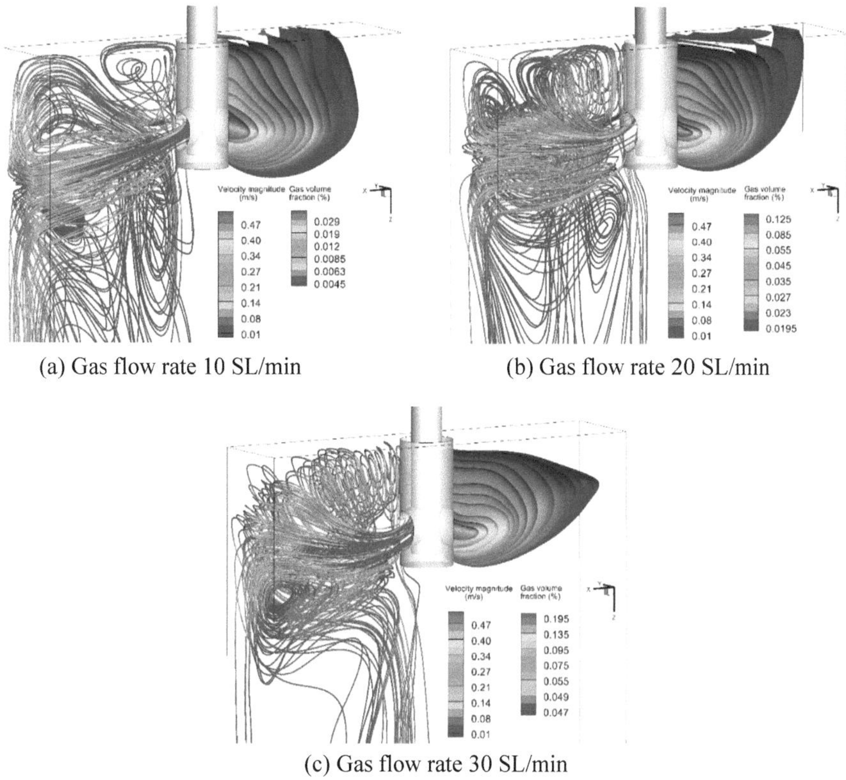

Fig. 4 Streamlines of liquid and gas volume fraction inside the mold with different gas flow rates

Accordingly, the minimizing gas flow rates are recommended to promote double roll flow pattern when considering nozzle clogging.

Figure 5 shows the flow pattern identification under different gas flow rate and casting speed. The results indicate that the flow pattern is single roll at the condition of low gas flow rate and casting speed. With the gas flow rate increasing, the flow pattern changes to the complex flow even at high casting speed. Although the complex roll flow pattern is relatively stable, the liquid level fluctuations become acutely which is harmful for slab quality such as the entrainment of the slag droplets. Therefore, the optimal on-site operation parameter is located in the bottom right hand corner of Fig. 5.

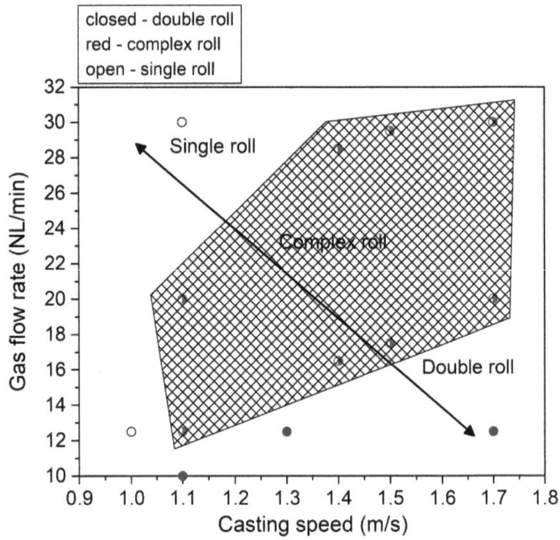

Fig. 5 Flow pattern identification under different gas flow rate and casting speed

Influence of Submergence Depth on the Flow Pattern

For a given casting speed, gas fraction and slab width, the influence of submergence depth on flow pattern is shown in Fig. 6. The bent jet to hit the top surface more easy for shallow submergence depth in Fig. 6(a). With increasing submergence depth, the vertical distance from the jet to the top surface increases which makes the bent jet to hit the top surface become difficult, as shown in Fig. 6(b) and (c). In summary, increasing the submergence depth will promote and stabilise the double roll flow formation which in turn will improve slab quality.

The relationship between different casting speed and submergence depth is plotted in Fig. 7. The flow pattern changes from single-roll to double roll flow with increasing submergence depth. However, the submergence depth almost has no remarkable effect on the flow pattern at high casting speed condition.

Conclusions

The present work is concerned specifically with a study of the effect of continuous casting process parameters on flow pattern in the continuous casting strand. The main conclusions can be drawn as follows:

(1) Increasing the casting speed may produce significant changes in the flow pattern. The flow pattern tends towards double roll when the liquid flow rate is near critical value.

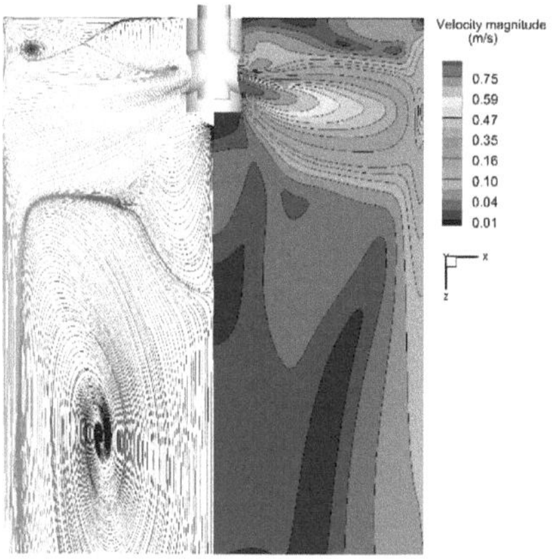

(a) Submergence depth 80 mm

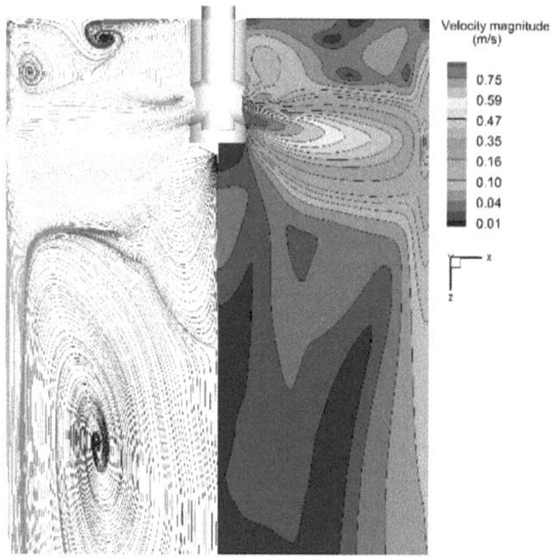

(b) Submergence depth 150 mm

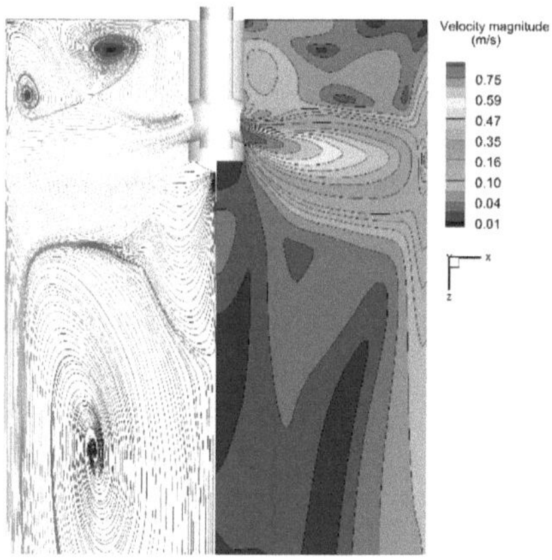

(c) Submergence depth 190 mm

Fig. 6 Streamlines of liquid inside the mold with different submergence depths

(2) The flow pattern changes from double roll to a complex flow, then to single roll pattern with increasing gas fraction.
(3) The flow pattern has more tendency to form double roll when nozzle submergence depth becomes deeper. However, the effect of submergence depth on the flow pattern is unremarkable at high casting speed.
(4) The flow pattern is identified under different gas flow rate and casting speed, and the optimal on-site operation parameters are higher casting speed and lower gas flow rate for narrow mold width.

Fig. 7 Relationship between different casting speed and submergence depth

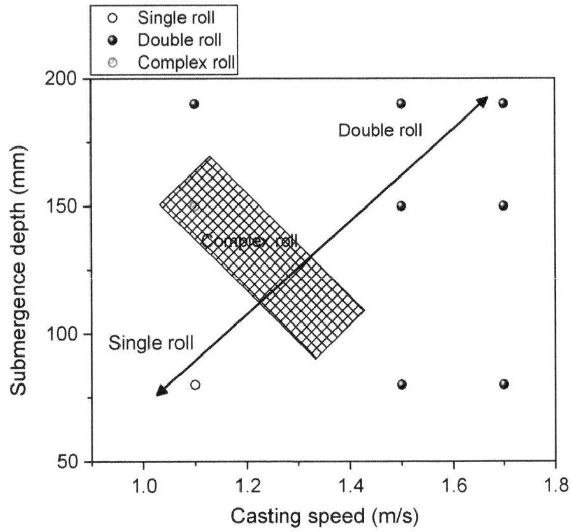

Acknowledgements The authors are grateful for the support from the National Science Foundation China (Grant No. U1860206), the National Key R&D Program of China (2017YFB0304001). Beijing Key Laboratory of Green Recycling and Extraction of Metals (GREM) and the High Quality Steel Consortium (HQSC) and Green Process Metallurgy and Modeling (GPM^2) at the School of Metallurgical and Ecological Engineering at University of Science and Technology Beijing (USTB), China.

References

1. Nakato H, Ozawa M, Kinoshita K, Habu Y, Emi T (1984) Factors affecting the formation of shell and longitudinal cracks in mold during high speed continuous casting of slabs. Trans ISIJ 24(11):957–965
2. Mcdavid RM, Thomas BG (1996) Flow and thermal behavior of the top surface flux/powder layers in continuous casting molds. Metall Mater Trans B. 27(4):672–685
3. Emling WH, Waugaman TA, Feldbauer SL, Cramb AW (1994) Subsurface mold slag entrainment in ultra low carbon steels. In: Steelmaking conference proceedings
4. Knoepke J, Hubbard M, Kelly J, Kittridge R, Lucas J (1994) Pencil blister reductions at inland steel company. In: Steelmaking conference proceedings
5. Kasai N, Mizukami H, Mutou A (2003) State of segregation with bubble in continuously cast slab of ultra low carbon steel. Tetsu-to-Hagane 89(11):1120–1127
6. Thomas BG, Huang X, Sussman RC (1994) Simulation of argon gas flow effects in a continuous slab caster. Metall Mater Trans B 25(4):527–547
7. Kubo N, Ishii T, Kubota J, Aramaki N (2002) Two-phase flow numerical simulation of molten steel and argon gas in a continuous casting mold. ISIJ Int 42(11):1251–1258
8. Yuan Q, Thomas BG, Vanka SP (2004) Study of transient flow and particle transport in continuous steel caster molds: Part I. Fluid flow. Metall Mater Trans B. 35(4):685–702
9. Liu R, Thomas BG, Sengupta J, Chung SD, Trinh MK (2014) Measurements of molten steel surface velocity and effect of stopper-rod movement on transient multiphase fluid flow in continuous casting. ISIJ Int 54(10):2314–2323

10. Ávila-Ortiz Y, Morales MD, Cedillo-Hernández V, Delgado-Pureco J (2018) Mathematical modeling of argon bubbling effects on fluid flow patterns of liquid steel in a slab mold. Steel Res Int 89(5)
11. Morales RD, Ortiz-Avila Y (2017) Fundamentals of scale up procedures to model liquid steel-argon flows in slab molds. In: The international conference on modeling and simulation of metallurgical processes in steelmaking
12. Andrzejewski P, Kohler KU, Pluschkell W (1992) Model investigations on the fluid flow in continuous casting moulds of wide dimensions. Steel Res 63(6):242–246

Flow Control in the Model of a Continuous Caster by Using Contactless Inductive Flow Tomography

I. Glavinić, S. Abouelazayem, M. Ratajczak, D. Schurmann, S. Eckert, F. Stefani, J. Hlava and T. Wondrak

Abstract The global flow pattern of liquid metal in the slab mold of a continuous caster is difficult to control, as it cannot be measured in real-time by conventional methods. Contactless inductive flow tomography (CIFT) can easily provide real-time information about the flow structure (double or single roll) and the angle of the jet coming out of the submerged entry nozzle (SEN) just from the raw sensor data. Furthermore, by solving the underlying linear inverse problem, the full velocity field can be reconstructed. This paper discusses the possibility of applying CIFT for flow pattern recognition in continuous casting, which is then used for setting an electromagnetic brake in order to control the angle of the fluid jet. The control loop will be implemented and developed for the Mini-LIMMCAST model of a continuous caster at Helmholtz-Zentrum Dresden-Rossendorf (HZDR).

Keywords Continuous casting · Flow control · Inductive measurement techniques Electromagnetic brake

Introduction

The idea of continuous casting traces back to 1843 when J. Laing patented a machine for producing pipes of low melting metallic alloys [1]. It took more than a century and a half to reach the point where 96% of world's crude steel production stems from continuous casting [2]. Usually, a continuous caster comprises of a tundish that acts as buffer storage of liquid metal, from which the metal flows through a submerged entry nozzle (SEN) to the water-cooled mold. A stopper rod or a sliding gate is typically used for controlling the flow rate of steel and therefore the casting speed. Simultaneously with the process of improvement of casting technologies and

I. Glavinić (✉) · M. Ratajczak · D. Schurmann · S. Eckert · F. Stefani · T. Wondrak
Helmholtz-Zentrum Dresden-Rossendorf, Bautzner Landstraße 400, Dresden, Germany
e-mail: i.glavinic@hzdr.de

S. Abouelazayem · J. Hlava
Technical University of Liberec, Studentská 1402/2, Liberec, Czech Republic

G. Lambotte et al. (eds.), *Materials Processing Fundamentals 2019*, The Minerals, Metals & Materials Series, https://doi.org/10.1007/978-3-030-05728-2_5

measurement techniques, the control systems and algorithms advanced as well, but because of the opaqueness and the high temperature of the melt [which excludes, e.g., the use of Ultrasound Doppler Velocimetry (UDV)], contactless flow measuring techniques are highly desirable. Even a rough knowledge of the velocity distribution and flow pattern in the mold would be useful for ensuring the quality of the end product [3]. Along with more traditional methods, such as nail board dipping or refractory paddles for determining the velocity at the meniscus [3], Lorentz force velocimetry and contactless inductive flow tomography (CIFT) have also been discussed. The CIFT technology, which was developed during the last two decades, has tremendous potential for flow reconstruction of conductive fluids not only in the field of continuous casting but also in Czochralski crystal growth [4]. Here we present a short overview of CIFT, the laboratory model (demonstrator) of the continuous caster, and the initial proposal for the control loop, which will be realized in scope of the TOMOCON (Smart **Tomo**graphic Sensors for Advanced Industrial **Con**trol) project.

The paper is structured as follows. The first section gives a short summary of the theory of CIFT, the second section provides an overall approach to the control methodologies, and the final section provides the description of the Mini-LIMMCAST facility.

Contactless Inductive Flow Tomography

In the presence of a magnetic field **B**, a fluid with electrical conductivity σ, moving with the velocity **v**, will induce an electric field which, in turn, will give rise to an electric current. According to Ohm's law, the induced current density is

$$\mathbf{j} = \sigma(\mathbf{B} \times \mathbf{v} - \nabla\varphi). \tag{1}$$

This induced current will generate its own magnetic field **b** according to Biot–Savart's law:

$$\mathbf{b}(\mathbf{r}) = \frac{\mu_0}{4\pi} \iiint\limits_{V} \mathbf{j} \times \frac{\mathbf{r} - \mathbf{r}'}{|\mathbf{r} - \mathbf{r}'|^3} dV. \tag{2}$$

Here, μ_0 is the magnetic permeability of the vacuum which is a good approximation of the permeability of the liquid metal and $\mathbf{r}'$ denotes the position vector in the volume.

Because of conservation of charges,

$$\nabla \cdot \mathbf{j} = 0, \tag{3}$$

we can obtain the Poisson equation for the electrical potential

$$\nabla^2 \varphi = \nabla(\mathbf{v} \times \mathbf{B}) \tag{4}$$

and end up with an integral equation system that couples the velocity field **v** with the induced magnetic field **b** for a specific point **r** outside the fluid volume:

$$\begin{aligned}\mathbf{b}(\mathbf{r}) = {} & \frac{\mu_0 \sigma}{4\pi} \iiint\limits_V \left[\mathbf{v}(\mathbf{r}') \times \mathbf{B}(\mathbf{r}')\right] \times \frac{\mathbf{r} - \mathbf{r}'}{|\mathbf{r} - \mathbf{r}'|^3} dV' \\ & - \frac{\mu_0 \sigma}{4\pi} \oiint\limits_S \varphi(\mathbf{s}')\mathbf{n}(\mathbf{s}') \times \frac{\mathbf{r}' - \mathbf{s}'}{|\mathbf{r} - \mathbf{s}'|^3} dS',\end{aligned} \tag{5}$$

$$\begin{aligned}\varphi(\mathbf{s}) = {} & \frac{1}{2\pi} \iiint\limits_V \left[\mathbf{v}(\mathbf{r}') \times \mathbf{B}(\mathbf{r}')\right] \times \frac{\mathbf{s} - \mathbf{r}'}{|\mathbf{s} - \mathbf{r}'|^3} dV' \\ & - \frac{1}{2\pi} \oiint\limits_S \varphi(\mathbf{s}')\mathbf{n}(\mathbf{s}') \times \frac{\mathbf{s}' - \mathbf{s}'}{|\mathbf{s} - \mathbf{s}'|^3} dS'.\end{aligned} \tag{6}$$

$\mathbf{n}(\mathbf{s}')$ represents the normal vector on the position $\mathbf{s}'$, and dV and dS denote the volume and surface elements, respectively.

The magnetic field **B** under the integral Eqs. (5) and (6) is, in general, the sum of the applied (excitation) magnetic field $\mathbf{B}_0$ and the flow-induced magnetic field **b**. The ratio between the applied and induced magnetic field is governed by the magnetic Reynolds number

$$R_m = v l \mu_0 \sigma. \tag{7}$$

For most industrial applications the magnetic Reynolds number is below 1, which excludes the possibility of magnetic self-excitation. In the case of continuous casting, we can assume typical values of $v \approx 0.1$ m/s, $l \approx 1$ m and $\sigma \approx 7 \cdot 10^5$ S/m, from which we obtain $R_m \approx 0.1$. In this case, it is viable to replace the total magnetic field **B** with $\mathbf{B}_0$ in the Eqs. (5) and (6), which gives us a linear equation system [5].

The non-uniqueness of the resulting linear inverse problem was discussed in [6]. Yet, a unique solution can still be derived utilizing Tikhonov regularization together with the L-curve method. For further literature about the inverse problem, refer to [7, 8].

Interestingly, instead of reconstructing the entire velocity field, useful information can already be extruded from the raw measurement data of the induced magnetic field, particularly if the flow pattern in the mold is a single or double roll. If it is indeed the desired double roll, the impingement point of the jet along the narrow wall of the mold can be easily observed. For a double roll flow structure shown in Fig. 1, the flow-induced magnetic field takes a characteristic S shape where the zero-crossing point coincides with impingement point and the flow is split in the

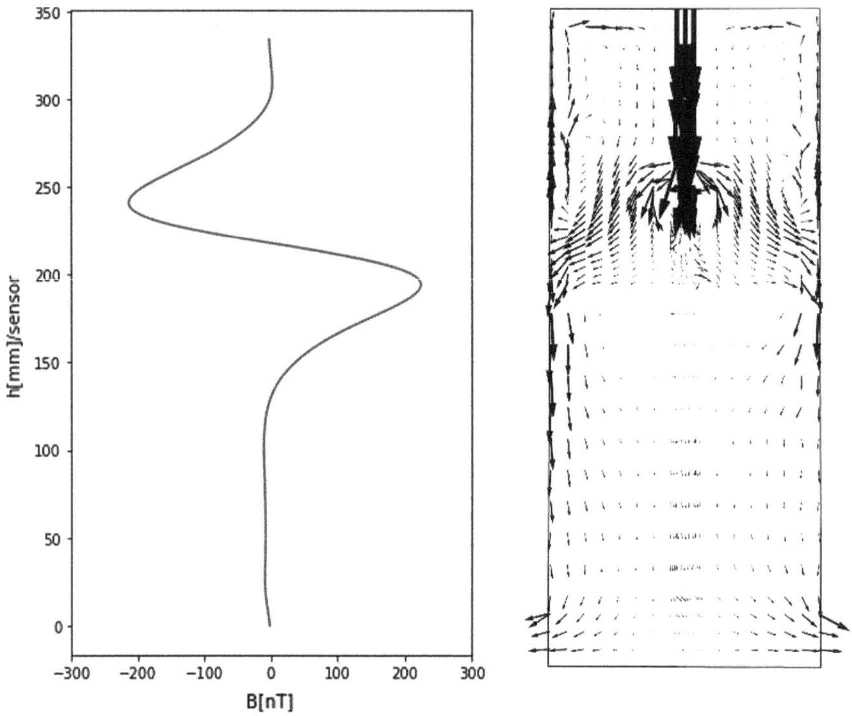

Fig. 1 Simulated sensor array measurements on the left side of the mold (left), reconstructed velocity field in the middle plane of the mold (right)

upper and lower roll. While not shown here, the magnetic field induced by the single roll flow, defined by the liquid metal meniscus velocities directed outward, lacks the *S* shape curve as the jet is not hitting the narrow wall with the downward angle but with an upward angle [9].

The simulated induced magnetic field is calculated from the velocity field generated by CFD simulation, and then solving Eq. (5), where the magnetic field $\mathbf{B}$ is the magnetic field $\mathbf{B}_0$ generated by a rectangular coil. A simulated measurement at the narrow face of the mold is shown in Fig. 1 (right). From this virtual measurement, the inverse problem is solved, and we obtain the reconstructed velocity field in the mold as shown in Fig. 1 (left).

Figure 2 shows the schematic representation of the CIFT measurement system in the Mini-LIMMCAST facility. Mini-LIMMCAST is a laboratory model of a continuous slab caster and can be equipped with various measuring techniques such as UDV and CIFT. The CIFT measurement system consists of two excitation coils generating the magnetic field strength of approximately 2 mT in the vertical direction. Seven gradiometric pick-up coils [10] on each side of the mold measure the

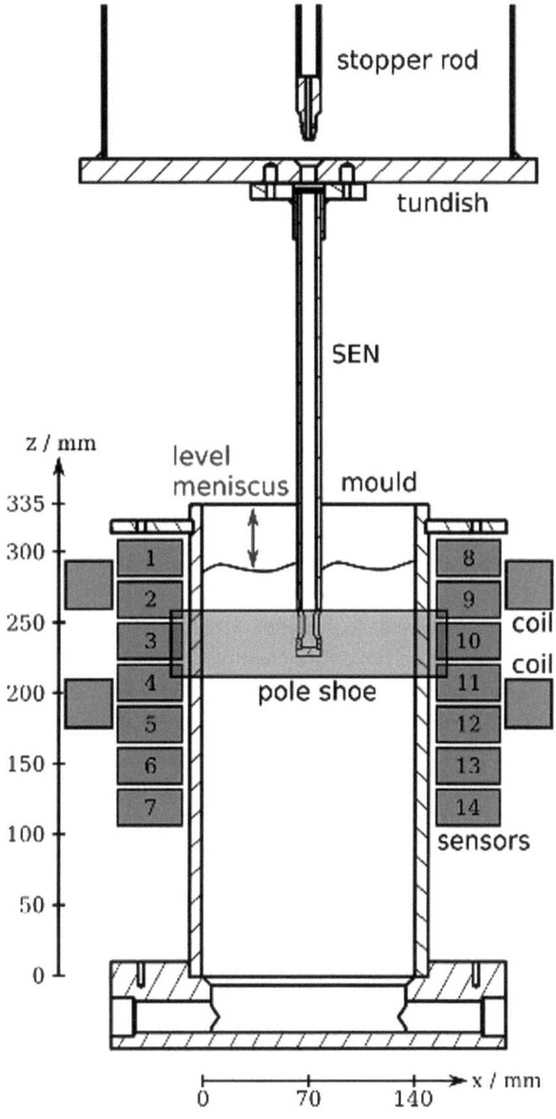

Fig. 2 Schematic sketch of the mini-LIMMCAST mold

flow-induced magnetic field. The flow velocity can be regulated by the position of the stopper rod. By positioning the pick-up coils on either side of the mold, we can observe irregularities in the normally symmetric double roll flow, and obtain a more precise reconstruction of the velocity field.

Control of Liquid Metal Flow

It is well known that the flow pattern in the mold has a severe impact on the quality of the final product [11]. The single roll flow is characterized by high meniscus speeds which increase the likelihood of entrapping slag or casting powder from the surface. The double roll flow profile is generally more suitable for steel slab casting. The flow structure can be controlled by adjustable parameters such as the immersion depth of the nozzle, the nozzle geometry (which can be only changed in between the production cycles), or by using electromagnetic actuators, such as electromagnetic brakes (EMBr) or electromagnetic stirrers (EMS) [12]. One such electromagnetic brake is a part of our demonstrator installation: it creates a uniform magnetic field perpendicular to the wide wall approximately at the height of the SEN's outlet and acting along the entire length of the mold (Fig. 2).

The effect of this electromagnetic actuator on the fluid flow has been modelled and analyzed frequently [13, 14]. A conductive fluid moving in the magnetic field of the EMBr will experience Lorentz forces that brake the fluid. Both experimental and numerical simulations have shown that the electromagnetic brake modifies the exit angle of the jet [15] and affects the fluid flow below the meniscus [16]. It is obvious that a concise control of these actuators during casting requires a robust and reliable flow measurement technique, which is able to monitor the actual flow structure in the mold during the entire casting sequence. Because of the lack of contactless flow measurement techniques, only a few papers exist which investigate the design of a closed control loop for electromagnetic actuators for continuous casting of steel. Dekemele et al. investigate the feasibility of a control loop based on measuring the velocity below the meniscus at one single location using numerical simulations. A refractory paddle was proposed as a measuring device, which might be difficult to utilize during the normal operation of the caster [17]. Therefore, we intend to utilize CIFT which allows us to reconstruct the entire flow structure in the mold. In a first attempt, the angle of the fluid jet exiting the SEN is controlled by varying the electromagnetic field of the EMBr.

One of the main challenges of implementing a tomography-based control loop is the integration of the data from the tomographic sensors in a manner which would enable the controller to calculate a suitable control action based on this data. The nature of the data provided by tomographic sensors is both spatial and temporal. However, classical control theory depends on the concept of lumped parameter systems. Therefore, various data analysis methods need to be considered to select the method which is most suitable for our process. The data analysis methods applied to process tomography can be categorized under two methods: raw measurement data, and the visualization of the process in the form of 2D or 3D images [18]. In this paper, we will utilize the magnetic flux profile as seen in Fig. 1 as the raw data for our control loop. This will allow us to avoid the delay needed for the reconstruction of the full velocity profile, as well as the degree of uncertainty. This approach has two advantages. First, it will allow us to avoid the delay arising because of the time needed for the reconstruction of the full velocity image. Second, it will eliminate the errors in the computed values that are induced by measurement and modelling inaccuracies, which may be quite significant because of the ill-posedness of the inverse problem. As mentioned above, the zero-crossing point of the *S* shape of the magnetic flux correlates with the jet impingement point at the narrow wall, therefore the jet angle can be easily calculated from this point. In the first step, instead of using the reconstructed velocity profile, we plan to use the induced magnetic field as input for the controller. The corresponding reduction of complexity from a distributed parameter system to a simpler lumped parameter system allows using simple PID controllers to properly adjust the flow, e.g. by setting the brake current to adjust the angle of the jet or compensate jet oscillations.

Additionally, the control strategy will be developed further to incorporate the full velocity field obtained by CIFT. This will serve to provide richer information for the controller. Control strategies based on distributed parameter systems need to be considered in this case, as conventional control techniques such as PID controllers will no longer be viable for the control objective. One of the main challenges here is the incorporation of the tomographic data into the control loop. Due to the above-mentioned issues with the inverse problem, specific velocity field reconstruction strategies have to be tested in order to facilitate the development of an implementable controller. In particular, we will consider the state estimation-based approach. This approach starts from state space modelling and is widely used in the system and control theory. Among its other advantages, it is a natural representation of the discretized distributed parameter models. The state space model of a dynamical system consists of the evolution model for the time-varying state variables and the observation model that relates the observations (in this case tomography data) to the state variables. Standard techniques like Kalman or extended Kalman filtering can then be used to obtain estimates of the state variables [19]. Many advanced control methods, such as Model-Predictive Control (MPC), are also often formulated using state space models, hence this approach will also streamline their application to the control of the continuous casting process, with the objective of achieving the optimal flow pattern in the mold.

Laboratory Setup

For the purpose of designing the control loop, experiments are currently carried out on a laboratory model of the continuous caster (Mini-LIMMCAST). Mini-LIMMCAST [20] has been recently equipped with the 1:5 scale of a typical mold and SEN. The cross section of the mold is 300×35 mm^2, the SEN outer diameter is 21 mm, the inner diameter 12 mm, and the height of the outlet ports of 13 mm. As the model of the liquid steel, eutectic alloy of Gallium-Indium-Tin (GaInSn) is being used, which is nontoxic, and liquid at room temperature. Figure 3 (left) shows the EMBr with the mold, and Fig. 3 (right) shows the mold filled with GaInSn.

Mini-LIMMCAST is equipped with a water-cooled electromagnetic brake that can produce a magnetic field up to 400 mT with the maximum current of 600 A. To

Fig. 3 EMBr (left), mold with GaInSn (right)

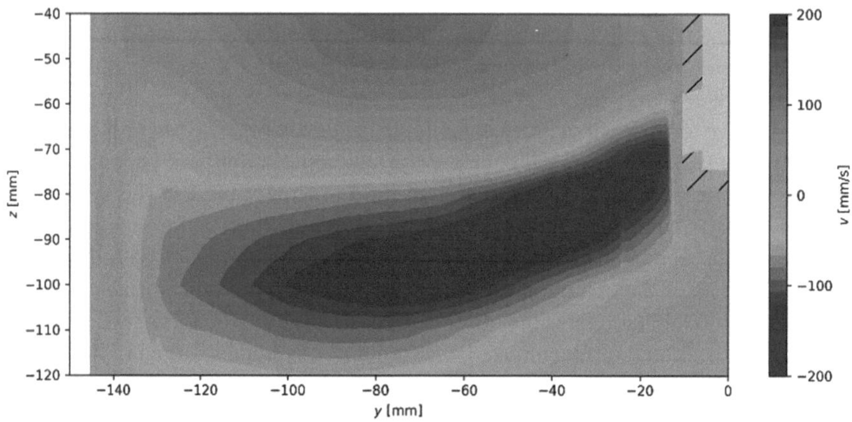

Fig. 4 Horizontal components of the velocity in the mold, UDV measurements

find the optimal position of the brake in relation to the SEN, the vertical position of the brake can be easily adjusted by using the movable table. Additionally, the flow rate of the liquid metal can be adjusted by the height of the stopper rod, while the level of the liquid metal can be set by changing the height of the overflow.

Figure 4 provides a result of one of the first measurements on the new setup and represents the time average of the horizontal components of the velocities in the mold.

The measuring system for CIFT consists of two excitation coils of 6–8 windings with the current of 30 A, giving an excitation field of approximately 2 mT in the vertical direction. The sensor system comprises 14 gradiometric coils, 7 on each narrow face of the mold, that measure the magnetic field induced by the flow of the liquid metal. The sketch of the CIFT system is shown in Fig. 5.

Fig. 5 Drawing of the CIFT excitation coil and sensor array

In order to gather enough data for the system identification of the controller, a series of measurements with different flow rates, heights of the brake, and strength of the magnetic field, are presently conducted using UDV and CIFT techniques. The measurements will be carried out for cases with conductive and nonconductive walls, and in the stationary and transient regime. These numerous experiments will not only serve to gather model data for the controller but also foster a better understanding of the jet behavior, and of the flow structure in the mold of the continuous caster.

Conclusions

By using the real-time measurements of CIFT, it is possible to identify the flow in the mold of a continuous caster. Knowing the flow patterns in the mold enables the application of different control mechanisms in order to influence the flow structure by adjusting the strength of the EMBr. First, we will focus on the angle of the jet as a process variable to control. Later, we plan to extend our research to the application of advanced control methodologies, such as model-predictive control, depending on the full reconstructed 3D velocity field, and on the two-phase flow in the SEN. The complexity of the flow structure and the lack of adjustable parameters present a challenge for the controller design. The feasibility of this approach is being tested and will be implemented in the model of a continuous caster.

Acknowledgements This project has received funding from the European Union's Horizon 2020 research and innovation program under the Marie Skłodowska-Curie grant agreement No 764902 (TOMOCON—www.tomocon.eu).

References

1. Laing J (1843) US Patent 3023
2. World Steel Association (2017) Steel statistical yearbook 2017
3. Dauby PH (2012) Continuous casting: make better steel and more of it! Rev Métall 109:113–136. https://doi.org/10.1051/metal/2012011
4. Wondrak T, Pal J, Stefani F et al (2018) Visualization of global flow structure in a modified Rayleigh-Bénard setup using contactless inductive flow tomography. Flow Meas Instrum 62:269–280
5. Ratajczak M, Gundrum T, Stefani F, Wondrak T (2014) Contactless inductive flow tomography: brief history and recent developments in its application to continuous casting. J Sens 2014:1–9. https://doi.org/10.1155/2014/739161
6. Stefani F, Gerbeth G (2000) On the uniqueness of velocity reconstruction in conducting fluids from measurements of induced electromagnetic fields. Inverse Prob 16:1. https://doi.org/10.1088/0266-5611/16/1/301
7. Stefani F, Gundrum T, Gerbeth G (2004) Contactless inductive flow tomography. Phys Rev E 70. https://doi.org/10.1103/PhysRevE.70.056306
8. Wondrak T, Stefani F, Gundrum T, Gerbeth G (2009) Some methodological improvements of the contactless inductive flow tomography. Int J Appl Electromagnet Mech 30:255–264. https://doi.org/10.3233/JAE-2009-1026
9. Wondrak T, Galindo V, Gerbeth G et al (2010) Contactless inductive flow tomography for a model of continuous steel casting. Meas Sci Technol 21:045402. https://doi.org/10.1088/0957-0233/21/4/045402
10. Ratajczak M, Wondrak T, Stefani F (2016) A gradiometric version of contactless inductive flow tomography: theory and first applications. Philos Trans A Math Phys Eng Sci 374. https://doi.org/10.1098/rsta.2015.0330
11. Zhang L, Thomas BG (2003) State of the art in evaluation and control of steel cleanliness. ISIJ Int 43:271–291. https://doi.org/10.2355/isijinternational.43.271
12. Gerber HL (1997) Electromagnetic processing of liquid steel. IEEE Trans Ind Appl 33:801–806. https://doi.org/10.1109/28.585873
13. Haiqi Y, Baofeng W, Huiqin L, Jianchao L (2008) Influence of electromagnetic brake on flow field of liquid steel in the slab continuous casting mold. J Mater Process Technol 202:179–187. https://doi.org/10.1016/j.jmatprotec.2007.08.054
14. Cukierski K, Thomas BG (2008) Flow control with local electromagnetic braking in continuous casting of steel slabs. Metall Mater Trans B 39:94–107. https://doi.org/10.1007/s11663-007-9109-3
15. Timmel K, Miao X, Lucas D et al (2010) Experimental and numerical modelling of the steel flow in a continuous casting mould under the influence of a transverse DC magnetic field. Magnetohydrodynamics 46:437–448
16. Harada H, Toh T, Ishii T et al (2001) Effect of magnetic field conditions on the electromagnetic braking efficiency. ISIJ Int 41:1236–1244. https://doi.org/10.2355/isijinternational.41.1236
17. Dekemele K, Ionescu C-M, De Doncker M, De Keyser R (2016) Closed loop control of an electromagnetic stirrer in the continuous casting process. 2016 European Control Conference (ECC). IEEE, Aalborg, pp 61–66
18. Romanowski A, Grudzien K, Williams RA (2005) A review of data analysis methods for electrical industrial process tomography applications. 4th World Congress on Industrial Process Tomography 916–921
19. Seppänen A, Voutilainen A, Kaipio JP (2009) State estimation in process tomography—reconstruction of velocity fields using EIT. Inverse Prob 25(8):085009
20. Timmel K, Eckert S, Gerbeth G et al (2010) Experimental modeling of the continuous casting process of steel using low melting point metal alloys—the LIMMCAST program. ISIJ Int 50:1134–1141. https://doi.org/10.2355/isijinternational.50.1134

Optimization of the Flow Behavior of Molten Steel in Ultrahigh-Speed Billet Continuous Casting Mold

Pei Xu, Dengfu Chen, Shixin Wu, Hengsong Yu, MuJun Long, Sheng Yu and Huamei Duan

Abstract Ultrahigh casting speed is an important tendency to improve the efficiency of continuous casting. A three-dimensional mathematical model and a hydraulic physical model on the billet mold with 160 × 160 mm cross section were established to investigate the flow behavior of molten steel with different SEN conditions and optimize the parameters of SEN at the casting speed of 6.0 m/min. Results indicate that when the immersion depth and the inner diameter of the SEN are 180 and 50 mm, respectively, the flow field and the surface velocity distribution in the mold are the most appropriate that the penetration depth of the stream is about 700 mm and the maximum surface velocity is 0.05 m/s. With the optimum parameters of SEN, the slag covers uniformly and keeps appropriately active, and no slag entrainment happens. Moreover, the differences are very slight between the results of the numerical and physical simulation, which can verify each other.

Keywords Ultrahigh-speed continuous casting · Fluid flow
Numerical simulation · Hydraulic simulation · SEN parameters

P. Xu · D. Chen (✉) · S. Wu · H. Yu · M. Long · S. Yu · H. Duan
College of Materials Science and Engineering, Chongqing University, Chongqing 400044, China
e-mail: chendfu@cqu.edu.cn

P. Xu
e-mail: xupeibest@126.com

M. Long
e-mail: longmujun@cqu.edu.cn

P. Xu · D. Chen · S. Wu · H. Yu · M. Long · S. Yu · H. Duan
Chongqing Key Laboratory of Vanadium-Titanium Metallurgy and New Materials, Chongqing University, Chongqing 400044, People's Republic of China

G. Lambotte et al. (eds.), *Materials Processing Fundamentals 2019*, The Minerals, Metals & Materials Series, https://doi.org/10.1007/978-3-030-05728-2_6

Introduction

In recent years, the entire steel industry is in a slump, and its cost of production is relatively high and efficiency is low. For improving this condition, High efficiency continuous casting is put forward, in which high-speed casting is an effective means. Especially to the billet produced by multi-flow continuous casting, high-speed casting can greatly increase productivity and decrease cost. At present, the maximum casting speed generally is about 3.0–4.0 m/min [1–3], but ultrahigh speed hardly is researched.

Compared with conventional casting speed, there exists an obvious difference in ultra high-speed continuous casting, especially to the mold. As the heart of the entire continuous casting process, a series of complex physicochemical coupling behaviors will occur inside it, and the flow behaviors of mold molten steel are crucial because it will affect the transport of inclusions, multi-phase flow phenomena and the effect heat transfer. To the flow behaviors of mold molten steel, some scholars, for example, B. G. Thomas, J. E. Mika, etc. [4–6] conduct relevant numerical simulation about the effect of the SEN (submerged entry nozzle), solidified shell and blowing, etc. on the flow behaviors of mold molten steel by establishing a three-dimensional mathematical model and obtain good calculation results, and these make people further understand the flow characteristics of molten steel. Others scholars, such as Miaoyong Zhu, Dengfu Chen, etc. [7–11] research the influence of different casting speed, the SEN structure and parameter on the flow behaviors of mold molten steel by numerical simulation, and also conduct corresponding physical hydraulic experiment to verify numerical simulation. Although these studies are very effective, they are basically concentrated on the conventional casting speed, and the research on the flow field of ultrahigh-speed continuous casting mold is still lacking.

Therefore, to the fluid flow behaviors of the billet mold with 160 × 160 mm cross section at the casting speed of 6.0 m/min, this paper conducts numerical simulation and physical hydraulic experiment, and both consequents can verify each other. So that the understanding of the flow field in the mold of ultra high-speed billet continuous casting has been strengthened, which is of certain significance to the realization of ultra high-speed billet continuous casting that its range is above 5.0 m/min generally.

Model and Method

Mathematical Model

(a) Assumptions

In the mathematical model, the liquid steel in mold was assumed to be a three-dimensional, steady-state, incompressible Newtonian fluid and regard as homogenous phase. The strand bulging, oscillation, mold taper, air gap, meniscus fluctuation and flux slag were neglected.

(b) Fluid and Turbulence Model

The liquid steel flow in the liquid pool was achieved through solving the continuity and momentum equations. The popular K-ε turbulent model [7] was chosen. The equations are

$$\rho \frac{\partial(\rho u_j)}{\partial x_i} = 0 \quad (1)$$

$$\rho \frac{\partial u_i u_j}{\partial x_j} = \frac{\partial}{\partial x_j}\left[\mu_{eff}\left(\frac{\partial u_i}{\partial u_x} + \frac{\partial u_j}{\partial x_i}\right)\right] - \frac{\partial P}{\partial x_i} + S_P \quad (2)$$

where u_i and u_j is the velocity component, x_i and x_j is the spatial coordinate, ρ is the density of liquid steel, P is the pressure, μ_{eff} is the effective viscosity, and S_P is the sink term of velocity.

(c) Solidification and Heat Transfer Model

The governing equation for heat transfer and solidification is

$$\rho u_i \frac{\partial H}{\partial x_i} = \frac{\partial}{\partial x_i}\left(\lambda_{eff}\frac{\partial T}{\partial x_i}\right) \quad (3)$$

where λ_{eff} is the effective thermal conductivity, T is the temperature, and H is the enthalpy of steel, which is the sum of the sensible enthalpy and latent heat.

(d) Computational Domain

Due to the symmetry of fluid flow in the mold, only a 1/2 of the strand was modelled to minimize computation. The casting parameters are shown in Table 1. Figure 1 is the mesh of the strand. The strand was meshed using about 390,000 hexahedral cells.

Table 1 The casting parameters of mold

Operating parameters	Values
Mold section	160 mm × 160 mm
Mold length	1000 mm
Mold radius	8000 mm
Clearance height	70 mm
Computational length	1930 mm
SEN type	Straight through nozzle
Casting speed	6.0 m/min
Superheat	13 K

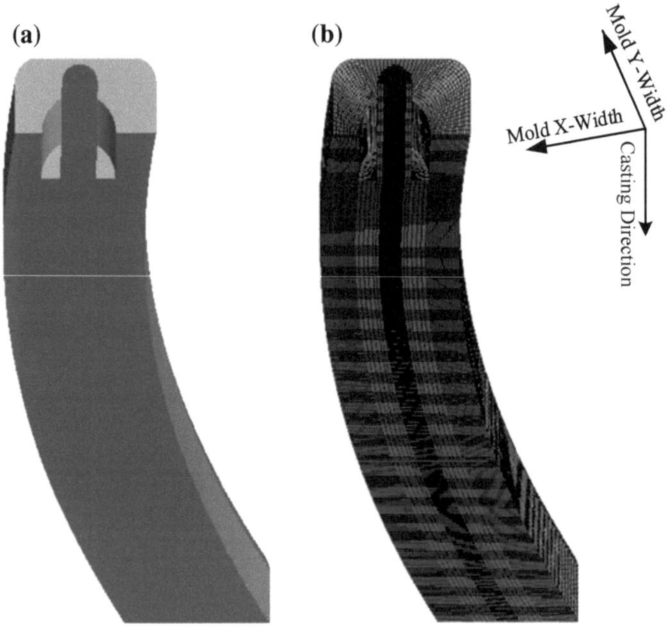

Fig. 1 The schematic diagram of mold model **a** the geometrical model, **b** the mesh model

(e) Boundary Conditions

At the inlet of SEN, the velocity-inlet boundary condition was used. The inlet velocity was calculated through the mass conservation based on the casting speed. The inlet value of turbulent kinetic energy and the rate of turbulent energy dissipation were 10^{-5} [9], respectively. The pressure-outlet boundary condition was applied at strand exit.

The top free surface was set as zero shear stationary wall and heat insulation. At the symmetry plane, the normal velocity components and normal gradients of all other variables were assumed to be zero. At the strand surface, the heat transfer boundary condition was employed the heat flux, which decreased along the casting direction.

(f) Physical Properties

In the simulation, the physical properties of steel are presented in Table 2.

Physical Model

Physical simulation is a simulation method based on similarity principle. In the process of fluid flow simulation, it is impossible to make the model completely

Table 2 The physical properties of steel

Physical properties	Values
Density, kg/m^3	7200
Specific heat, J $(kg\ K)^{-1}$	657
Thermal conductivity coefficient, W $(m\ K)^{-1}$	58.4545–0.0165 × T(K)
Viscosity, kg $(m\ s)^{-1}$	0.0062
Latent heat, J/kg	264,500
Solidus temperature, K	1732
Liquidus temperature, K	1790

similar to the actual process because of the complexity of the actual process. In general, only considering the dominating similarity principle. For the flow behavior of molten steel in mold, geometric similarity and dynamic similarity are two basic similarity conditions and need considering.

In order to ensure dynamic similarity, the Reynolds Number and Froude Number between the prototype of mold and physical hydraulic model are required to be equal, respectively. However, it is very difficult to satisfy that the two similar criteria of the prototype and the model are equal at the same time. Considering the experimental condition, the fluid flow state in the model is in the second self-modeling region. Therefore, the Reynolds Number can be neglect, and only that ensuring the gravity-related Froude Number equal is required to the dynamic similarity.

When the Froude Number is equal, geometric similarity can select any proportion. Based on the need of precise simulation and observation of the fluid flow in the physical hydraulic mold model, the larger the proportion of geometric similarity is, the better simulation effect is. This paper adopts 1:1 similarity proportion between the prototype and the model. Figure 2 is the physical model of the billet mold.

Optimization Method

Via adjusting SEN relevant parameters, this paper accomplishes flow behavior optimization of molten steel in 6.0 m/min casting speed billet mold. The specific process is as follows.

(1) Selecting the SEN of inner diameter 40 mm, numerical simulation simulates molten steel flow in billet mold when the SEN immersion depth is 120, 140, 160, 180 and 200 mm, respectively, and analyzing simulation results. Afterwards gaining the best SEN immersion depth H that optimized flow field corresponds.
(2) Selecting the SEN of immersion depth H, numerical simulation simulates molten steel flow in billet mold when the SEN inner diameter is 40, 45, 50 and

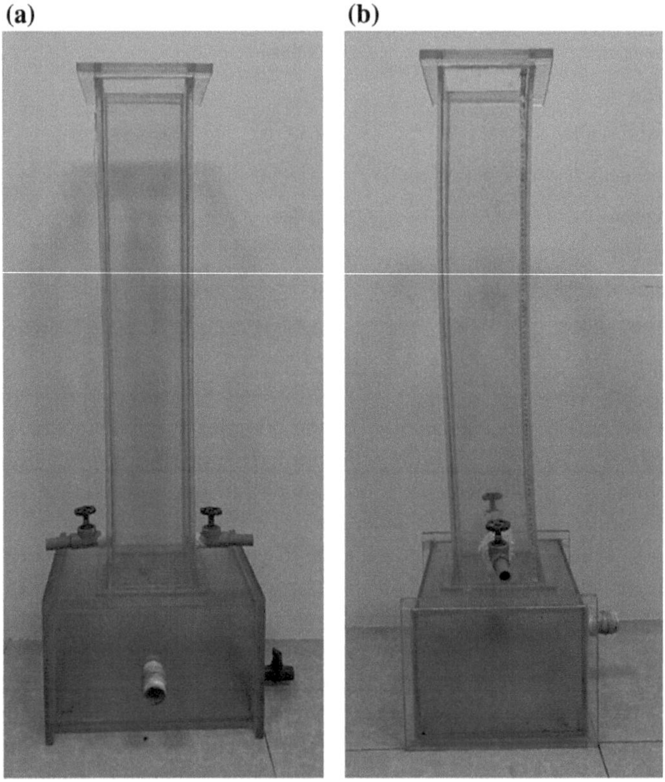

Fig. 2 The physical model of the billet mold **a** front view, **b** side view

55 mm, respectively, and analyzing simulation results. Afterward gaining the best SEN inner diameter *R* that optimized flow field corresponds.

(3) According to numerical simulation, the best SEN parameter, conducting physical hydraulic experiment and then comparing the corresponding results with numerical simulation ones.

Results

(1) Selecting the SEN of inner diameter 40 mm, numerical simulation simulates molten steel flow in billet mold when the SEN immersion depth is 120, 140, 160, 180 and 200 mm, respectively. Figure 3 is a schematic diagram that shows the flow distribution of mold central symmetry plane. In Fig. 3a, this paper thinks the distance from the isoline's bottom that its velocity value equal to the casting speed to meniscus is defined the stream impact depth. Figure 4 shows the impact depth of five SEN immersion depths. Except for the flow field of

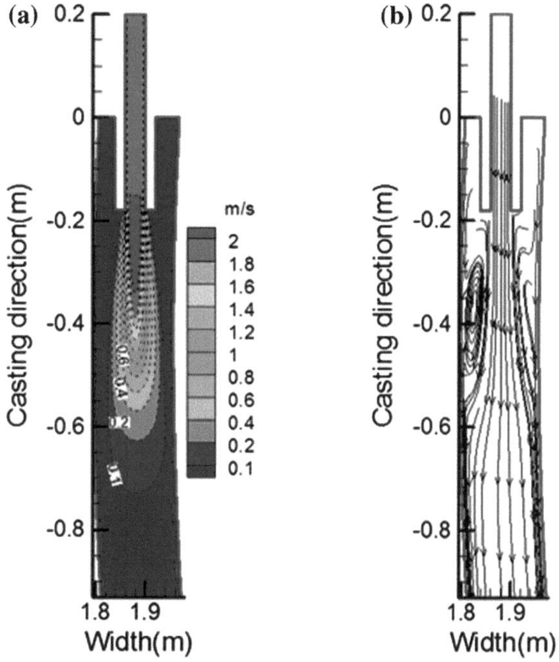

Fig. 3 Flow field of central symmetry plane **a** cloud picture, **b** streamline picture

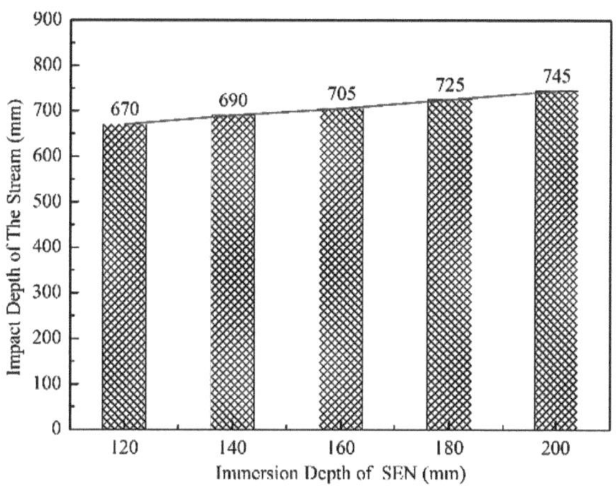

Fig. 4 Impact depth of five SEN immersion depths

mold central symmetry plane, liquid surface velocity field is an also important optimized aspect. Figure 5 shows the mold liquid surface velocity distribution of five SEN immersion depths.

(2) Fixing the SEN immersion depth 180 mm, numerical simulation simulates molten steel flow in billet mold when the SEN inner diameter is 40, 45, 50 and 55 mm, respectively. Figure 6 shows mold liquid surface velocity distribution of four SEN inner diameters.

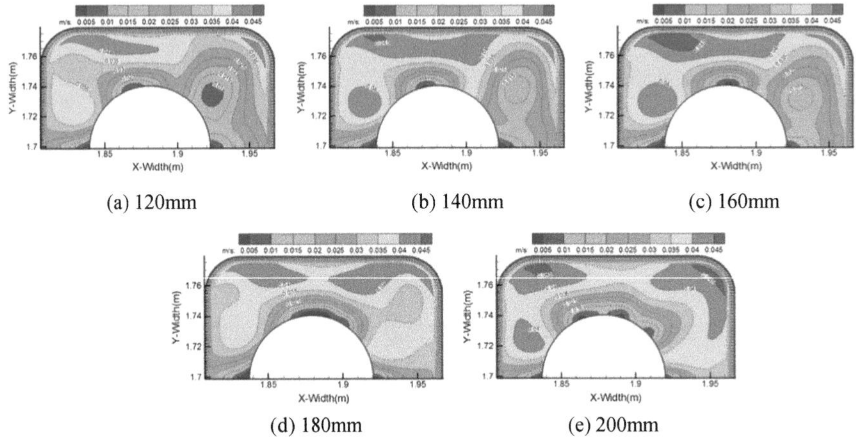

Fig. 5 Liquid surface velocity field of five SEN immersion depths

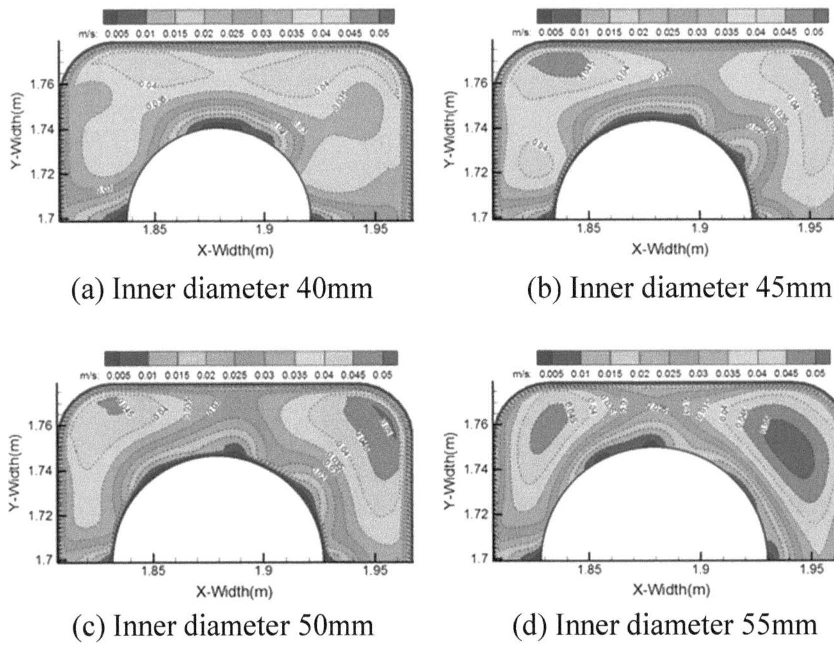

Fig. 6 Liquid surface velocity field of four SEN inner diameters

(3) By adopting the SEN of inner diameter 50 mm and immersion depth 180 mm, the physical hydraulic experiment was conducted, and then compared with mathematical simulation. Figure 7 is mold flow field comparison between the physical simulation and the mathematical simulation. To study the fluid flow of mold liq-

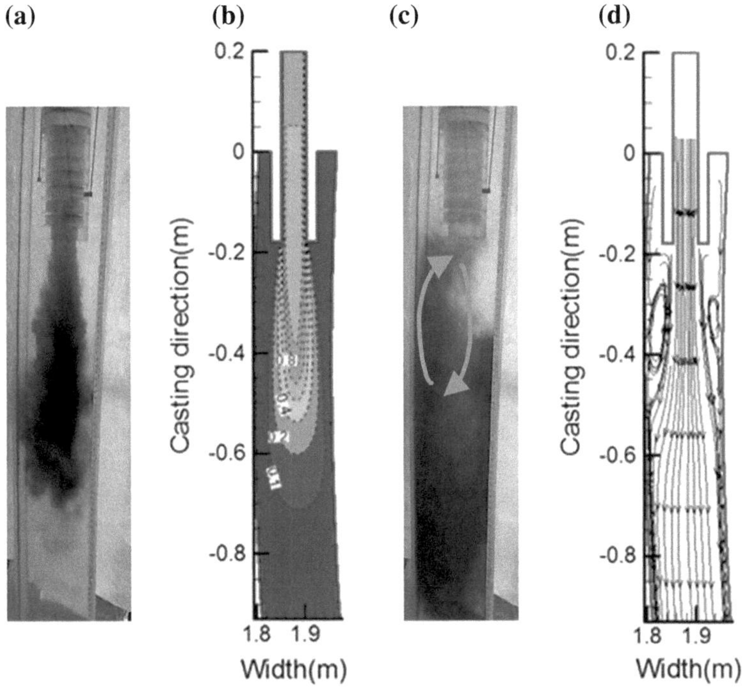

Fig. 7 Mold flow field comparison between the physical simulation and the mathematical. **a** Physical simulation, **b** cloud picture, **c** physical simulation, **d** streamline picture

(a) Overhead view (b) Upward view

Fig. 8 Liquid slag distribution on mold liquid surface

uid surface, the liquid slag physical simulation experiment was conducted under the best SEN parameter. Figure 8 shows liquid slag distribution on mold liquid surface.

Discussion

The Study on Immersion Depth of SEN

Figure 3a shows that flow fields of different immersion depths are similar and that the velocity of the main stream becomes weak gradually along the casting direction. Figure 3b shows the molten steel flow forms circuiting zone around inner and outer arcs, and obviously the range and strength of outer arc's one is bigger than inner arc's. It is because of billet mold's arc structure so that the stream is closer to inner arc and the fluid flow room around it gets smaller than outer arc. In Fig. 4, as the SEN immersion depth becomes deep from 120 to 200 mm, the stream impact depth increases gradually and increment is 75 mm. Due to the immersion depth will influence inclusions floatation and mold meniscus fluctuation, so it must keep an appropriate value, otherwise goes against producing high-quality products. To optimize liquid surface velocity field, its uniform, symmetry, velocity gradient and maximum velocity value are mainly considered In Fig. 5. The maximum velocity of five SEN immersion depths is all about 0.04–0.045 m/s, and it can let mold liquid surface keep appropriately active, contributing to flux slag melting. When immersion depths are 120, 140 and 160 mm, their uniform and symmetry are bad and there exists local liquid surface velocity to be relatively small around the inner arc, going against flux slag distribution. When being in 180 and 200 mm, the symmetry gets better than ones, but the local velocity around SEN is small and its gradient is great, and they can result in slag-bonding and influence product quality. On the whole, the velocity field of 180 mm SEN immersion depth is the most appropriate. Based on above-mentioned analyses, the best immersion depth that optimized flow field corresponds is 180 mm.

The Study on Inner Diameter of SEN

Fixing the SEN immersion depth 180 mm, numerical simulation simulates molten steel flow in billet mold and the basic fluid flow rules of central symmetry are similar to Section "The Study on Immersion Depth of SEN". Adopting the same impact depth definition, the stream impact depths of four kinds of SEN inner diameters are about 720 mm, and there is a little difference among them. In Fig. 6 the maximum velocity of four SEN inner diameters is all about 0.04–0.05 m/s, and it can let mold liquid surface keep appropriately active, contributing to flux slag melting. When inner diameters are 40 and 45 mm, their uniform are bad and there exists local liquid surface velocity to be relatively small around the SEN, and they can result in slag-bonding and influence product quality. When being in 55 mm, its velocity gradient is great, going against flux slag distribution. On the whole, the velocity field of 50 mm SEN inner diameter is the most appropriate. Based on above-mentioned analyses, the best inner diameter that optimized flow field corresponds is 50 mm.

The Physical Hydraulic Simulation Experiment

In the physical hydraulic experiment, mold flow field was visual by adding blue ink into water. Figure 7a, b indicates that mold flow field is similar between two simulation methods that the stream from SEN port spreads around gradually along the casting direction. In Fig. 7a, this paper thinks the distance from the stream bottom that it spreads to just touch the mold shell to meniscus is defined the stream impact depth. The impact depths of two simulations are all about 700 mm, and the mathematical is slightly larger but it is within the allowable range of error. Figure 7c, d shows that after the fluid pours into billet mold, circuiting zone will form around inner and outer arc. However, the circuiting zone around inner is little and not easy to observe. Moreover, the effect of the fluid mixture is fine and the flow field is uniform under the best SEN parameter. In Fig. 8, mechanical pump oil was used to simulate flux slag. Via observing 5 min, the liquid slag on the mold liquid surface is not only uniform distribution and there is no slag entrapment, but also can keep properly active.

Conclusions

In this work, a numerical simulation and a physical hydraulic simulation were conducted to make sure the best SEN parameter by optimizing billet mold flow field. Conclusions are as follows:

(1) As the SEN immersion depth becomes deep from 120 to 200 mm, the stream impact depth increases gradually and increment is 75 mm. The maximum velocity of five SEN immersion depths is all about 0.04–0.045 m/s. and the best immersion depth that optimized flow field corresponds is 180 mm.
(2) The stream impact depth of four kinds of SEN inner diameters is about 720 mm. The maximum velocity of four SEN inner diameters is all about 0.04–0.05 m/s. and the best inner diameter that optimized flow field corresponds is 50 mm.
(3) The impact depths of two simulations are all about 700 mm, and their difference is within the allowable range of error. The liquid slag on the mold liquid surface is not only uniform distribution and there is no slag entrapment but also can keep properly active.

Acknowledgements The work is financially supported by the Natural Science Foundation of China, Project No. 51374260 and 51504048.

References

1. Chow C, Samarasekera IV (2013) High speed continuous casting of steel billets: Part 1: General overview. Ironmaking Steelmaking 29(1):53–60
2. Chow C, Samarasekera IV, Walker BN, Lockhart G (2013) High speed continuous casting of steel billets: Part 2: Mould heat transfer and mould design. Ironmaking Steelmaking 29(1):61–69
3. Li C, Thomas BG (2002) Maximum casting speed for continuous cast steel billets based on sub-mold bulging computation. Paper presented at 85th Steelmaking Conference, Nashville, 10–13 March, 2002
4. Thomas BG, Mika LJ, Najjar FM (1990) Simulation of fluid flow inside a continuous slab-casting machine. Metall Trans B 21(2):387–400
5. Lait JE, Brimacombe JK, Weinberg F (1974) Mathematical modeling of heat flow in the continuous casting of steel. Ironmaking Steelmaking 1(2):90–97
6. Ho YH, Chen CH, Hwang WS (2007) Analysis of molten steel flow in slab continuous caster mold. ISIJ Int 34(3):255–264
7. Zhu M (1997) Numerical simulation of the coupled molten steel flow and heat transfer in continuous casting tundishes. Acta Metall Sin 33(9):933–938
8. Ren B, Zhu M, Wang H, Chen Y (2008) 3D numerical simulation of electromagnetic field and flow field in bloom continuous casting mold with electromagnetic stirring. Acta Metall Sin 44(4):507–512
9. Zheng S (2006) Water model study on removing inclusion in a ladle with argon injected through nozzle and porous plug. Acta Metall Sin 42(11):1143–1148
10. Jin X, Chen DF, Xie X, Shen J, Long M (2013) Investigation on water model for fluid flow in slab continuous casting mold with consideration of solidified process. Steel Res Int 84(1):31–39
11. Zhang L, Yang S, Cai K, Jiying LI, Wan X, Thomas BG (2007) Investigation of fluid flow and steel cleanliness in the continuous casting strand. Metall Mater Trans B 38(1):63–83

Part II
Steel—Microstructure and Properties

A New Alloy System Having Autogenous Grain Pinning at High Temperature

Tihe Zhou, Hatem S. Zurob and Ronald J. O'Malley

Abstract This contribution proposes a new alloy in which a small volume fraction of austenite particles is used to pin ferrite grain growth at high temperatures. During the reheating process, when the temperature is higher than 1200 °C, the coarsening of austenite particles is driven by volume-diffusion-controlled behaviour and ferrite grain growth is dominated by the pinning effect of austenite particles. At low temperature (<1280 °C), grain growth occurred at a rate which is proportional to the particle coarsening rate; while at high temperature (>1280 °C), grain growth is much lower than that expected without pinning. During the solidification process, austenite particles nucleate along ferrite grain boundaries and retard grain growth. Grain growth can be completely arrested with more austenite particle precipitates. This new alloy can be applied to control grain coarsening in the thin slab casting direct rolling process, grain size control in the HAZ of welds and grain growth resistance at high temperature.

Keywords High temperature · Particle coarsening · Particle pinning Applications

Introduction

It has been well established that, of all strengthening mechanisms, only finer grains can improve both strength and toughness in microalloyed steels [1, 2]. In industrial processes, one of the most common methods to refine grain size is to use carbides/nitrides of Ti, Nb and V to limit grain growth at high temperatures [3, 4].

T. Zhou (✉) · H. S. Zurob
Department of Materials Science and Engineering, McMaster University, 1280 Main Street West, Hamilton, ON L8S 4L7, Canada
e-mail: tom.zhou@stelco.com

R. J. O'Malley
Department of Materials Science and Engineering, Missouri University of Science and Technology, 1400 N. Bishop Ave, Rolla, MO 65409-0330, USA

G. Lambotte et al. (eds.), *Materials Processing Fundamentals 2019*, The Minerals, Metals & Materials Series, https://doi.org/10.1007/978-3-030-05728-2_7

However, when the steel is held at high temperature for a long time, these precipitates become ineffective at pinning grain growth for two reasons: First, these nano-scale fine particles coarsen rapidly because of their small size and as a consequence the pinning force drops very rapidly. Second, most of these fine particles will dissolve at very high temperatures. In some cases, it is possible to use TiN particles that do not dissolve at high temperatures. However, these particles precipitate in the liquid and coarsen quickly to a large size which limits their effectiveness at pinning grain growth. In this paper, a new steel system which can automatically pin the delta grain growth by using a small volume fraction of austenite particles at high temperature is proposed. During the reheating process, the austenite particle coarsening kinetics and the delta-ferrite grain growth kinetics were investigated. In order to verify the effectiveness of this alloy in inhibiting high temperature grain growth under realistic casting conditions, solidification experiments were performed to reproduce the as-cast microstructure. These microstructures were then used as starting microstructures for studying grain growth kinetics during subsequent cooling with different cooling rates. The alloy system which has autogenous grain pinning at high temperature and the possible applications are discussed.

Materials and Experimental Procedure

To study the coarsening behaviour of the delta/austenite two-phase microstructure at high temperature, 1.5 wt% Al was introduced into a low carbon steel (Table 1). The addition of Al can stabilize delta-ferrite down to room temperature as shown in Fig. 1. A two-phase mixture of delta-ferrite and austenite will exist at temperatures between 1310 °C and the eutectoid temperature [5, 6]. The new alloy was prepared by induction melting at CANMET Materials Technology Lab (Hamilton, Canada); the as-received microstructure is delta-ferrite with grain size of approximately 85 μm after pilot mill hot rolling to the thickness of 10 mm.

A high temperature tube furnace was used to investigate the austenite particles coarsening kinetics and the delta grain growth kinetics, all samples were heated to different temperatures (1200, 1280, 1295, 1305 and 1310 °C) for different holding times. After each heat treatment, the samples were quenched into ice water. To simulate the solidification process, the materials from CanMet was re-melted using an induction furnace. A copper bar on a steel rod was then dipped into the molten metal to initiate solidification. Once the dipping bars were removed from the melt, the solidified shells were cooled down to room temperature at different rates using air, forced

Table 1 Chemical composition of the new alloy used in this investigation (wt%)

wt%	C	Mn	Si	Al	Ti	Nb	N
Fe-1.5 Al% new alloy	0.051	1.00	0.36	1.5	0	0	0
API X80	0.060	1.65	0.25	0.025	0.012	0.034	0.005

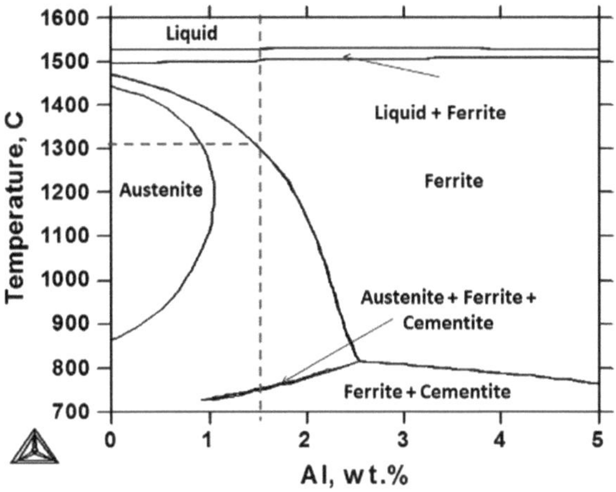

Fig. 1 Phase diagram of Fe-1.5% Al-0.051% C-1.0% Mn-0.36% Si new alloy predicted by Thermo-Calc by using TCFE6 database [7]

air and water quenching. All the samples were then sectioned and prepared using standard metallographic techniques. The microstructure was studied using optical microscopy. Image analysis was performed using Clemex PE5.0 software. The average austenite phase particle size and the delta-ferrite grain size were measured based on the equivalent circular diameter.

Results

Reheating Process

Figure 2 is an example of the microstructure evolution during the reheating process at 1200, 1280, 1305 and 1310 °C. The dark particles are the austenite phase which transformed to martensite when quenched to room temperature and the matrix is delta-ferrite. The austenite area fraction was measured for each reheating condition. It was found that the austenite area fraction did not change with holding time for the conditions investigated. This suggests that the equilibrium volume fraction of austenite is achieved rapidly. The measured austenite area is listed in Table 2.

The measured austenite particle size and the ferrite grain size, as a function of reheating temperature and holding time, are summarized in Figs. 3a and b. As expected, the austenite coarsening rate increases with increasing temperature. The evolution of ferrite grain size is more complicated due to the strong pinning effect exerted by the austenite particles. The grains grow very slowly at 1200 °C: at shorter holding times, the grain size is almost constant, while at longer times grain growth is observed. When the reheating temperature is increased to 1280, 1295 and 1305 °C,

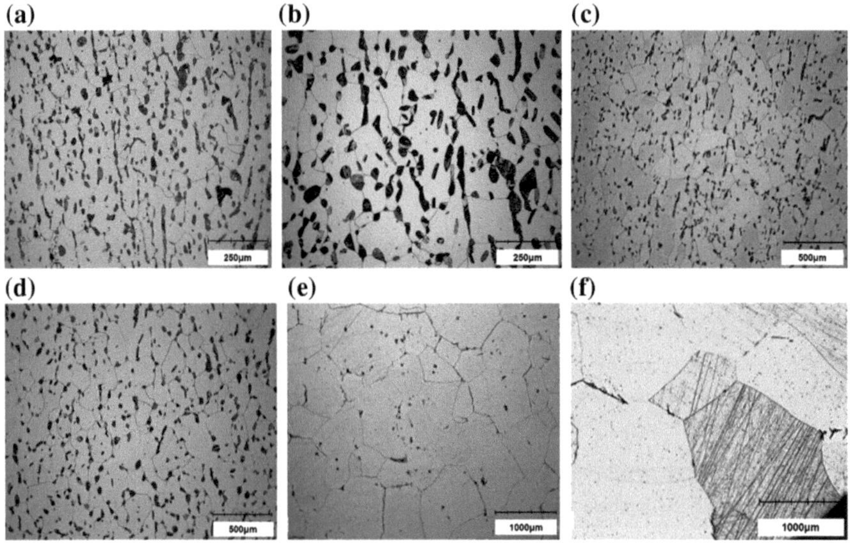

Fig. 2 Fe–Al new alloy microstructure obtained after **a** 15 min at 1200 °C, **b** 180 min at 1200 °C, **c** 5 min at 1280 °C, **d** 60 min at 1280 °C, **e** 15 min at 1305 °C, **f** 15 min at 1310 °C

Table 2 Measured austenite area fraction (%) at different reheating temperature

Temperature (°C)	1310	1305	1295	1280	1200
Experimental measurement	0	1.3	6.7	9.6	21.0

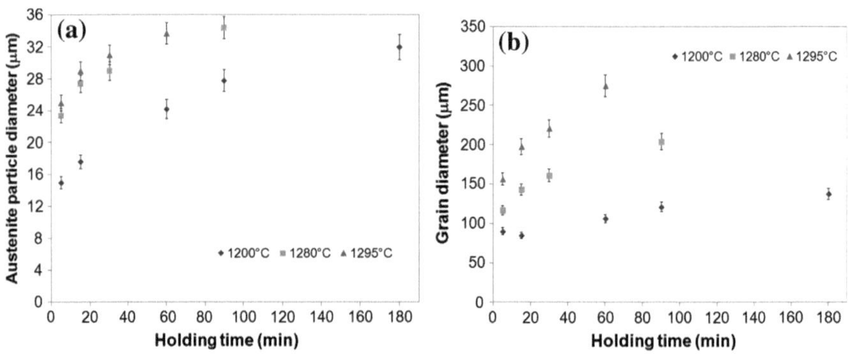

Fig. 3 **a** The measured austenite particle size as a function of temperature and holding time, **b** The measured delta grain size as a function of temperature and holding time

grain growth occurs for all of the times investigated. When the material is reheated to 1310 °C, grain growth rate increases dramatically due to the austenite particle volume fraction approaches zero (Fig. 2f).

Solidification Process

Figure 4 shows the microstructure evolution of new alloy shells that solidified on the steel bar that was dipped into the melt. Figure 4a corresponds to air cooling of the solidified shell down to room temperature, while Fig. 4b and c correspond to forced air and water cooling, respectively. The dark particles are the austenite phase which transformed to martensite and/or pearlite on cooling to room temperature and the matrix is delta-ferrite. The thermal profile was measured during solidification as shown in Fig. 4d. In addition, the ingot which resulted from the solidification of the new alloy melt within the furnace was analyzed. The evolution of the grain size as a function of position from the surface of the ingot is shown in Fig. 4e. The essentially constant grain size as a function of depth suggests that the delta-ferrite grains were pinned by the austenite particles which are uniformly distributed within the ingot. The microstructure of the industrially cast, 85 mm, slab of APIX70 steel from thin slab casting direct rolling process (TSCDR) is shown in Fig. 4f. At the surface of the APIX70 slab, the austenite grain size is about 58 μm, while at the centre of the slab, the austenite grain size is as large as 1342 μm. The industrial as-cast microstructure is non-uniform with extremely large grains at the center.

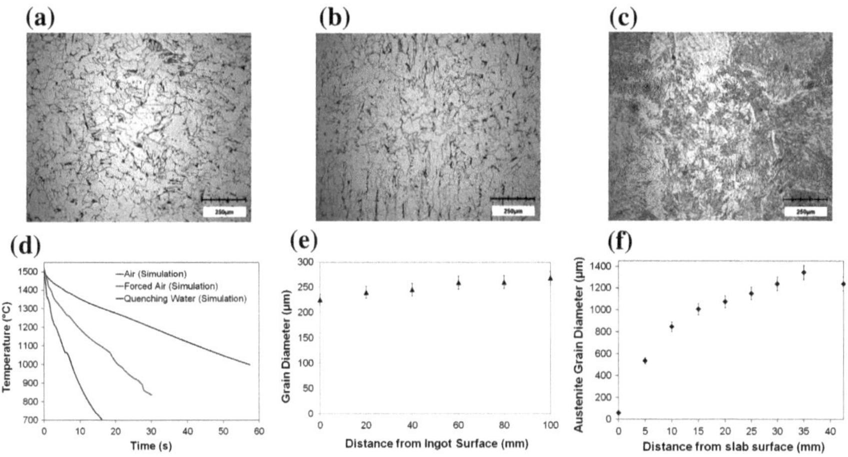

Fig. 4 Microstructure evolution of solidified shells of new Fe–Al alloy on a steel dipping bar with different cooling rates using **a** air, **b** forced air, **c** water quenching, **d** the thermal profile during solidification with different cooling rates, **e** measured delta grain size with the ingot distance from the ingot surface to center, and **f** measured austenite grain size with distance from API X70 slab surface to center

Discussion

Reheating Process

In the present new alloy, austenite particles can pin the grain growth of delta-ferrite over a wide temperature range. Experimental results suggest that delta grain growth is strongly retarded up to 1305 °C (Fig. 2e). Quantitative optical metallography clearly shows that the austenite particle volume fraction was constant during holding at a given reheating temperature. As such, the experimental observations correspond to the stable distribution of second-phase particles. The theory for coarsening of stable distribution of second-phase particles was developed by Greenwood [8], Lifshitz and Slyozov [9] and Wagner [10] which is often referred to as the LSW theory. These approaches rely on a diffusion analysis in which the rate of change of the diameter of each particle is independent of the position of other particles. The principal prediction of the LSW model is shown below:

$$\left(\overline{r}^{n} - \overline{r}_{0}^{n}\right) = kt \tag{1}$$

where $\overline{r}_o$ is the initial average austenite particle radius, $\overline{r}$ is the average particle size, t is time and k is the proportionality constant. When coarsening is controlled by bulk diffusion, the value of n is 3 and when it is controlled by grain boundary diffusion control, the exponent n is equal to 4 [8–10]. Figure 5 shows the variation of average austenite particle radius with time at different reheating temperature 1200, 1280 and 1295 °C. In order to ensure that the kinetics correspond to the coarsening of a stable distribution of second-phase (austenite) particles, the initial time, t_o, used for these plots was 5 min at 1200, 1280, and 1295 °C. The results are reasonably fitted using n = 3. Consequently, the rate controlling step could be the bulk diffusion which dominates for the high temperatures and relatively large grain sizes.

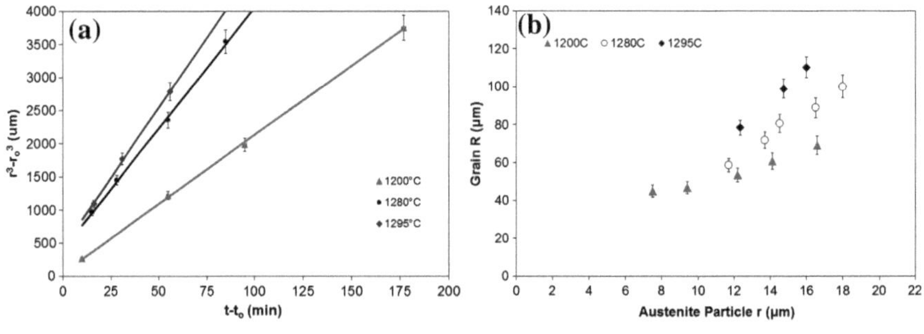

Fig. 5 **a** Plot of $(r^n - r_0^n)$ versus $(t - t_0)$ for 1200, 1280, and 1295 °C, n = 3, and **b** The relationship between delta grain size and austenite particle radius at different temperature

Second-phase particles have been used extensively to control the kinetics of grain growth. Considering the effects of boundary–particle correlation, the ferrite grain growth rate is related to the effective driving pressure ΔP [5, 6].

$$\frac{d\overline{R_\delta}}{dt} = M\Delta P = M(P - P_{ZC}) = M\left(\frac{\alpha\gamma_{gb}}{\overline{R_\delta}} - \frac{1,2F_v^{2/3}}{\overline{r}}\right) \tag{2}$$

where α is a geometric constant near unity, $\overline{R_\delta}$ is the mean radius of an individual grain and M is the grain boundary mobility, γ_{gb} denotes the grain boundary interfacial energy and F_v is austenite particle volume fraction. The relationship between ferrite grain size and austenite particle radius at different temperatures is shown in Fig. 5b. When the austenite particle pinning pressure is only slightly larger than the driving force for grain growth, grain growth will be pinned initially, but it can later start to occur due to the coarsening of the austenite particles. The growth of the ferrite grains would then occur at a rate which is controlled by the rate of change of the austenite particle size [11]:

$$\frac{d\overline{R_\delta}}{dt} = k\frac{d\overline{r}}{dt} \tag{3}$$

At reheating temperature of 1200 °C, the delta-ferrite grain growth was pinned for the first 15 min. For a longer holding time, ferrite grain growth started to take place as shown in Figs. 2b and 5b. When the austenite particle pinning pressure is smaller than the driving pressure, delta grain growth would occur under a net driving force of $P - P_{ZC}$. As such, grain growth will occur at a rate which is significantly smaller than what would be expected in the absence of pinning. When the reheating temperature was increased to 1310 °C, the austenite particle volume fraction approached zero (Fig. 2f). As a consequence of the elimination of particle pinning, the ferrite grain size increases at a rate which is much larger than that reported in the presence of particle pinning at a slightly lower temperature of 1295 °C. This confirmed the concept of using a small volume fraction of austenite to pin ferrite grain growth over a wide temperature range.

All of the experiments performed in the above discussion involved reheating samples with a very homogenous microstructure that was obtained by rolling and homogenization. Under industrial conditions, such as TSCDR conditions and welding, the ferrite and austenite grain growth both occur in a heterogeneous microstructure produced by solidification. In order to verify the effectiveness of the new alloy in inhibiting high temperature grain growth under realistic starting conditions, some work was performed to reproduce the as-cast microstructures.

Solidification Process

Solidification starts when the dipping bar is submerged into the melt. Under equilibrium cooling conditions, the new Fe–Al steel will solidify as delta-ferrite and austenite precipitation occurs in the solid state at about 1310 °C. However, industrial cooling conditions will lead to non-equilibrium solidification and some austenite will form during solidification. Figure 4a, b and c confirm the presence of austenite precipitates along the delta-ferrite grain boundaries. In these figures, the dark phase is austenite which transformed to pearlite or/and martensite on cooling. The small grains along the grain boundaries may also correspond to austenite grains which transformed to ferrite during cooling to room temperature. Using the assumption of no diffusion of the substitutional elements in the solid, the SCHEIL module of Thermo-Calc [7] was used to analyze the phases formed during non-equilibrium cooling of the new steel. In this calculation, carbon back-diffusion in the solid is allowed to take place. As shown in Fig. 6a, the liquid will start to solidify as delta-ferrite at 1528 °C. Once the temperature drops below 1455 °C, the austenite phase will precipitate. The predicted austenite mole fraction with temperature is shown in Fig. 6b. At 1445 °C, the calculated volume fraction of austenite is about 0.5%. These particles are expected to form along the delta-ferrite grain boundaries as shown in Fig. 4a, b and c. The austenite particles will exert a pinning pressure on the delta-ferrite grains and as the temperature decreases, the volume fractions of these particles will increase leading to complete pinning of the delta-ferrite grain growth.

In order to quantify the analysis, a simple non-isothermal grain growth model is utilized to capture the evolution of the delta-ferrite grain size in the Fe–Al new steel [12].

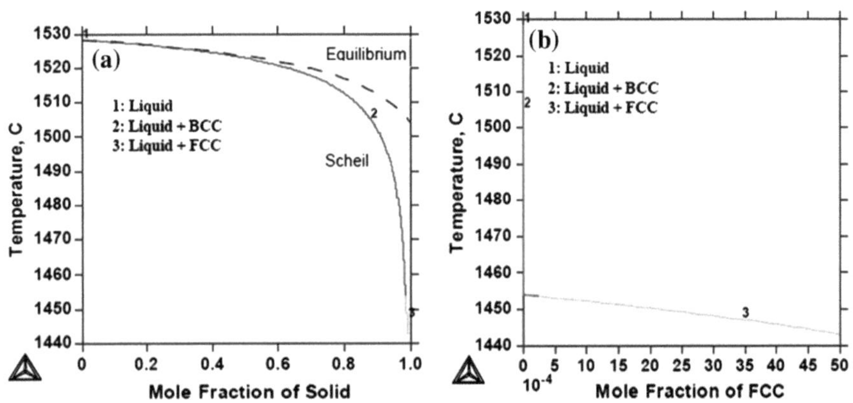

Fig. 6 ThermoCalc using SCHEIL model predicted non-equilibrium solidification **a** mole fraction of solid, **b** mole fraction of austenite particle

$$\overline{R}^2 = \overline{R}_o^2 + 4\gamma_{gb} \int_0^t \alpha M(t)dt \tag{4}$$

where $\overline{R}$ is the mean radius of an individual delta-ferrite grain, $\overline{R}_o$ is the initial grain radius, which will be assumed to be 7.5 μm according to Fig. 4c, and γ_{gb} denotes the grain boundary energy per unit of area, a reasonable value for which is 0.8 Jm^{-2}. Finally, *M(t)* is the mobility of the delta-ferrite grain boundaries [12]:

$$\alpha M(t) = \frac{0.7075}{T(t)} \times exp\left(\frac{-20{,}995.43}{T(t)}\right) \tag{5}$$

In this equation, *T(t)* is an expression for the temperature as a function of time which was obtained experimentally from the data recorded, using a thermocouple, during solidification (Fig. 4d).

The predicted delta-ferrite grain size, in the absence of pinning (solid line), is compared to the experimental data for the new Fe–Al steel for different cooling rates in Fig. 7 [13]. Evidently, the model overestimates the delta-ferrite grain size because particle pinning was not taken into account. In order to capture the effect of particle pinning, it is assumed that the delta-ferrite grains are completely pinned by the austenite particles as soon as the austenite phase appears. The temperature at which the austenite forms was estimated from the Scheil model and the predicted delta-ferrite grain size evolution was plotted as the dot-dash curves in Fig. 7. The very good agreement between the model predictions, in the presence of pinning, and the experimental data confirms the effectiveness of new alloy in inhibiting high temperature grain growth under post-solidification cooling conditions.

Applications

Application to Thin Slab Casting Direct Rolling Process

During TSCDR process grain growth occurs during the cooling of the solidified slab as well as isothermal holding in the soaking furnace. Figure 8a shows the temperature profile at the surface as well as those at 5 and 10 mm below the surface of an 85 mm, APIX70 slab, cast at a speed of 3.4 m/min at Essar Steel Algoma Inc. using the DSPC process. The temperature profiles are calculated using the CON1D Slab Casting Heat Transfer Model [14]. The figure also includes the recorded thermal profile obtained from the laboratory solidification experiments using the new alloy. The simulated cooling rate, using water quenching, is close to the cooling rates predicted at 5 mm and 10 mm below the surface of the industrially cast 85 mm slab. One can, therefore, compare the average grain sizes obtained from the dipping tests to the grain sizes measured at 0, 5 and 10 mm below the surface of the industrial slab. In addition, the cooling rate for the furnace cooled Fe–Al steel ingot was extremely slow and could

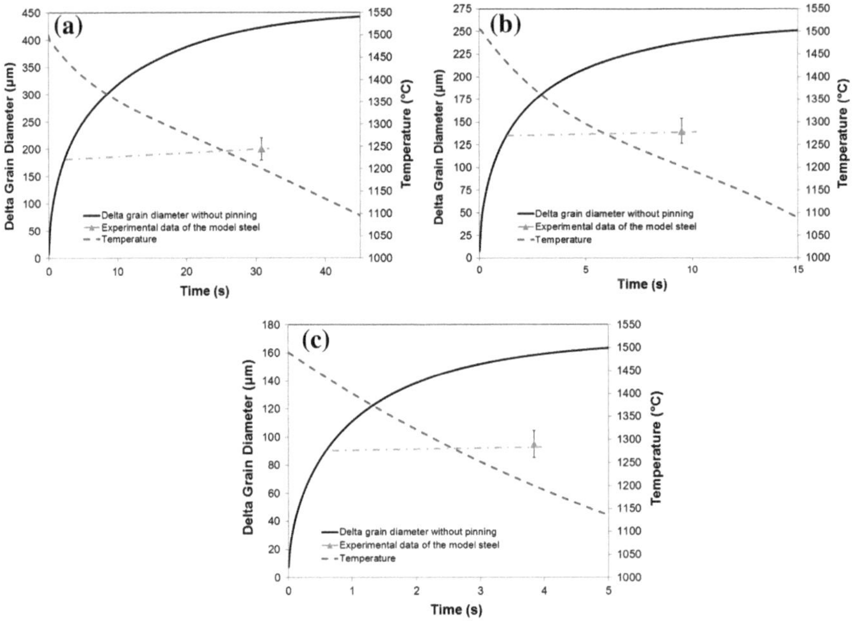

Fig. 7 Comparison of the predicted delta grain size without pinning and experimental data of new Fe–Al alloy solidified with different cooling rates **a** air, **b** forced air and **c** water quenching

be compared to that at the centrer of the industrially cast slab. These comparisons are shown in Fig. 8b, in which, the grain size prior to entry into the soaking furnace was obtained by directly measuring the prior austenite grain size as a function of distance from the surface of the slab (Fig. 4f). The data points for the new steel were positioned by matching the cooling rates in the dipping test to the position at which these cooling rates would be observed within the slab. Based on this comparison, if the new steel was cast in the form of an 85 mm slab, the expected grain size at the center would be 280 μm, compared to 1340 μm for APIX70. This clearly demonstrates the potential advantage of the new alloy. One could also compare the grain sizes within the APIX70 slab after exiting the soaking furnace, to those expected in the new alloy. As a result, the grain size of the new alloy is essentially unchanged as a result of soaking which can prevent excessive grain growth prior to the onset of thermomechanical processing.

Application to Welding Process

During welding, the base metal lying adjacent to the fused zone will be subjected to one or more high temperature thermal cycles. These thermal cycles will introduce significant changes in the microstructure and mechanical properties within the heat

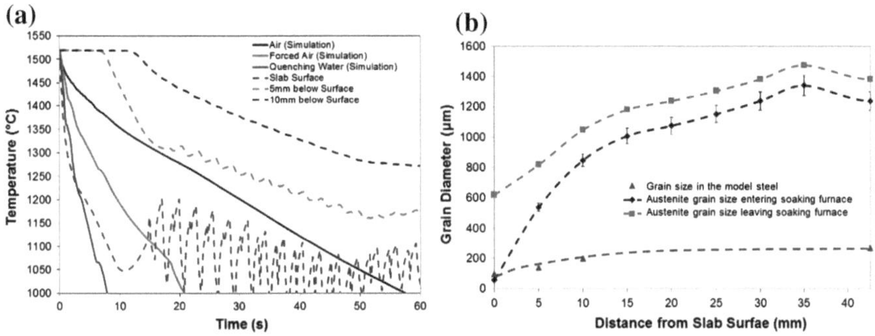

Fig. 8 **a** Comparison of CON1D predicted temperature profiles on the surface, 5 and 10 mm below the APIX70 85 mm slab and the recorded thermal profile during simulation process, **b** comparison of grain size evolution with slab distance using TSCDR process to produce APIX70 and Fe–Al% new alloy

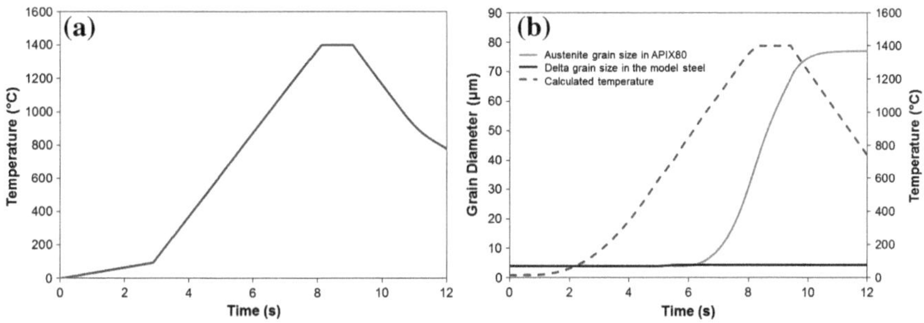

Fig. 9 **a** A heating profile for APIX80 during a welding trial, **b** Comparison of model predicted austenite grain size in API X80 and delta grain size in new Fe–Al steel during welding process

affected zone (HAZ). In conventional low alloy steels, the fine niobium, titanium, or vanadium carbonitrides that inhibit austenite grain growth will dissolve or coarsen within the HAZ. As a result, excessively large austenite grains may develop within the HAZ. The proposed new Fe–Al steel can prevent austenite coarsening in the HAZ. Figure 9a shows the heating profile measured during the welding of APIX80 steel, whose chemistry is shown in Table 1 [15]. The steel was heated at a rate of 250 K/s to a peak temperature of 1400 °C followed by cooling at 250 K/s. In order to model grain growth during this heating cycle, the microalloyed APIX80 steel was assumed transforms to austenite at 720 °C. The austenite grain growth was modelled by integrating Eq. (2) from t_1 to t, where t_1 is the time corresponding to the completion of the alpha ferrite/pearlite to austenite phase transformation during heating. The product of α and austenite grain boundaries mobility is 0.3 times of Turnbull estimate [12, 13]:

$$\alpha M(t) = \frac{0.1920}{T(t)} \times exp\left(\frac{-20{,}837.14}{T(t)}\right) \tag{6}$$

where *T(t)* is obtained from the heating profile of Fig. 9a. The calculated grain size evolution for APIX80 is shown in Fig. 9b. An initial austenite grain size of 4 μm was assumed in the above calculation. For comparison, the ferrite grain size in the new alloy is not expected to change during the above heating cycle due to the strong pinning effect exerted by the austenite particles.

Steel that Resists Grain Growth at High Temperature

As discussed in section "Reheating Process", the delta-ferrite grain growth in the new steel is dominated by the pinning effect of the austenite particles. At high temperatures, the delta-ferrite grains grow at a rate which is controlled by the rate of coarsening of the austenite particles. As a result, the growth of the delta-ferrite grains occurs at a very slow rate due to the fact that austenite particles are large (micron scale) and consequently have a small driving force for coarsening. Thus, the proposed new alloy is a promising candidate for high temperature applications which require stable grain size. In Fig. 10, the experimental grain growth kinetics in this two-phase material is compared with the predicted grain growth kinetics in the absence of pinning at 1200 and 1295 °C. The temperature profile, shown on the secondary axis, was obtained experimentally form the data recorded by the thermocouple attached to the specimen. The solid points refer to the delta-ferrite grain diameter which was experimentally measured in the new Fe–Al alloy. For comparison, the ferrite grain size in the absence of particle pinning was calculated using Eq. (2) and the ferrite grain boundary mobility was taken from reference 12 and 13. When these two materials are heated to 1295 °C for 5 min, the materials without particle pinning is predicted to have a grain size of 1100 μm, while, the new alloy is expect to have a grain size of 157 μm.

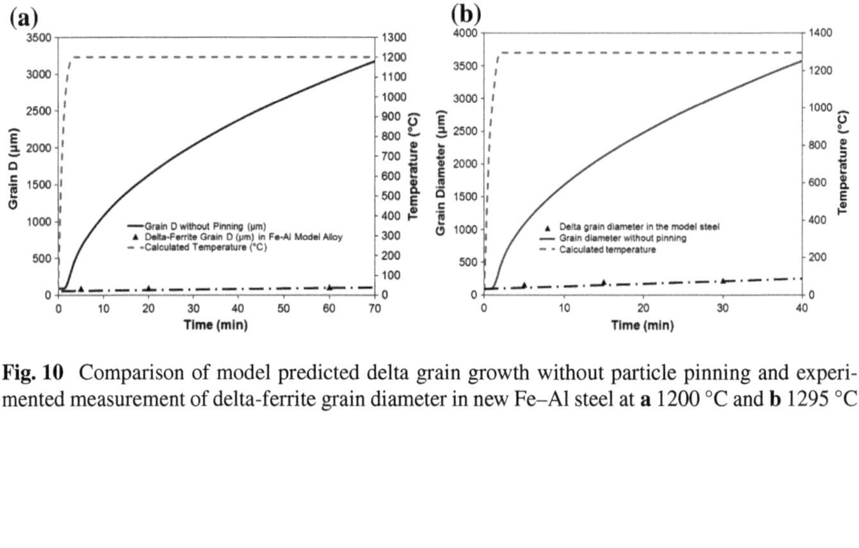

Fig. 10 Comparison of model predicted delta grain growth without particle pinning and experimented measurement of delta-ferrite grain diameter in new Fe–Al steel at **a** 1200 °C and **b** 1295 °C

Conclusions

(1) In this new alloy, austenite particles can pin the grain growth of delta-ferrite over a wide temperature range, delta grain growth is strongly retarded up to 1305 °C. When the reheating temperature is higher than 1200 °C, the coarsening of the austenite particles was bulk diffusion controlled. When the pinning pressure is equal to the driving force for grain growth, grain growth proceeds at a rate controlled by the coarsening kinetics of the austenite particles, Once the pinning pressure is smaller than the driving force, grain growth occurs with a grain growth is much lower than that expected without pinning.

(2) Solidification and cooling rate simulations confirmed that the proposed Fe–Al new alloy can resist grain growth under industrial solidification and cooling conditions. During non-equilibrium solidification, the austenite phase precipitates along the delta-ferrite grain boundaries. The austenite particles pin the delta-ferrite grain growth during cooling and soaking. As a result, a finer and more uniform as-cast macrostructure can be obtained.

(3) The proposed new alloy has the potential to control grain size and prevent excessive grain growth during the process of TSCDR. During welding, the novel alloy can greatly limit coarsening in the HAZ. More generally, this material can be used to prevent grain growth at high temperature due to the presence of a stable distribution of pinning particles.

References

1. Gladman T (1997) The physical metallurgy of microalloyed steel. Institute of Metals, London
2. Baker TN (2016) Microalloyed steels. Ironmak & Steelma 43(4):264–307
3. Hillert M (1988) Inhibition of grain growth by second-phase particles. Acta Metall 36:3177–3181
4. Zhou T, Overby D, Badgley P, Martin-Root C, Wang X, Liang SL, Zurob H (2018) Study of processing, microstructure and mechanical properties of hot rolled ultra high strength steel. Ironmak & Steelmak. https://doi.org/10.1080/03019233.2018.1468652
5. Zhou T, Zurob H, O'Malley RJ, Rehman K (2015) Model Fe–Al steel with exceptional resistance to high temperature coarsening. Part I: coarsening mechanism and particle pinning effects. Metall Mater Trans A 41A:178–189
6. Zhou TH, Gheribi AE, Zurob HS (2013) Austenite particle coarsening and delta-ferrite grain growth in model Fe–Al alloy. Can Metall Q 52(1):90–97
7. Available from Thermo-Calc Software. www.thermocalc.com
8. Greenwood GW (1956) The growth of dispersed precipitates in solutions. Acta Metall 4:243–248
9. Lifshitz IM, Slyozov VV (1961) The kinetics of precipitation from supersaturated solid solutions. J Phys Chem Solids 19:35–50
10. Wagner C (1961) Theorie der Alterung von Niederschlaegen durch Umloesen (Ostwald reifung). Z Eleclraehem 65:581–591
11. Hillert M (1965) On the theory of normal and abnormal grain growth. Acta Metall 13:227–238
12. Zhou T, O'Malley RJ, Zurob HS (2010) Study of grain-growth kinetics in delta-ferrite and austenite with application to thin-slab cast direct-rolling microalloyed steels. Metall Mater Trans A 41A:2112–2120

13. Zhou T, Zhang P, O'Malley RJ, Zurob H, Subramanian M (2015) Model Fe–Al steel with exceptional resistance to high temperature coarsening. Part II: Experimental validation and applications. Metall Mater Trans A 41A:190–198
14. Yand Meng, Thomas BG (2003) Heat-transfer and solidification model of continuous slab casting: CON1D. Metall Mater Trans B 34B:685–705
15. Banerjee K, Militzer M, Perez M, Wang X (2010) Nonisothermal austenite grain growth kinetics in a microalloyed X80 linepipe steel. Metall Mater Trans A 41A:3161–3172

Effect of Casting Temperature on the Surface Finish of Grey Iron Castings

Izudin Dugic

Abstract One of the most common surface defects in sand casting of grey cast iron is caused by metal penetration into the sand mould. Metal penetration is a surface condition in which metal or metal oxides have filled the voids between sand grains to various depth without displacing them, thus yielding a phase of sand grains surrounded by metal and frequently by mould–metal reaction products. The penetration is often so severe that casting components are beyond the point of economical rework and must be scrapped. This experimental work has focused on reducing metal penetration on casting component on a production scale. The casting component produced has strongly affected by sand sintering metal penetration. A series of simulations were performed with the casting simulation program MagmaSoft® in order to investigate the solidification characteristics as well as the porosity formation in the casting component.

Keywords Grey cast iron · Metal penetration · Casting simulation · Solidification Surface finish

Introduction

One of the most important factors affecting the surface finish of cast components in the grey iron foundry is metal penetration. Draper and Gaindhar [1] proposed the general definition of metal penetration, accepted by the foundry industries, is the condition in which cast metal has entered into the pore spaces of the mould sand and core beyond the mid-point of the surface layer of sand grains. Metal penetration is more prevalent at casting "hot-spots" and in heavy section areas or "transition zones" from thin to heavy metal section. The defect is tightly bonded to the casting surface

I. Dugic (✉)
Faculty of Technology, Department of Mechanical Engineering, Linnaeus University, 35195 Vaxjo, Sweden
e-mail: Izudin.dugic@lnu.se

, , https://doi.org/10.1007/978-3-030-05728-2_8

and can only be removed by extensive chipping and grinding. The metal penetration defect can occur over a small area of casting or spread over the entire surface.

Numerous names have been given to the various types of penetration described in the published literature [2–5]. Stefanescu et al. [6] has presented a detailed summary of the significant findings from this review.

The expansion penetration was first identified by Levelink and Julien [2], who called it exudation penetration. They indicated that the expansion occurring during eutectic solidification may result in exudation of eutectic in locations where a solidifying metal shell does not obstruct it. The expansion penetration depends upon the metallurgical characteristics of the solidifying metals and alloys. Levelink and Julien have demonstrated that expansion penetration is most common when the carbon equivalent is high. This has been confirmed in an earlier work by Dugic et al. [7].

The theory of exudation or metal expansion penetration proposed by Levelink and Julien [2] was developed further. Diószegi and Dugic [8] suggested a new description for the metal expansion penetration mechanisms, considering the nucleation and growth of both the primary austenite grains and the eutectic cells.

Metal penetrations have been discussed by many authors and lot of works have been done in the industries to understand how to avoid metal penetration [9–12]. To give an increased knowledge of the metal expansion penetration and the influence from different factors, a series of test castings were performed at the XY Foundry.

Experimental

Alloys and Other Materials

The experiments in this work were carried out at the XY Foundry with a Seiatsu production line. The XY Foundry produces several different components for pumps in the industry. In the experimental works, a typical casting component was selected, Fig. 1, and six castings were mounted on the pattern plate, illustrated in Fig. 2. The weight of each component is 6.0 kg, the weight of the gating system is 7.6 kg and the total weight is 43.6 kg.

The metal was melted in a high-frequency furnace with a charge composition of 35% recycled metal, 20% pig iron and 45% steel. Five series of casting experiments with different casting temperature were carried out. After melting, the melt is transported in a 1-ton pouring ladle to a production line.

The inoculant used in these experiments was a ferro-silicon-strontium type. The chemical composition of the used inoculant is shown in Table 1. The inoculant had a particle size between 1.0 and 6.0 mm. For all experimental series, the inoculation was made in the stream when pouring the melt from the melting furnace to the pouring ladle. The amount of inoculant was 0.15% for all experiments, and time between inoculation and casting in the mould flasks was 2 min.

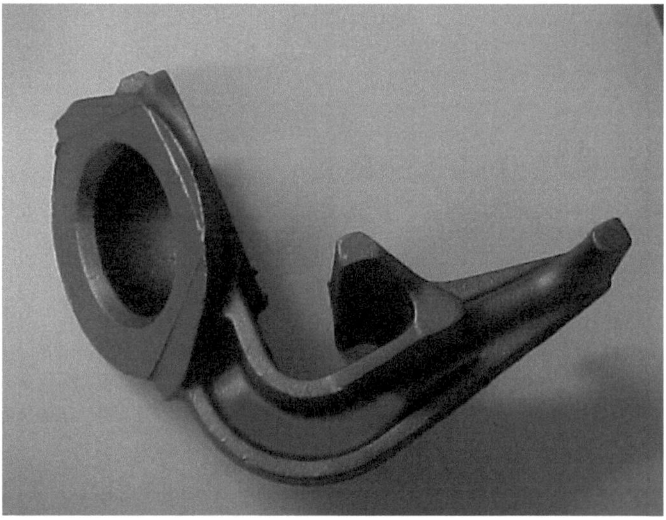

Fig. 1 The casting component

Fig. 2 The castings lower part

The chemical compositions were determined by the light emission spectrometer—ARL 3460 from cast coin specimens. The coins were cast immediately before the melt was poured into the moulds. The chemical compositions of the base melt for each casting and pouring temperature are shown in Table 2.

Table 1 The chemical composition of the inoculant used

Inoculant	% Si	% Ca (max.)	% Al (max.)	% Sr
A	75 ± 3	0.10	0.50	0.80 ± 0.20

Table 2 The chemical composition of the base melts and pouring temperatures

Experiment	Element in wt%							C^{a}_{equ}	Pouring temp °C
	C	Si	S	P	Mn	Cr	Cu		
I	3.34	1.99	0.089	0.100	0.79	0.06	0.50	4.037	1375
II	3.35	2.00	0.089	0.100	0.79	0.05	0.48	4.050	1390
III	3.33	2.01	0.087	0.099	0.80	0.05	0.51	4.033	1400
IV	3.35	1.99	0.088	0.098	0.78	0.06	0.50	4.046	1415
V	3.34	2.00	0.087	0.097	0.79	0.05	0.49	4.039	1430

[a]C_{equ} = %C + 1/3 (%Si + %P)

Table 3 The sand analyze used in the experiments

Sand	Experiment				
	I	II	III	IV	V
Temperature, °C	45	42	44	46	45
Bentonite, %	6.70	7.15	6.85	6.75	6.90
Carbon, %	3.00	2.90	2.85	3.10	3.05
Moisture content, %	2.61	3.20	2.70	2.75	2.80
Compaction pressure, %	29.50	30.00	29.75	29.68	30.10

The castings were made in green sand moulds. The sand analysis for each casting is shown in Table 3. The core used in the experiments are manufactured with silica sand and a water-based phenolic resin system cured by the use of CO_2 gas by Laempe machine LFB20. The silica sand with particle size (MK) of 0.21 mm was mixed with a resin content of 2.1% based on the weight of the sand.

The mould hardness were measured at two critical places, where metal penetration is often seen to occur, with an Electronic Mould Strength Tester Type PFP. These locations are shown in Fig. 3 and the results of the average hardness value for all experiments are shown in Table 4.

For each of the different pouring temperature, 15 flasks were moulded and cast. This yields a total of 75 flasks, and consequently 450 castings, thereby providing a sound statistical basis for evaluation of the effect of pouring temperatures on metal penetration.

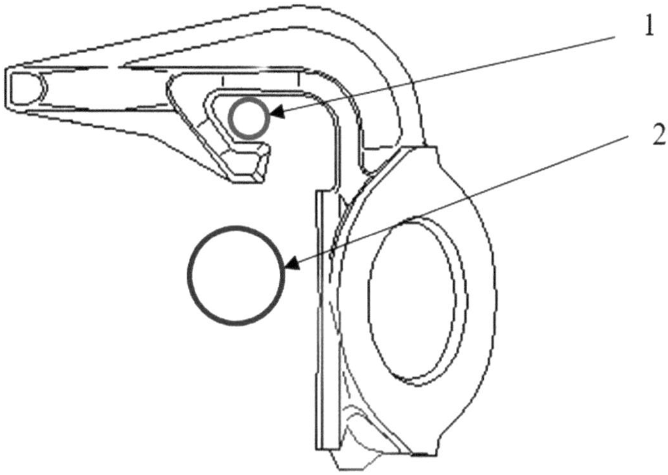

Fig. 3 The casting component—the arrows show the areas for measuring of mould hardness

Table 4 The average mould hardness value for all experiments

Experiment	Mould hardness, N/cm^2	
	Area 1	Area 2
I	14.40	26.90
II	15.80	29.70
III	14.30	24.70
IV	14.50	25.30
V	14.70	27.10

Results and Discussion

Casting Defects

After sandblasting, all castings were investigated by an ocular inspection. A scale from 0 to 4 to classify the degree of penetration was used (0 is no penetration, 1 is low degree of penetration and 4 is a high degree of penetration). Data from the penetrated area from the experiments are shown in Table 5.

In the experiment with low pouring temperature (1375 °C), another type of casting defect had seen to be formed, the cold shuts. Totally six castings of 6.67% had the phenomena.

Table 5 The results of metal penetration investigation

Experiment	Penetration grade, %				
	0	1	2	3	4
I	50.00(45)	25.56(23)	14.44(13)	6.67(6)	3.33(3)
II	45.55(41)	26.66(24)	16.67(15)	6.67(6)	4.45(4)
III	33.33(30)	22.22(20)	21.11(19)	11.11(10)	12.23(11)
IV	32.22(29)	20.00(18)	16.67(15)	13.33(12)	17.78(16)
V	22.22(20)	16.67(15)	18.88(17)	16.67(15)	25.56(23)

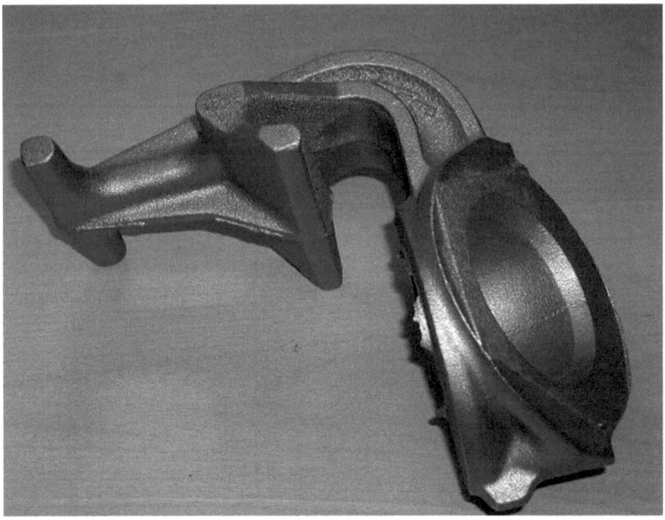

Fig. 4 The casting component defected with metal penetration

Figures 4 and 5 show the casting components defected with the surface defect metal penetration.

Casting Simulation

A series of simulations were performed with the casting simulation program MagmaSoft® using the add-on module MagmaIron®, especially developed for cast iron simulation. The casting component with its gating system was modeled for the simulation. In the simulation, the mould filling sequence as well as the solidification sequence was considered. Figures 6 and 7 show 3D result from the simulation.

Fig. 5 The casting component defected with metal penetration

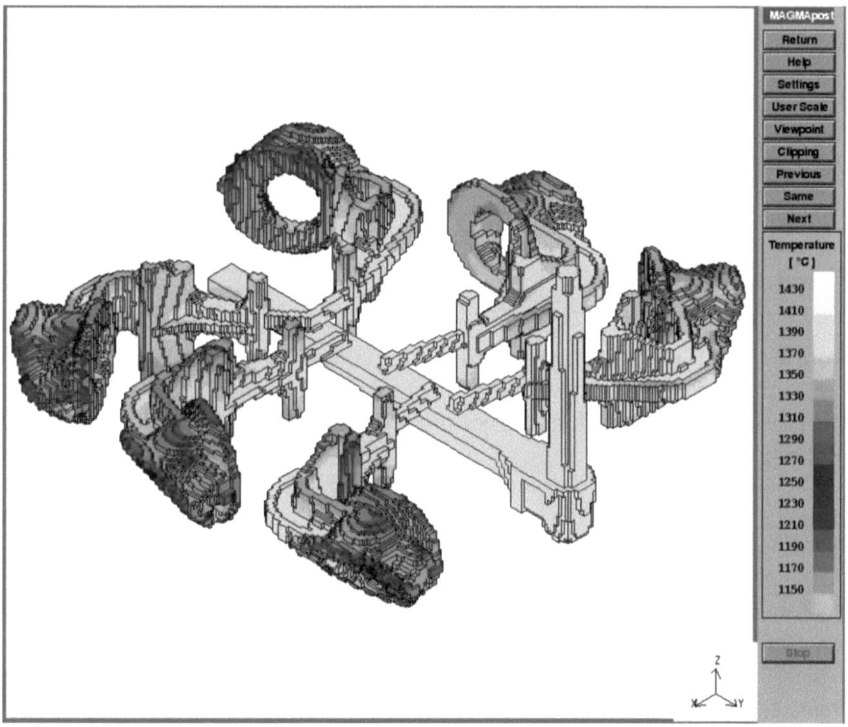

Fig. 6 3-D temperature gradient results from the simulation when 100% of the mould is filled

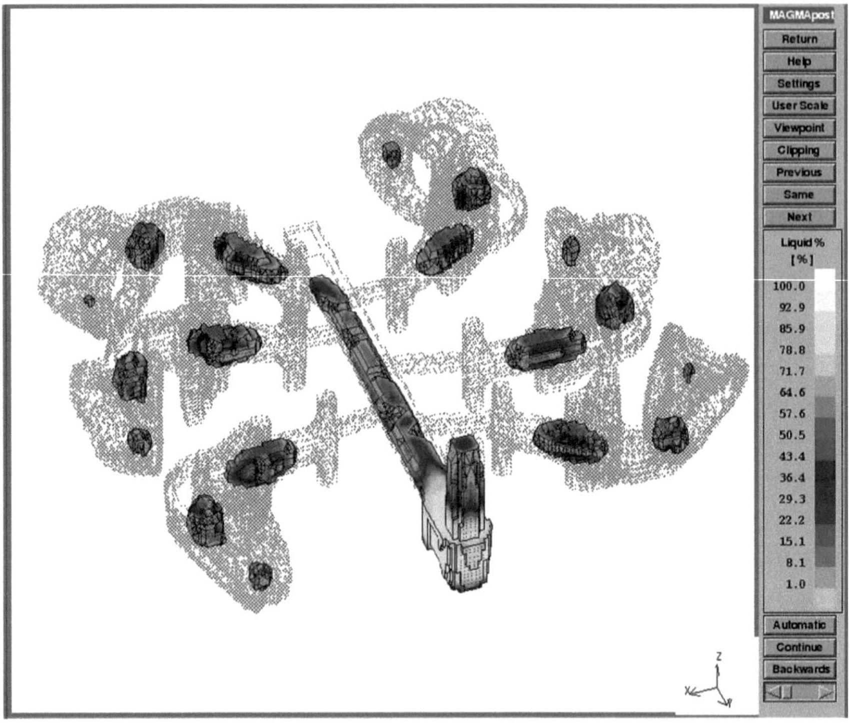

Fig. 7 3D fraction liquid results from the simulation when 10% liquid remains in total, using X-ray option

Conclusions

From the results of the ocular inspection of the components, it can be concluded that the pouring temperature plays an important role in the formation of surface defects. The casting show two different types of surface defects, namely, metal expansion penetration and cold shuts.

The influence of the casting temperature showed that a lower temperature reduces the metal penetration, but the cold shuts occur.

The best results were obtained using pouring temperature 1390 °C, which is only 4.45%. No cold shuts were seen on these castings.

From the simulation no distinct differences could be observed between castings considering the mould filling and the solidification path. By simulation it is possible to detect the areas where porosities are likely to be formed. The amount of porosity predicted is more or less identical for all castings. Simulation results show that the expansion penetration generally occurs in the same regions for each casting.

From lower mould hardness, at least two effects can be observed. The first and probably the most important one is that the volume change differs when the mould is heated up. Second, the thermal properties will be lower and this will affect the heat flow from the casting. These both effects may together result in much greater tendency to metal penetration locally. This effect is very difficult to predict by the casting simulation.

Acknowledgements The author would like to thank Linnaeus University, Faculty of Technology, Department of Mechanical Engineering, Växjö, Sweden.

References

1. Draper AB, Gaindhar JL (1977) Metal penetration—a critical literature review. AFS Trans 85:163–199
2. Levelink HG, Julien FPMA (1973) Penetration and shrinkage by interaction of solidifying cast iron and casting mould—Part 2. AFS Cast Met Res J 9(2):105–109
3. Thorpe PJ (1971) Avoidance of metal penetration and sand burn-on in iron castings. Brit Foundryman 64:38–396
4. Kagawa A, Kiguchi S, Osada M (1995) Volumetric change in freezing cast irons. Trans Japan Foundrymen's Soc 14:18–23
5. Svoboda JM, Geiger GH (1996) Mechanisms of metal penetration in foundry molds. AFS Trans 77:281
6. Stefanescu DM, Piwonka TS, Giese S, Lane A (1996) Cast iron penetration in sand moulds, Part I: Physics of pentration defects and penetration model. AFS Trans 104:1233–1248
7. Dugic I, Svensson IL (1999) The effect of chemical composition on the metal expansion penetration in grey cast iron. Research report 99:1, ISSN 1404-0018, Division of Component Technology, The School of Engineering, Jönköping University, Sweden
8. Diószegi A, Dugic I (2006) The mechanisms of metal expansions penetration in grey cast iron. In: Conference Proceeding ISCP8, Beijing, China
9. Stefanescu DM, Giese S (1996) Cast iron penetration in sand molds, Part II: Experimental evaluation of some main parameters responsible for penetration. ASF Trans 104:1249–1257
10. Srivatsan TS, Sudarshan TS (1999) The influence of phosphorus on shrinkage porosity in cast irons. Mater Lett 41(4):186–191
11. Moosavian TC, Archibald J (1998) Methods for improving iron casting quality. In: One hundred second annual meeting of the American Foundrymen's Society, Atlanta, GA, USA, pp 419–425
12. Dugic I, Svensson IL (1999) An investigation of the effect of inoculants on the metal expansion penetration in grey iron. Int J Cast Metal Res 11(5):333–338

Carbide Precipitation of TBM Cutter Ring Steel During Tempering

Shaoying Li, Hanjie Guo, Mingtao Mao and Xiao Shi

Abstract In this comparative study, the carbides of TBM cutter ring steel at different tempering temperatures of 530 and 560 °C, were studied using ASPEX inclusion automatic analyzer and EPMA, respectively. The results show that the number of carbides increased by about 180% at the tempering temperature of 530 °C. The inclusions of test steels were characterized through the carbides and carbides with the core of Al_2O_3. The thermodynamics results indicate that Al_2O_3 inclusions were generated in the liquid phase, and carbides started to form in the solid–liquid two-phase region. Al_2O_3 inclusion promoted the formation of carbides through serving as preferred nucleation sites. A lower temperature in the solid phase increases the difference value of actual solubility product and equilibrium solubility product, thus it is beneficial to the formation of carbide. The thermodynamic calculations are in accordance with the experimental results.

Keywords Carbide · Precipitation · Tempering · TBM cutter ring

Introduction

TBM (Tunnel Boring Machine) is widely used in tunnel excavation at various environments such as cities, seabeds and mountains. The high chromium content cold work die steel which is high hardness and toughness [1] is mostly used for excavating tools, especially TBM (Tunnel Boring Machine) cutter rings.

There are many failure modes of the cutter ring, for example, normal wear, partial wear, breakage, and so on. But the replacement of normal wear cutter ring accounts

S. Li · H. Guo (✉) · M. Mao · X. Shi
School of Metallurgical and Ecological Engineering, University of Science and Technology Beijing, Beijing 100083, China
e-mail: guohanjie@ustb.edu.cn

S. Li · H. Guo · M. Mao · X. Shi
Beijing Key Laboratory of Special Melting and Preparation of High-End Metal Materials, Beijing 100083, China

G. Lambotte et al. (eds.), *Materials Processing Fundamentals 2019*, The Minerals, Metals & Materials Series, https://doi.org/10.1007/978-3-030-05728-2_9

for 80% of total replacement. Therefore, it is especially important to improve the wear resistance of materials [2]. At present, methods for improving wear resistance include producing a metal matrix composite (MMC) by spraying carbide on a metal surface and inlaying a hard alloy outside the steel matrix [3, 4]. Due to the high production cost, it is not widely used.

For the steel material matrix, the control of carbide precipitation is a reasonable means to improve wear resistance [5, 6]. In the heat treatment process of tool steel, the methods of controlling carbide precipitation proposed by most scholars were mainly reflected in austenitizing temperature control [7, 8], deep cryogenic treatment [9, 10], spheroidization [11], control of hot pressing conditions for carbide fracture [12, 13] and cold speed control after austenitization [14], and so on.

However, systematic studies on the effects of tempering temperature on carbide precipitation and wear resistance have rarely been reported. In this paper, X90CrMoV is taken as an example to study the influence of tempering temperature on carbide precipitation control from the perspective of precipitation thermodynamics. The analysis of carbide size, quantity and morphology are carried out by means of automatic analysis system of inclusions and electron probe.

Experimental Design

Material

The selected tool steel grade, X90CrMoV, was provided by Pan Gang Group China. The chemical composition is presented in Table 1. It should be noticed that all the alloying elements, except Si, are strongly carbide forming ones. In the case of the cold work Cr–Mo–V tool steel, the higher Cr and C percentages lead, during solidification, to the formation of characteristic, eutectic, coarse chromium carbides.

Heat Treatment

The specimens were divided into two groups; each one of them was then subjected to proper heat treatment that led to a final hardness of 58 and 63 HRC. In all cases, the heat treatment procedure followed was that recommended by the materials' supplier and the whole thermal cycle included spheroidizing annealing, austenitizing and

Table 1 Commercial names and chemical compositions (wt%) of the examined tool steel

Commercial name	Fe	C	Si	Mn	Cr	Mo	V
X90CrMoV	Bal.	1.2	1.0	<0.6	8.5	2.50	0.50

Table 2 Heat treatment conditions applied for hardening of the X90CrMoV

Heat treatment stage	Temperature (°C)/time (min)	Temperature (°C)/time (min)
Spheroidizing annealing	800–900/40–60	800–900/40–60
Austenitizing	1040/30	1040/30
First tempering	560/90	530/90
Second tempering	560/90	530/90
Final hardness (HRC)	58	63

two tempering stages. During austenitizing, the polished surfaces of specimens were protected against decarburization employing oil baths that prevented contact with air. The actual conditions are presented in Table 2.

Microstructural Characterization

To study the microstructures, the specimens after tempering were metallographically ground, polished and etched with 4% nital solution. The microstructures of the etched samples were observed under OM (Leica, Wetzlar, Germany). The microstructures of the unetched samples were observed under EPMA (EPMA-1720H, JEOL, Japan) as well as SEM with the function of the EDS (Carl-Zeiss, Oberkochen, Germany) in secondary electron mode. An automated inclusion analysis system (INCA, Oxford Instruments, Oxfordshire, United Kingdom) was employed to measure the size distribution and number of the precipitation particles. INCA Steel could detect carbides with the minimum size of 0.587 μm on the surface of a specimen of 4 mm × 4 mm, and measure the equivalent diameter of each carbide in the field of view.

Results

Morphology of Carbides or the Mixture of Carbides and Oxides

(1) The morphology of carbides by OM

The microstructures of etched specimens that were tempered at different temperatures are shown in Fig. 1.

The observation shows that the microstructure consists of tempered martensite, retained austenite, primary carbides in a matrix of martensite, and a small amount of secondary carbides in a matrix of martensite. Based on the size of carbides, the carbide particles can be classified as primary carbides (>5 μm) and secondary carbides (<5 μm). The classification of carbides and selection of their size limits have been detailed in previous studies [15].

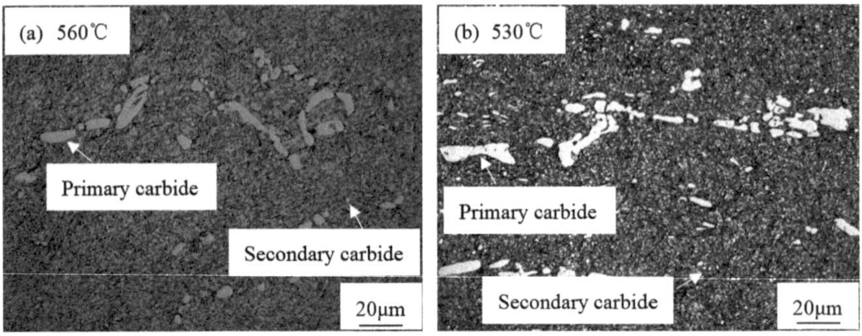

Fig. 1 Metallographic observation of the microstructures of specimens: **a** tempering at 560 °C, **b** tempering at 530 °C

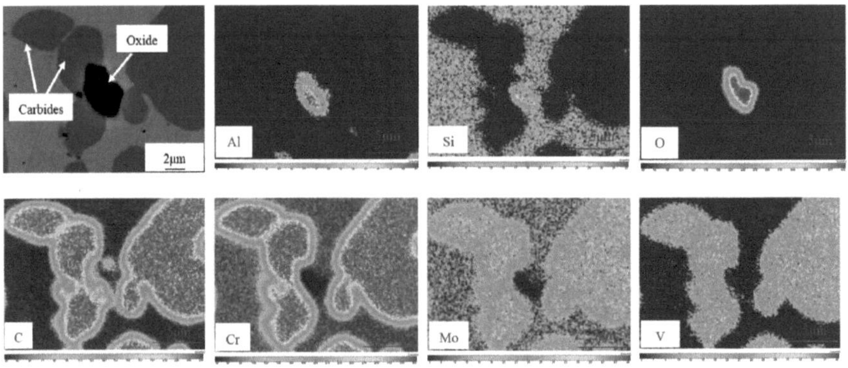

Fig. 2 Elemental maps of simple A

(2) The morphology of carbides nucleated oxides by EPMA

In order to further observe the distribution of elements in carbides and oxides, EPMA was carried out for sample A and B. The EPMA measurements showed strong segregation of alloying elements such as aluminum, silicon, oxygen, carbon, chromium, molybdenum and vanadium at the oxide and carbide. The elemental segregation has taken place at the boundaries between oxide and carbide. The results of the elemental plots revealed that elemental partitioning of simple A has been as same as simple B. According to Fig. 2, the oxide was Al_2O_3/SiO_2 and the carbides were (Fe, Cr)-rich Cr_7C_3 (converted to molar ratio by mass percentage according to EDS) for both simple A and B. Meanwhile, there was small amounts of molybdenum and vanadium in the carbide. Carbide nucleated oxide could be attributed to the precipitation thermodynamics. This part will be discussed in Section "Discussion".

(3) Size and distribution

The details of the size distribution and number of the carbides and carbides nucleated on oxides measured by INCA Steel are given in Fig. 3a, b. As shown in Fig. 3, the

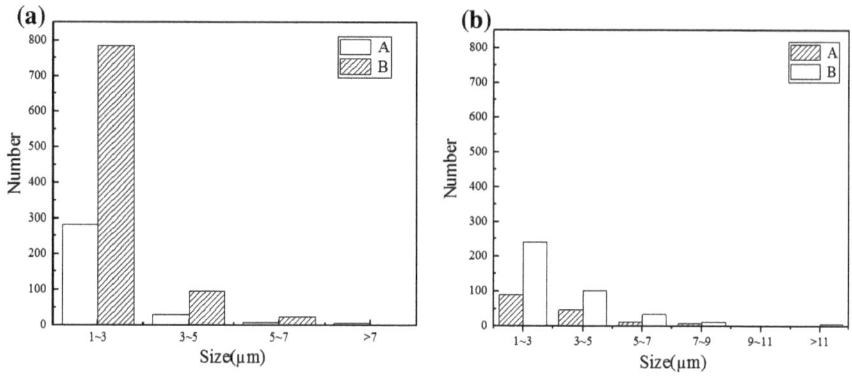

Fig. 3 Number distribution of various types of sizes **a** carbides **b** carbides and oxides

columns of different heights meant the number of carbides or carbides nucleated on oxides in the dimension range indicated by the x-axis. As can be seen from Fig. 3a, the size between 1 and 3 μm was the largest number for carbides, when the tempering temperatures were both 560 °C (833 K) and 530 °C (803 K). In terms of the same size, the number of carbides at tempering temperature 530 °C (803 K) was more than the tempering temperature 560 °C (833 K), especially the size less than 5 μm. The above regularity was as same as carbides nucleated on oxides, as shown in Fig. 3b. Therefore, it could be inferred that low tempering temperature was beneficial to enlarge the number of the carbides.

Discussion

From Section "Morphology of Carbides or the Mixture of Carbides and Oxides", it can be concluded that precipitate types after tempering are mainly Cr_7C_3, $Cr_{23}C_6$. Experimental results in Fig. 2 show the morphology of carbide nucleated oxide. The initial generation temperatures of carbides and oxide inclusions will be calculated by thermodynamic to explain this phenomenon. The solidification process of molten steel can be divided into three stages, namely, liquid phase, solid–liquid dual-phase, and solid phase. The liquidus temperature (T_L) and solidus temperature (T_S) of 8% Cr alloy can be calculated by the Empirical formula [16], respectively.

$$\begin{aligned} T_L = {} & 1809 - 83w(\text{C}) - 7.8w(\text{Si}) - 5w(\text{Mn}) - 32w(\text{P}) - 31.5w(\text{S}) \\ & - 1.5w(\text{Cr}) - 2w(\text{Mo}) - 2w(\text{V}) - 3.6w(\text{Al}) - 18w(\text{Ti}) \end{aligned} \tag{1}$$

$$\begin{aligned} T_S = {} & 1809 - 344w(\text{C}) - 12.3w(\text{Si}) - 6.8w(\text{Mn}) - 124.5w(\text{P}) \\ & - 183.5w(\text{S}) - 1.4w(\text{Cr}) - 4.1w(\text{Al}) - 4.3w(\text{Ni}) \end{aligned} \tag{2}$$

where w is the percentage of mass concentration for the element.

Formation of Oxides and Carbides in the Liquid Phase

Depending on the composition of molten steel, the possible forms of oxides in the steel include Cr_2O_3, Al_2O_3, MnO, SiO_2, V_2O_3 and MoO_2. The Gibbs free energy of formation of oxides in the liquid steel is obtained through following chemical reaction, as shown in the literature [17]. In order to make the reaction (1–7) spontaneously proceed to the right, the condition to be reached is $\Delta G<0$.

$$\Delta G = \Delta G^{\theta} + \mathrm{RT}\ln\frac{a_{M_xO}}{a^{x}_{[\mathrm{M}]}a_{[\mathrm{O}]}} = \Delta G^{\theta} + \mathrm{RT}\ln\frac{a_{M_xO}}{f^{x}_{[M]}[\%\mathrm{M}]^{x}f_{[O]}[\%\mathrm{O}]} \tag{3}$$

where $a_{M_xO}, a_{[M]}, a_{[O]}$ are the activities of oxides M_xO, M and O, respectively; [%M] and [%O] are the mass fractions of metal M and oxygen in molten steel, respectively; $f_{[\mathrm{M}]}$ and $f_{[\mathrm{O}]}$ are the activity coefficients of the metal elements M and oxygen, respectively, calculated by the Wagner model [18].

The Wagner model can be characterized by a single interaction coefficient, in which the activity coefficient expression of each element is

$$\lg f_k = \sum_{j=2}^{n} e_k^j[\%j] \tag{4}$$

where e_k^j is the first interaction coefficient between the solute elements, k is any solute component of the multi-phase, and j is the 2, 3, …, n solute component. The first interaction coefficient of each element in the molten steel for O, Cr, Al, Mn, Si, V, Mo can be found in literature [18], the activity coefficient of each element was obtained which are $f_{\mathrm{C}} = 1.02, f_{\mathrm{O}} = 0.06, f_{\mathrm{Cr}} = 0.71, f_{\mathrm{Al}} = 1.45, f_{\mathrm{Mn}} = 0.82, f_{\mathrm{Si}} = 2.14, f_{\mathrm{V}} = 0.35$ and $f_{\mathrm{Mo}} = 0.76$.

In the liquid phase (T > 1682 K), each activity coefficient and concentration are substituted into the formula (3), and the Gibbs free energy of the oxide is generated. The relationship diagram of $\Delta G \sim T$ is shown in Fig. 4a. It can be seen that only Al_2O_3 is formed in the liquid phase, and other oxides cannot be formed in the liquid phase.

When the actual solubility product of the precipitate-forming element is larger than the equilibrium solubility product, carbide can be formed. The solubility product of Cr_7C_3 or $Cr_{23}C_6$ in liquid steel is deduced as follows [19]:

$$23[\mathrm{Cr}] + 6[\mathrm{C}] = \mathrm{Cr_{23}C_6(s)} \quad \Delta G^{\theta} = -887{,}890 + 1284.48\mathrm{T} \tag{5}$$

$$7[\mathrm{Cr}] + 3[\mathrm{C}] = \mathrm{Cr_7C_3(s)} \quad \Delta G^{\theta} = -356{,}120 + 417.6\mathrm{T} \tag{6}$$

According to the principle of precipitation thermodynamics, the equilibrium solubility product of M_xC_y in liquid steel can be expressed as

$$\ln\left([\%\mathrm{M}]^x[\%\mathrm{C}]^y\right) = B + \frac{A}{T} - x\ln f_M - y\ln f_C \tag{7}$$

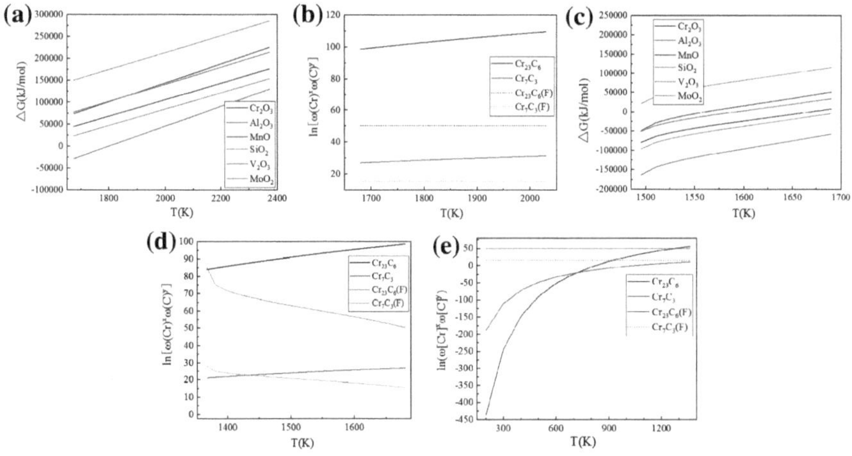

Fig. 4 Precipitation thermodynamic calculation results **a** oxides in liquid **b** carbides in liquid **c** oxides in solid–liquid dual-phase **d** carbides in solid–liquid dual-phase **e** carbides in solid phase

According to formula (5)–(7), the solubility products of $Cr_{23}C_6$ and Cr_7C_3 in liquid steel are obtained as follows:

$$\ln\left([\%\text{Cr}]^{23}[\%\text{C}]^{6}\right) = -1284.48 + \frac{887{,}890}{T} - 23\ \ln f_{Cr} - 6\ \ln f_C \quad (8)$$

$$\ln\left([\%\text{Cr}]^{7}[\%\text{C}]^{3}\right) = -417.6 + \frac{356{,}120}{T} - 7\ \ln f_{Cr} - 3\ \ln f_C \quad (9)$$

Put $f_C = 1.02$, $f_{Cr} = 0.71$ into formulas (8) and (9), the relationship between $\ln([\%\text{Cr}]^x[\%\text{Cr}]^y)$ and T is shown in Fig. 4(b). It can be seen that $Cr_{23}C_6$ and Cr_7C_3 cannot be formed in liquid steel.

Formation of Oxides and Carbides in Solid–Liquid Dual-Phase

Due to the element segregation in the solidification front, the Scheil formula is introduced in the solidification process as follows:

$$[\%\text{N}] = \frac{[\%\text{N}_0]}{\text{P}_\text{s}(\text{K}_\text{N} - 1) + 1} \quad (10)$$

$$[\%\text{M}] = [\%\text{M}_0](1 - \text{P}_\text{s})^{(K_M - 1)} \quad (11)$$

where [%M] and [%N] are the mass fraction of the metal element M and the non-metal element N in the solid–liquid phase during solidification; $[\%\text{M}_0]$ and $[\%\text{N}_0]$ are the mass fraction of the metal element M and the non-metal element N in the liquid steel before solidification; K_M and K_N are the equilibrium solute partition coefficient for M and N, respectively.

$$P_s = \frac{(T_m - T_s)(T_l - T)}{(T_l - T_s)(T_m - T)} \quad (12)$$

In the literature [18], the equilibrium partition coefficients of solute elements oxygen, carbon, vanadium, chromium, molybdenum and aluminum in the solidification process are 0.02, 0.17, 0.90, 1, 0.80, 0.6; T is the temperature of the system during solidification, K; T_m—the melting point of pure iron, 1809 K; T_l, T_s—liquidus and solidus temperature of steel, 1682 K, 1365 K.

Put the oxygen concentration and the metal element concentration calculated by the Scheil into the formula (3), and the relationship of oxides between ΔG and T can be obtained, as shown in Fig. 4c. In the same way, the relationship of carbides between $ln([\%Cr]^x[\%C]^y)$ and T is shown in Fig. 4d. It can be seen from Fig. 4c that when the temperature is in the solid–liquid two-phase region, the Gibbs free energy of Al_2O_3 is much smaller than other oxides, so Al_2O_3 is easily formed. Due to the segregation of solidification, the concentration of the liquid phase in the solid–liquid two phases is greater than the solid phase.

The equilibrium solubility product curves of Cr_3C_7 and $Cr_{23}C_6$ in the liquid phase of the solidification front have intersections with the actual solubility product curves. With the decrease in temperature, the actual solubility product becomes greater than the equilibrium solubility product, and the precipitation thermodynamic condition is reached. The intersection points are the initial precipitation temperatures of precipitates in the solid–liquid two-phase zone, and the theoretical precipitation temperature of Cr_3C_7 and $Cr_{23}C_6$ are 1425 and 1375 K, respectively, which can be judged by the intersection point.

Formation of Carbides in Solid Phase

Since the Gibbs free energy of element dissolution in solid-phase ferrite has not been studied, empirical data in austenite is still used in the calculation process. During the solidification process, X90CrMoV directly enters the austenite zone and has no solute in the high temperature ferrite zone according to Fig. 3. According to literature [18], the chemical reaction of Cr forming carbides in austenite is shown in the formula (13) and (14).

$$23[Cr]_\gamma + 6[C]_\gamma = Cr_{23}C_6(s) \quad \Delta G^\theta = -959{,}797.2 + 1172.7T \quad (13)$$

$$7[Cr]_\gamma + 3[C]_\gamma = Cr_7C_3(s) \quad \Delta G^\theta = -389{,}247.6 + 376.98T \quad (14)$$

Combined with the solubility product formula, the relationship between the solubility product and the temperature can be obtained for $Cr_{23}C_6$ and Cr_7C_3, as shown in Fig. 4e. As can be seen from Fig. 4, $Cr_{23}C_6$ and Cr_7C_3 begin to precipitate.

The temperature is lower and the difference between the actual solubility product and the equilibrium solubility product is larger. The carbide is easier to be formed. This is consistent with the statistical results in Section "Morphology of Carbides or the Mixture of Carbides and Oxides", i.e. the amount of carbides and carbides nucleated oxides in B is greater than A. The thermodynamic calculation results of the precipitation of oxides and carbides are in agreement with the observations at Section "Formation of Oxides and Carbides in Solid–Liquid Dual-Phase". Because Al_2O_3 precipitates in the liquid phase, Cr_3C_7 and $Cr_{23}C_6$ begin to precipitate in the solid–liquid two-phase region. Therefore, as the, the phase nucleation of $Al_2O_3$is precipitated first, and Cr_3C_7 and $Cr_{23}C_6$ are attached to the edge as the core.

Conclusions

(1) In terms of the same size, the number of carbides at tempering temperature 530 °C (803 K) was more than the tempering temperature 560 °C (833 K), especially the size less than 5 μm.
(2) There are carbides and carbides nucleated oxides of Al_2O_3 in X90CrMoV steel. Because the phase nucleation of Al_2O_3 is precipitated first, Cr_3C_7 and $Cr_{23}C_6$ are attached to the edge as the core.
(3) According to thermodynamic calculation results, the temperature is lower and the difference between the actual solubility product and the equilibrium solubility product is larger. The carbides are easier to be formed at a relatively low temperature.

Acknowledgements The authors are thankful for the support from the National Natural Science Foundation of China (Nos. U1560203 and 51274031), and the Beijing Key Laboratory of Special Melting and Preparation of High-End Metal Materials in the School of Metallurgical and Ecological Engineering of University of Science and Technology Beijing, China.

References

1. Abdul Rahim MASB, Minhat MB, Hussein NISB, Salleh MSB (2018) A comprehensive review on cold work of AISI D2 tool steel. Metall Res Technol 115(1):104
2. Wang L, Kang Y, Cai Z, Zhang Q, Zhao Y, Zhao H, Su P (2012) The energy method to predict disc cutter wear extent for hard rock TBMs. Tunn Undergr Space Technol 28:183–191
3. Zhang G, Xing J, Gao Y (2006) Impact wear resistance of WC/Hadfield steel composite and its interfacial characteristics. Wear 260(7):728–734
4. Nurminen J, Näkki J, Vuoristo P (2009) Microstructure and properties of hard and wear resistant MMC coatings deposited by laser cladding. Int J Refract Metal Hard Mater 27(2):472–478
5. Hetzner DW, Van Geertruyden W (2008) Crystallography and metallography of carbides in high alloy steels. Mater Charact 59(7):825–841
6. Nanesa HG, Boulgakoff J, Jahazi M (2016) Influence of prior cold deformation on microstructure evolution of AISI D2 tool steel after hardening heat treatment. J Manuf Process 22:115–119

7. Meng F, Tagashira K, Azuma R, Sohma H (1994) Role of eta-carbide precipitations in the wear resistance improvements of Fe-12Cr-Mo-V-1.4C tool steel by cryogenic treatment. ISIJ Int 34(2):205–210
8. Koneshlou M, Meshinchi Asl K, Khomamizadeh F (2011) Effect of cryogenic treatment on microstructure, mechanical and wear behaviors of AISI H13 hot work tool steel. Cryogenics 51(1):55–61
9. Zhou XF, Fang F, Jiang JQ, Zhu WL, Xu HX (2014) Refining carbide dimensions in AISI M2 high speed steel by increasing solidification rates and spheroidising heat treatment. Mater Sci Technol 30(1):116–122
10. Ghomashchi MR, Sellars C (1993) Microstructural changes in As-Cast M2 grade high speed steel during hot forging. Metall Mater Trans A Phys Metall Mater Sci 24(10):2171–2180
11. Bombac D, Fazarinc M, Saha Podder A, Kugler G (2013) Study of carbide evolution during thermo-mechanical processing of AISI D2 tool steel. J Mater Eng Perform 22(3):742–747
12. Luo Y, Guo H, Guo J (2018) Effect of cooling rate on the transformation characteristics and precipitation behaviour of carbides in AISI M42 high-speed steel. Ironmaking Steelmaking 2018:1–7
13. Ko D, Kim S, Kim B (2015) Influence of microstructure on galling resistance of cold-work tool steels with different chemical compositions when sliding against ultra-high-strength steel sheets under dry condition. Wear 338–339:362–371
14. Kim H, Kang J-Y, Son D, Lee T-H, Cho K-M (2015) Evolution of carbides in cold-work tool steels. Mater Charact 107:376–385
15. Ko D-C, Kim S-G, Kim B-M (2015) Influence of microstructure on galling resistance of cold-work tool steels with different chemical compositions when sliding against ultra-high-strength steel sheets under dry condition. Wear 338–339:362–371
16. Jiaqi W, Jirong H (1977) Solidification of metal and its control. China Machine Press, Beijing, pp 9–35
17. Chen J (2006) Steelmaking common chart data manual. Metallurgical Industry press, Beijing
18. Xigu H (2011) Principle of steel metallurgy. Metallurgical Industry Press, Beijing
19. Ning AG (2015) Investigation on nanoscale precipitates in hot-work die steel and comprehensive strengthening mechanism of steel, Ph.D. thesis. University of Science and Technology Beijing

Analysis of Large Inclusions in Crankshaft Steel by Ingot Casting

Qinghai Zhou, Jiongming Zhang and Yanbin Yin

Abstract By means of the anhydrous solution electrolysis, large inclusions in crankshaft steel samples are non-destructively extracted. Through optical microscopy and scanning electron microscopy, the three-dimensional morphology, composition, size distribution and quantity density of the large inclusions in the refining, teeming and steel ingot were revealed. The size of the inclusions obtained is 50–200 μm. The results showed that the main components of inclusions in ladle furnace (LF) sample, the heated billet and the rolled products were Al_2O_3 and SiO_2. After LF and Ruhrstahl Hereaeus (RH) refining, the main components were Al_2O_3, SiO_2, CaO, and a small amount of MgO. It is found that the secondary oxidation during the pouring process is the main source for large inclusions. The shapes are mostly irregular blocks. There are more inclusions at the top of rolled products, and more inclusions in head and less inclusions in tails. During the pouring process, the molten steel in the ingot mold circulates, and the molten steel near the wall of the ingot mold flows downward and entangled in the protective slag, then captured by the solidified shell near the wall of the ingot mold, which is also the reason why large inclusions are distributed on the surface layer.

Keywords Crankshaft steel · Large inclusions · Composition evolution Quantity distribution

Introduction

42CrMoA crankshaft produced by a steel plant has crack defects due to large inclusions. Research [1–4] shows that inclusions are the main source of stress cracks in crankshaft steel. The number of large inclusions is very small, and its distribution in steel is random, it has greatly affected the quality of the product. The composition,

Q. Zhou · J. Zhang (✉) · Y. Yin
State Key Laboratory of Advanced Metallurgy, University of Science and Technology Beijing, Beijing 100083, China
e-mail: jmz2203@sina.com

G. Lambotte et al. (eds.), *Materials Processing Fundamentals 2019*, The Minerals, Metals & Materials Series, https://doi.org/10.1007/978-3-030-05728-2_10

size, position and distribution of the large inclusions have not yet been carried out much more. This paper combines products and experimental results to analyze the evolution of inclusions in steel during refining and solidification process, and systematically studies the types, sizes, compositions and morphologies of inclusions in different parts of ingots. Finding out the position where non-metallic inclusions in the ingot are easily gathered, thereby guiding the control of large inclusions in the crankshaft steel.

Currently, the common methods for detecting inclusions are metallographic methods, acid dissolution methods, and electrolysis methods [5]. The sample is ground into a metallographic sample, and the number, composition and morphology of the microscopic inclusions in the steel are analyzed by means of an automatic scanning analysis system (INCASTEEL, Oxford). The maximum size of the inclusions can be detected to be about 20 μm. The inclusions extracted by the traditional large-scale electrolysis method are larger in size, generally larger than 100 μm, and the inclusion analysis with the size of 20–100 μm is a "blind zone" in the detection and analysis work [6, 7]. In this paper, through a non-aqueous electrolysis method, the full-size non-metallic inclusions in steel can be extracted intact, and the number, size and composition of inclusions are analyzed by light microscopy, scanning electron microscopy and energy spectrometer.

Experimental Method

The 42CrMoA crankshaft steel produced by a steel plant was selected as the research object. The main chemical composition is shown in Table 1. The production process of the steel is: Converter smelting → LF refining → RH refining → Molded bottom casting → Rough rolling blanking → Continuous rolling, the final material is billet with round of 160 mm. The specific process is as follows: molten iron and scrap are charged into a converter, and oxygen is blown into the molten for decarburization and dephosphorization. In the tapping process, alloy additive is added into the ladle, deoxidized by aluminum iron. Subsequently, the ladle is transferred to a LF refining station for heating, alloy adjustment and further refining. Then the ladle is transferred to the RH refining station, and the molten steel is degassed, the composition is finely adjusted and the inclusions are removed. The refining end temperature is 1560 °C, the refining time is ≥30 min, and the pouring method is poured in the ingot mold by the bottom injection method, and the pouring temperature is around 1530 °C.

Figure 1 is a schematic diagram of an anhydrous electrolysis experiment apparatus. The electrolyte is a saturated $FeCl_3$ ethanol solution prepared from anhydrous

Table 1 Chemical composition of 42CrMoA steel

C	Si	Mn	S	Cr	Al	Mo
0.3965	0.292	0.747	0.0003	1.129	0.0233	0.233

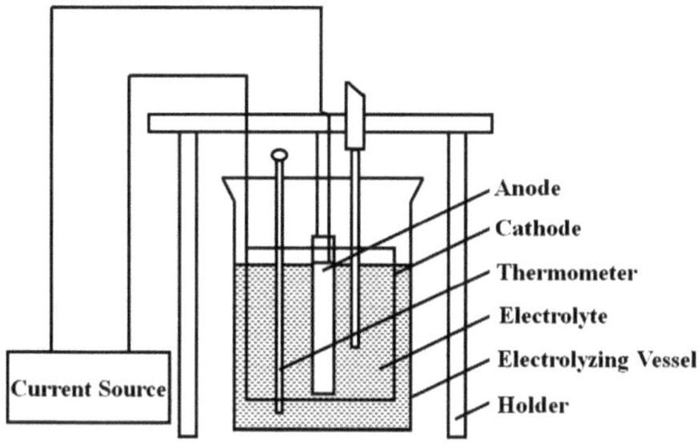

Fig. 1 Schematic diagram of anhydrous solution electrolysis device

$FeCl_3$ and high purity ethanol. The sample is used as the anode and the stainless steel rim is used as the cathode. Elemental Fe in the sample loses electrons turn to be ions and enter the electrolyte, and the inclusions do not change and precipitate into the anode mud. A 2000 mesh screen (mesh size of 5 μm) is used to insulate between the anode and the cathode to ensure that the deposited inclusions are always in the screen. The anode slime is collected and filtered through a filter to obtain inclusions after the electrolysis is completed. The extracted large inclusions were placed under a body mirror of 45 times magnification, and the inclusions were photographed using a DLC500 high-resolution digital camera. The ScopePhoto 3.0 image measurement software statistically analyzed the inclusions, and finally observed under a scanning electron microscope. The inclusions were topographically analyzed and their composition was analyzed by means of Energy Dispersive Spectrometer (EDS).

This steel is a high value-added steel, and due to the particularity of the production conditions, the steel cannot get samples directly from ingots, so it is considered to take samples in the rolled material. Through the three-dimensional finite element rolling model, the position of the sampling position of the rolled material is reversely positioned at the position of the ingot and before rolling: the sampling of the head corresponds to the connection position between the upper part of the ingot and the cap before the hot rolling, and the sampling in the middle corresponds to the heat. Before rolling, the upper part of the ingot is in the upper position, and the tail sampling corresponds to the lower part of the steel ingot before hot rolling. The sampling position of the edge of the cross section of the rolled material is reversed to the surface of the ingot, and the sampling position of the cross section of the rolled material is reversed to the one-fourth surface of the cross section of the ingot. As shown in Fig. 2, the upper cross-sectional dimension of the ingot is 719 mm * 887 mm, the lower cross-sectional dimension is 555 mm * 791 mm, and the height of the ingot is 2760 mm. The ingot was sent to a hot rolling mill at an initial rolling temperature of 1150 °C, an initial ingot speed of 0.1 m/s, a roll speed of 80 rad/min, and a final billet of 160 mm round billet after rolling.

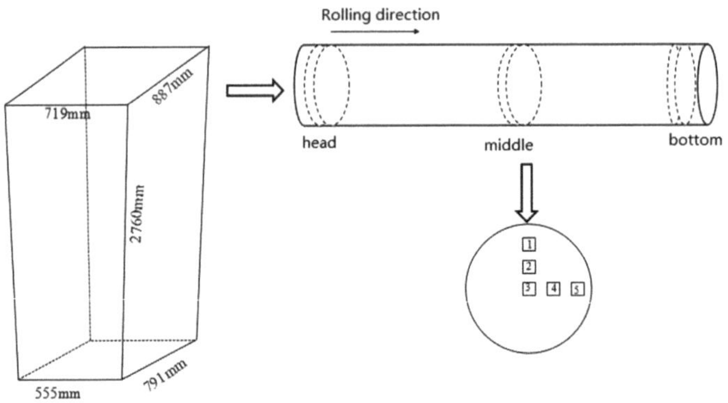

Fig. 2 Sampling schematic diagram

Discs samples were taken at three locations on the head, middle and tail of the rolled material. Five samples (20 mm * 20 mm * 20 mm) were obtained from each disc sample, with sample 3 at the center position, sample 2 and sample 4 at the center, sample 1 and sample 5 are 20 mm from the edge. Fifiteen samples were obtained from each ingot, and the samples were ground and polished into metallographic samples. The inclusions, composition and morphology of the inclusions in the steel were analyzed by the inclusion scanning automatic analysis system (INCASTEEL, Oxford). The minimum inclusion size to be analyzed is set to 2.5 μm and the scan area is set to 18 mm * 18 mm. After the automatic scanning analysis is completed, representative inclusions are repositioned and EDS spectrum analysis is performed to determine the composition.

Experimental Results

Result of Inclusions Before and After Refining

Figure 3 and Table 2 reflect the morphology and composition evolution of inclusions before LF refining, after LF refining, after RH refining, in the casting platform and in heated samples.

Result of Inclusion Behavior in Solidification Process

In order to fully understand the source of inclusions in crankshaft steel, based on the chemical composition of molten steel, the Equilib module of FactSage 7.0 software is used to calculate the precipitation of inclusions during solidification of molten

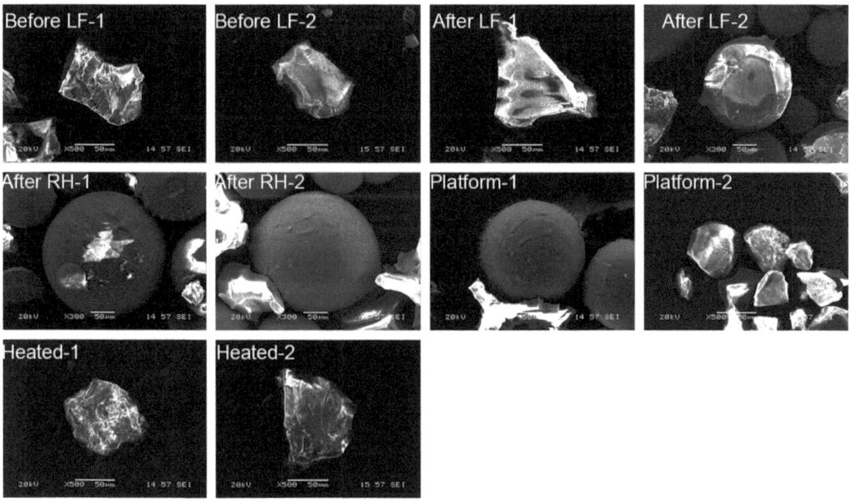

Fig. 3 The morphology and composition of inclusions in samples

Table 2 Composition statistics of inclusions in samples

	CaS	CaO	Al_2O_3	SiO_2	MgO	K_2O
Before LF 1	0.00	0.00	93.19	6.81	0.00	0.00
Before LF 2	0.00	0.00	100.00	0.00	0.00	0.00
After LF 1	0.97	52.44	34.72	6.47	5.39	0.00
After LF 2	1.95	52.97	29.28	7.70	8.10	0.00
After RH 1	0.00	46.14	31.67	8.75	13.44	0.00
After RH 2	0.00	42.19	31.62	7.27	18.23	0.00
Platform 1	0.00	0.00	15.92	76.75	1.81	3.40
Platform 2	0.00	0.00	10.26	84.98	1.44	3.32
Heated 1	0.00	0.00	96.44	0.00	0.00	0.00
Heated 2	0.00	0.00	98.72	0.00	0.00	0.00

steel. This module is mainly used to calculate the reaction in a given reaction while the balance of each species is based on the minimum Gibbs free energy of the system under constant temperature and constant pressure conditions. It is believed in FactSage that the dissolved metal atom M in the molten steel has a tendency to combine with the dissolved O atoms to form a dissolved MO complex. An association solution model was employed for the Gibbs free energy calculation of the combination. The results are shown in Fig. 4.

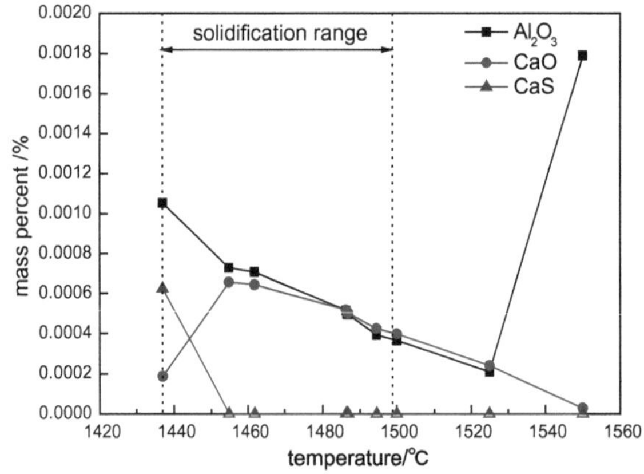

Fig. 4 The calculated result of the inclusion generation during the molten steel solidification

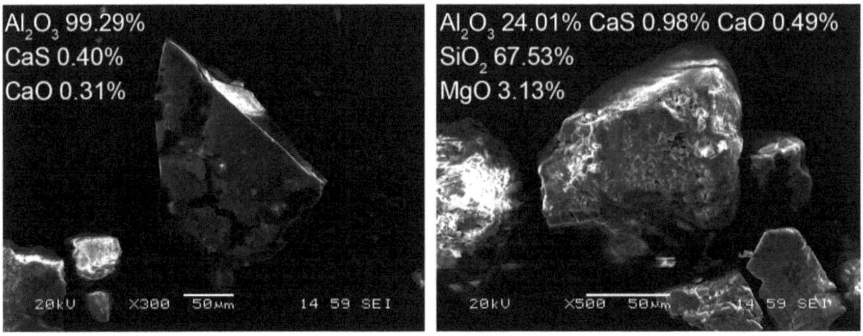

Fig. 5 Morphology and composition of typical inclusions in axle steel ingot

Result of Large Inclusions in Ingots

The chemical composition of the inclusions in the hot-rolled ingot was compared with the inclusion composition of the casting platform sample, and it was found that the contents of MgO and Al_2O_3 in the ingot increased and the CaO content decreased. In addition, during the solidification of molten steel, the mass fraction of Al_2O_3 and CaS in the ingot inclusions increased due to the precipitation of Al_2O_3 and CaS. The results are shown in Fig. 5.

Figure 6 is a histogram of the number density of large inclusions at different positions of 1# and 2# ingots. The number density of inclusions is the number of inclusions larger than 50 μm per unit cubic centimeter of steel obtained in the absence of electrolysis. The ordinate is the number of inclusions per unit cubic centimeter, and the abscissa is the distribution of large inclusion particle size intervals at different sampling positions of the head, middle and tail.

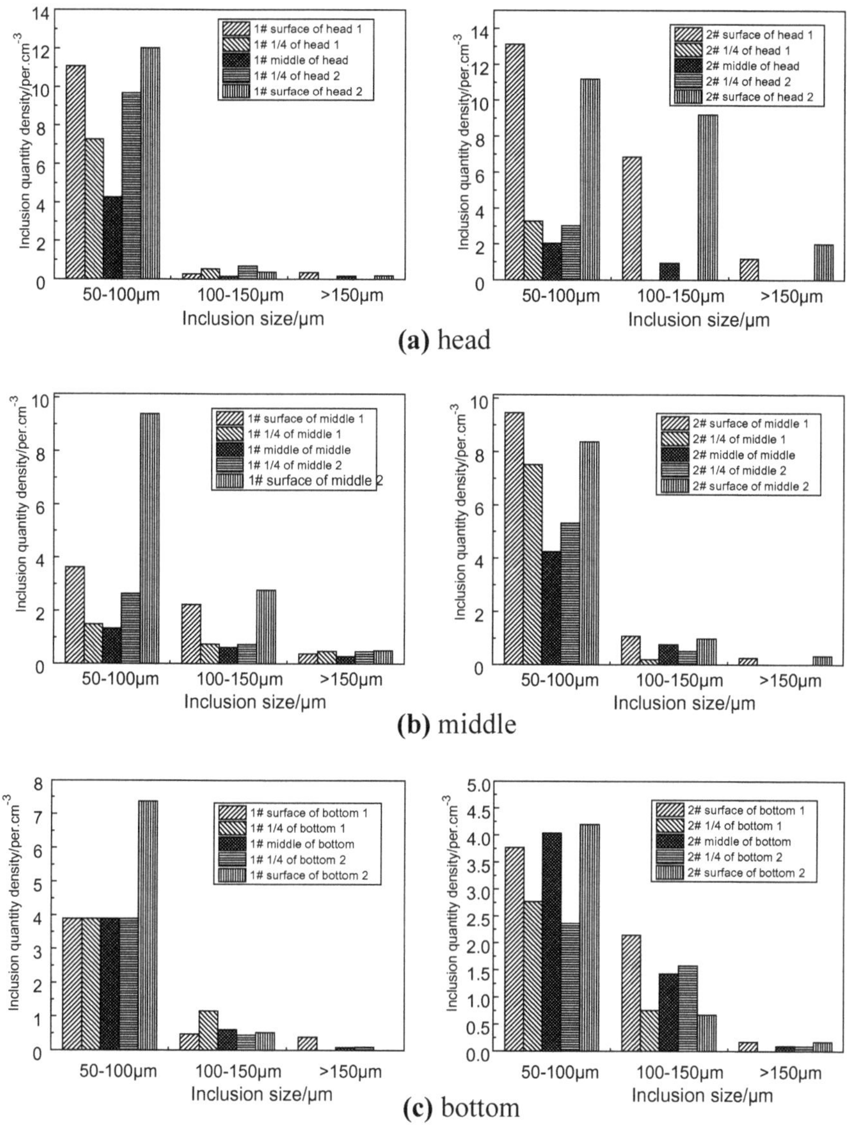

Fig. 6 Histogram of mass density for large inclusions at different positions

(a) is the position of the steel ingot head. The particle size of the electrolytic inclusions at the cross-sectional positions of the 1# sample is mainly 50–100 μm and the distribution law is that more in surface and less in center. Since the junction between the riser and the steel ingot is easy to form a circulation and the cooling strength is large here, it is easy to entrap the inclusion and capture, thereby forming an inclusion accumulation area. The 2# sample is between 50–100 and

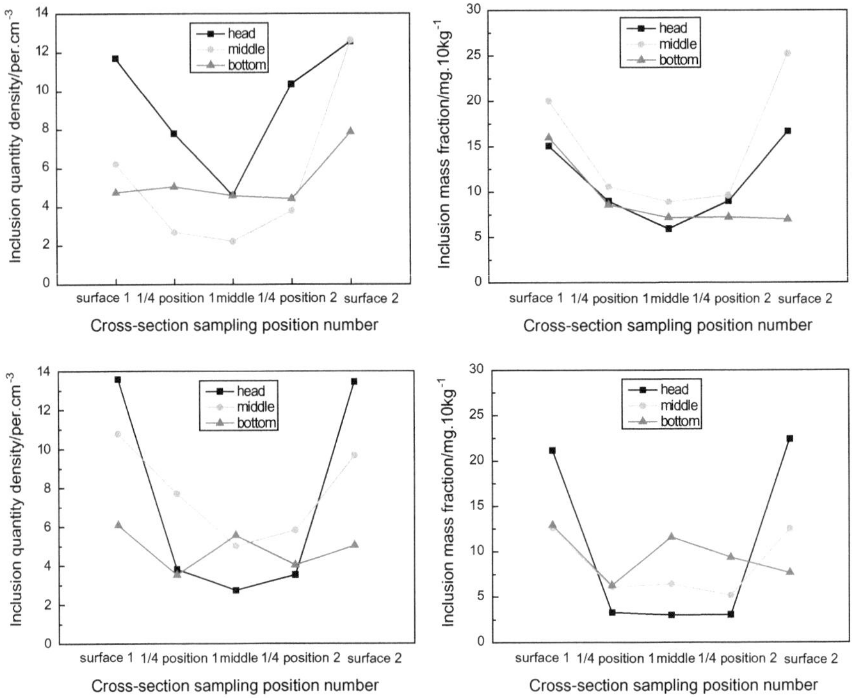

Fig. 7 Mass fraction distribution of large inclusions in crankshaft steel

100–150 μm, and also reflects the regular that there are less inclusions in center and more inclusions on surface. The number of inclusions larger than 150 μm is small.

(b) reflect the middle position of the ingot, the size of the electrolytic inclusions at the cross-sectional positions of the 1# and 2# samples is mainly concentrated at 50–100 μm. The distribution of large inclusions in the cross section is more inclusions in surface, and the number of inclusions larger than 150 μm is small.

(c) shown the position of the steel ingot tail, the size of the electrolytic inclusions at the five cross-sectional positions of the 1# and 2# sample layers is mainly concentrated at 50–100 μm and the large inclusions of the 100#-150 μm of the 2# sample also have some. The distribution law is different from the head and the middle, but basically meets the regular that more inclusions in surface and less inclusions in internal.

According to the number of inclusions in each group, the number density and mass fraction distribution of large inclusions electrolyzed in the two ingots can be obtained through integration, as shown in Fig. 7.

Discussion

It can be seen from Fig. 3 and Table 2 that before LF refining, most of the inclusions in the steel are irregular single-block or cluster-like aluminum deoxidation products Al_2O_3 inclusions; after LF refining, some Al_2O_3 inclusions in the sample turn to be a spherical CaO–Al_2O_3 system and a MgO–Al_2O_3 composite inclusion. After the RH refining, the composition of the inclusions became relatively stable, mainly CaO–Al_2O_3 system and CaO–MgO–Al_2O_3 composite inclusions, and the average composition of the components entered the lower melting point inclusion region. Most of the inclusions in casting platforms are irregular block aluminates inclusions, while the inclusions in heated samples are single-block Al_2O_3 inclusions.

During the solidification process of molten steel, as shown in Fig. 4, the dissolved oxygen and deoxidizing elements in the molten steel are supersaturated, the solubility of sulfur in the molten steel is lowered, and Al_2O_3, CaO and CaS can be precipitated to form a composite inclusion. Among them, CaO and Al_2O_3 precipitated first. As the temperature is lowered to about 1454 °C, the segregation of sulfur increases the sulfur content before solidification, and the single CaS is stably precipitated, and the mass fraction of Al_2O_3 and CaS in the inclusion increased.

Figure 5 shows the morphology and composition of typical inclusions under electron microscopy. As can be seen from the figure, the main components of the inclusions in the ingot are Al_2O_3, CaO, CaS and MgO, and a small amount of SiO_2. Among them, Al_2O_3 has the largest proportion, and the chemical composition of the inclusions in the gray and black parts is similar, and it is CaO–MgO–Al_2O_3 inclusions. The number of such inclusions is the largest, and the size is more than 50 μm. The large inclusions shown in the figure are generally irregular blocks with a large number of pores on the surface. Some inclusions have sharp corners at the outer contour, and the largest size is about 200 μm. Studies [8] have shown that the reason for the different shapes of inclusions is that the growth mechanism of inclusions in different size ranges is different. When the size of inclusions is small, the growth is controlled by diffusion and Brownian motion collision. When the size is larger, the steel is liquid, which is controlled by turbulence collision.

As shown in Fig. 6, The size of large inclusions is mostly in the range of 50–100 μm, and the number of inclusions that larger than 150 μm is small. The inclusions are mainly located at the surface of the ingot. From the overall distribution, the large inclusions are concentrated in the surface of the ingot, and the number density distribution tends to gradually increase from the center outward. The reason for the analysis is that the molten steel near the wall of the ingot mold flows downward, and entangled in the mold flux, and then caught by the solidified shell near the mold wall.

It can be seen from Fig. 7 that the comparison of the number density and mass fraction of inclusions in the crankshaft steel ingot, the distribution in the height direction of the ingot can clearly see that the number of large inclusions in the head is large, the number of inclusions in the middle is second, and the number of inclusions is less in the tail, and the distribution of inclusions in the height direction of the ingot is gradually reduced from head to tail. The amount of inclusions is not evenly distributed within a certain horizontal section within the ingot. The mass fraction distribution of large inclusions in the ingot, the head, the middle and the tail show the same pattern in the width direction: the edge surface > the wide side 1/4 position > center. Therefore, the number of large inclusions in the ingot is more in the lateral outer part and less in the inner part, and more inclusions in head and less inclusions in tails.

Conclusion

(1) The three-dimensional shape of the large inclusions of crankshaft steel after LF and RH refining is generally spherical, while the inclusions in the casting platform and the rolled material are mostly irregular blocks. The composition evolution of inclusions in the refining process is "$Al_2O_3 \rightarrow MgO–Al_2O_3 \rightarrow CaO–MgO–Al_2O_3 \rightarrow CaO–Al_2O_3$".
(2) The inclusions in the hot-rolled ingot of crankshaft steel are mainly $CaO–MgO–Al_2O_3$ and $CaO–Al_2O_3$ composite inclusions, and the inclusion components in different parts have no obvious change. The distribution of large inclusions is more in the lateral outer part and less in the inner part, and more inclusions in head and less inclusions in tails.
(3) During the refining process before pouring, large inclusions collide and polymerize are easily removed by floating. During the pouring process, the molten steel in the ingot mold circulates, and the molten steel near the wall of the ingot mold flows downward and is entangled in the protective slag, and is captured by the solidified shell near the wall of the ingot mold, which is also the reason why large inclusions are distributed on the surface layer.

References

1. Cheng J, Eriksson R, Jönsson P (2003) Determination of macroinclusions during clean steel production. Ironmaking Steelmaking 30(1):66–72
2. Miki Y, Kitaoka H, Sakuraya T et al (1992) Mechanism of separation of inclusions from molten steel stirred with rotating electro-magnetic field. Tetsu-to-Hagane 78:431–438
3. Ragnarsson L, Sichen D (2010) Inclusions generated during ingot casting of tool steel. Steel Res Int 81:40–47
4. Liu J-H, Zhuang C-L, Cui X-N et al (2014) Inclusion distribution in ingots investigated by dissection. J Iron Steel Re Int 21:660-665
5. Kaike C (2010) Continuous casting billet quality control. Metallurgical Industry Press
6. Dunmin B, Jiongming Z, Shunxi W (2015) Analysis of large particle inclusions in GCr15 bearing steel. In: National special steel annual meeting
7. Jin Y, Bo H, Jiongming Z (2011) Analysis of non-metallic inclusions in Q420 steel. Foundry Technol 32(1):43–45
8. Lifeng Z (2011) Inclusion in the casting process during the molding process. In: International symposium on clean steel production technology

Research on the L2 Control Model Technology of Double Cold Reduction During Continuous Annealing Process

Wei Guo, Hui Wang, Yanglong Li, Jie Wen, Meng Yu and Fengqin Wang

Abstract The double cold reduction (DCR) process of ultra-thin plate rolling, which is seldom based on the model calculation is generally carried out by manual input rolling instruction parameters. The L2 process control model plays an important role in any process control system. In this paper, the DCR process control of continuous annealing is taken as the study object, and the control principle and function of each control module in L2 control models are studied, and the significance of L2 model to process control and technology application is analyzed. Some suggestions for the improvement of the control model application are proposed, which can provide effective guidance for the high automatic control of the double cold reduction of the ultra-thin strip steel.

Keywords Double cold reduction · L2 control model · Rolling · Shape Unsteady state

Background

Generally, the double cold reduction (DCR) is one of the main processes to ensure the dimensional accuracy, shape, surface quality and stamping performance of cold rolled strip [1]. Domestic and foreign scholars have done a lot of research on DCR, including roll crown optimization [2], shape control [3], surface roughness control [4], color difference control [5], etc. At present, there are less than 40 sets of DCR for the production of high–quality, ultra-thin tin plate in the world. DCR technology is a common process for the production of tinplate with of thin specifications, and the thinnest target thickness can even be 0.12 mm [6]. Nevertheless, during DCR process, the degree of automation is not high at present. Adjusting or configuring the strategy table mostly depends on the experience of operators, lacking of theoretical basis and mathematical model as the support. The shape quality fluctuations and defects can

W. Guo (✉) · H. Wang · Y. Li · J. Wen · M. Yu · F. Wang
Shougang Research Institute of Technology, Beijing 100043, China
e-mail: guowei0514@126.com

G. Lambotte et al. (eds.), *Materials Processing Fundamentals 2019*, The Minerals, Metals & Materials Series, https://doi.org/10.1007/978-3-030-05728-2_11

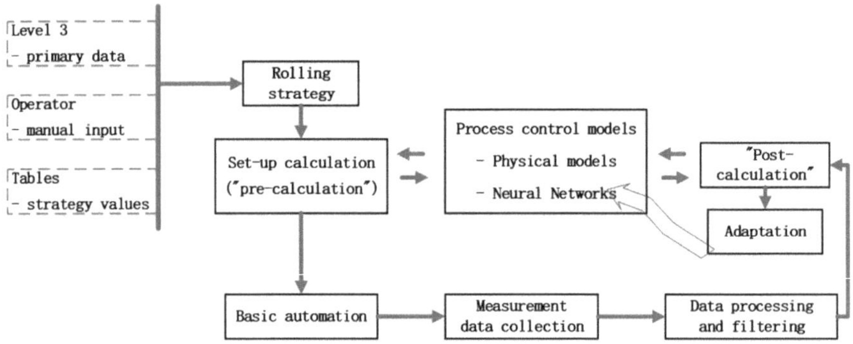

Fig. 1 Frame diagram of L2 process control system

only be adjusted by the experience of operators, and cannot be automatically adjusted by L2 control model. Such a situation has considerable limitations, which seriously restricts the development of automatic process control of cold rolling. Due to the growing complexity of the plants, operators need a sophisticated process automation to make the process transparent and to enable quick interventions.

L2 Process Control Model of DCR

L2 control model is the "brain" of the automatic process of DCR. It is responsible for the setup of various process parameters in the production process, and the accuracy of these process parameters directly affects product quality and production stability. The L2 control model of DCR includes two main parts: setup calculation and adaptation calculation. The setup calculation model is made up of pass schedule calculation, profile & flatness setup calculation, and wedge transition control model. The pass schedule calculation includes rolling physics model and the strip temperature model. The profile and flatness setup model includes roll bending model and roll temperature and wear model. In the adaptation part, the parameters of the roll force, yield stress, friction coefficient, flatness (roll gap contour), etc. ,are self-learning, and the adaptation of the yield stress and friction coefficient use neural network algorithms.

Figure 1 is the whole frame diagram of L2 process control system. Due to DCR, the strip temperature model and roll temperature and wear model are not introduced in this paper.

Pass Schedule Calculation and Rolling Physics Model

According to the strategy parameters (including operator input data, strategy table values and primary data), the pass schedule calculation system starts with the current entry thickness and the material characteristics of the coil and calculates the pass-specific setpoints such that the final strip thickness is achieved after the last pass. The specific plant limits are always considered. Various technological models are used to calculate the setpoints. The setpoints required for rolling operation are calculated using a process model. This process model is based on the theoretical principal of cold strip rolling and mathematical equations.

Rolling physics model is the "technological core" of DCR process automation system, which concludes four sub models: friction model, yield stress model, roll force model and temperature model. Since the temperature is constant at room temperature in this research, the temperature model cannot be studied in this paper.

Friction Model

The friction model in the current research is exponential model. The friction model's output variable is the Coulomb frictional coefficient. The model expression is as follows:

$$\mu = \mu_0 \cdot \left(1 + C_R \cdot \frac{R - R_0}{R_0}\right) \cdot \left(1 + \frac{c_W}{1 + \frac{L}{L_0}}\right) \cdot \alpha_{\text{baseC}} + \mathrm{d}\mu_{\mathrm{v}} \cdot \mathrm{e}^{-\frac{\mathrm{V}}{\mathrm{V0}}} \cdot \left(1 + C_R \cdot \frac{R - R_0}{R_0}\right) \cdot \left(1 + \frac{c_W}{1 + \frac{L}{L_0}}\right) \cdot \alpha_{\text{speedC}} \tag{1}$$

where, μ is the Coulomb frictional coefficient, μ_0 is the friction base value, v is the roll line speed, v_0 is the reference speed, $d\mu_v$ is the velocity coefficient, R is the roll surface roughness, R_0 is the reference roll surface roughness, C_R is the roughness coefficient, L is roll length, L_0 is the reference wear length, c_W is the wear coefficient. α_{baseC} and α_{speedC} are correction factors.

The variables of roll surface roughness, roll length and roll line speed have effects on the friction coefficient. The effects of these variables on the friction coefficient are as follows in Fig. 2: (when discussing the effect of any variable on the friction coefficient, the other variables remain consistent.)

The effect of roll surface roughness on friction coefficient is linearly related. The larger the roll surface roughness is, the bigger the friction coefficient. But the effect of roll surface roughness on the friction coefficient numerical value is not very great. When roll surface roughness increases from 0.5 to 3 μm, the friction coefficient increases by only 0.001.

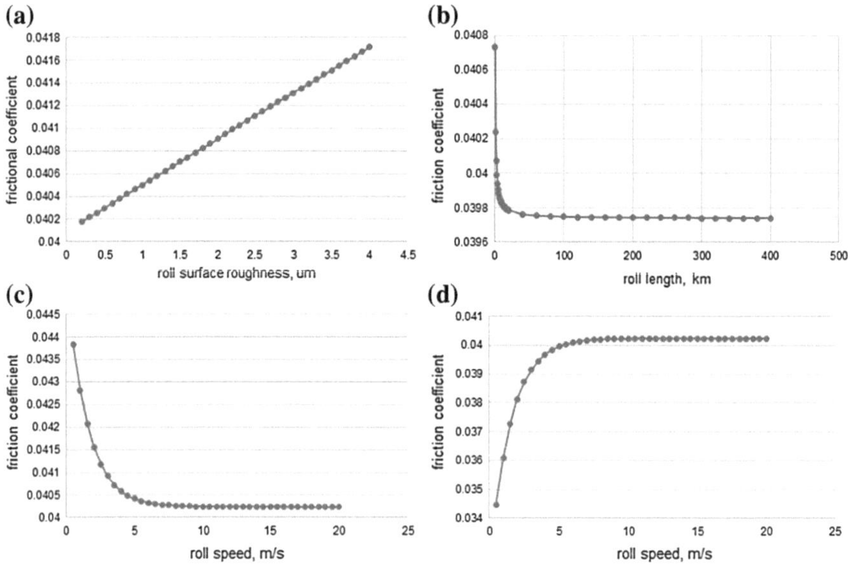

Fig. 2 Effects of variables on friction coefficient **a** effect of roll surface roughness on friction coefficient, **b** effect of roll length on friction coefficient, **c** effect of roll speed on friction coefficient when velocity coefficient is positive, **d** effect of roll speed on friction coefficient when velocity coefficient is negative

When the roll length is less than 20 km, the friction coefficient decreases significantly with the increase of roll length. When the roll length is more than 20 km, the friction coefficient hardly changes with the increase of rolling length.

The influence of positive and negative velocity coefficient on friction coefficient should be considered. One paper points out [7] that with the increase of rolling speed, the friction coefficient of dry temper rolling increases. The other paper [8] points out that similar to ordinary cold rolling, with the increase of rolling speed, the friction coefficient of wet temper rolling decreases. Thus, when the velocity coefficient is negative, friction coefficient increases with the increase of rolling speed, corresponding to dry temper rolling. When the velocity coefficient is positive, friction coefficient decreases with the increase of rolling speed, corresponding to wet temper rolling. When the rolling speed is more than 10 m/s, the friction coefficient hardly changes with the increase of rolling speed.

Yield Stress Model

Yield stress model is composed of thickness deformation dependent base function, strain rate correction, temperature correction and alloy correction.

Base function:

$$KfB = B0 + (B1 - B0) * eps^{B2} \tag{2}$$

where $eps = 1 - e^{-phi}$, $phi = \ln(HSM/h_i)$, $B0$, $B1$ and $B2$ are the base constant, HSM is the thickness before DCR, h_i is the exit thickness of the stand.

Strain rate correction function:

$$Kf(phi\ dot) = (phi\ dot/p\,\mathrm{Re}f)^{pA \times kf\ 0^{pB}} \tag{3}$$

where $phi\ dot$ is the strain rate, pA, pB and $p\,\mathrm{Re}f$ are strain rate constant, $kf\ 0$ is the initial yield stress constant.

Temperature correction function:

$$tempfactor = \frac{0.5 - a\tan\left(\frac{temp-C}{sigma}\right)/3.1416}{0.5 - a\tan\left(\frac{20-C}{sigma}\right)/3.1416}$$
$$sigma = (C - 20)/tempFS \tag{4}$$

where $temp$ is the temperature, C is the temperature constant, $tempFS$ is temperature slope factor.

Alloy correction function:

$$KfC = \frac{\sum_{i=0}^{3} e^{-\frac{(eps-E[i])^2}{sigmaE^2}} \times aKf[i]}{\sum_{i=0}^{3} e^{-\frac{(eps-E[i])^2}{sigmaE^2}}} \tag{5}$$

where $sigmaE$ is the reduction rate coefficient, $E[i]$ $(\mathrm{i} = 0 \sim 3)$ is reduction rate constant, $aKf[i]$ $(i = 0 \sim 3)$ is the yield stress adaptation coefficients.

Thus, the yield stress model is as follows:

$$kf = KfB * KfC * Kf(phi\ dot) * tempfactor \tag{6}$$

where kf is the yield stress.

The effects of the thickness deformation, strain rate, temperature and adaptation coefficients on yield stress are as follows in Fig. 3: (when discussing the effect of any variable on the yield stress, the other variables remain consistent.)

The yield stress increases with the increase of thickness deformation and strain rate. When the strain rate is high, the yield stress increases slowly.

The yield stress decreases linearly with the increase of temperature. The temperature is usually 0–100 °C during DCR of continuous annealing process. When the temperature is at this range, the effect of temperature increase on yield stress is not great, and the variation of yield stress is about 5 MPa.

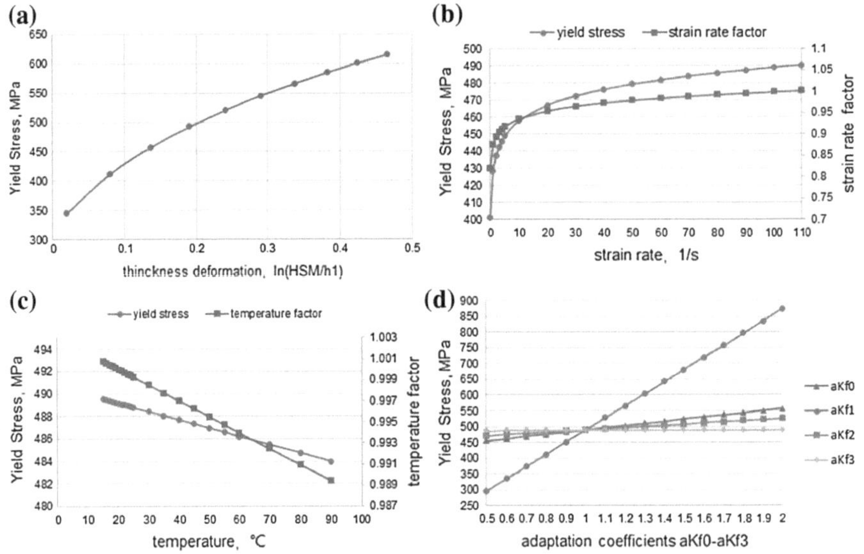

Fig. 3 Effects of variables on yield stress **a** effect of thickness deformation on yield stress, **b** effect of strain rate on yield stress, **c** effect of temperature on yield stress, **d** effect of adaptation coefficients on yield stress

The calculation of yield stress involves four neural network correction coefficients aKf0 ~ aKf3. The four neural network correction coefficients have great effect on the yield stress calculation results. While aKf0 ~ aKf3 have different influence on yield stress. aKf1 has the greatest influence on yield stress, that is, slight changes of aKf1 cause major changes in yield stress, and followed by aKf0, then aKf2 and finally aKf3. Changes of aKf3 only have a slight impact on yield stress. Thus, the adjustment of aKf0 ~ aKf3 should be made according to actual demand.

Roll Force Model

The biggest difference between DCR and ordinary cold rolling is that the reduction and the contact arc length are very small for DCR. The contact arc between the strip and the roll is no longer circular. Figure 4 shows the contact section of strip steel. The usual assumption of circular contour is no longer applicable to the prediction of rolling force of extremely thin strip. This model considers the impact of roll elastic flattening and recovery on roll force, based on the non-circular arc theory of Fleck, Johnson [9] and Sutcliffe [10] (FJS) to accurately predict rolling force. The contact between the strip and the roll is divided into the following areas: elastic compression zone, entry plastic zone, neutral zone, exit plastic zone and elastic recovery zone. The equations of the roll force model which describe the contact zones between the strip and the roll can refer to the references [10, 11].

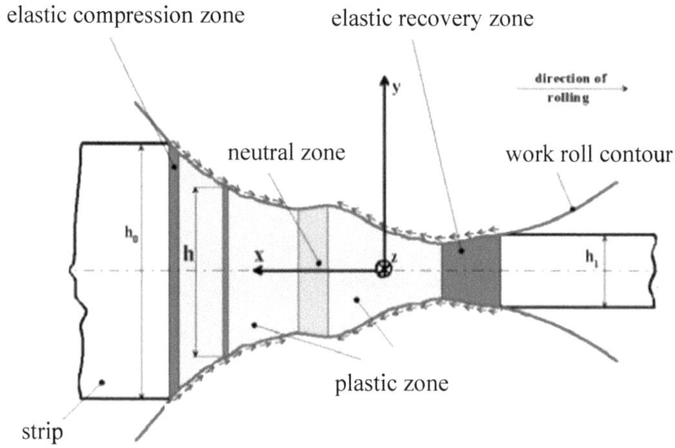

Fig. 4 The section model of strip steel based on elastoplastic theory

Roll force model is the core part of the whole L2 rolling setup model. The quantitative influence of the main factors on roll force is analyzed. The effects of reduction rate, tensile stress, friction coefficient, and roll radius on roll force for two different steel grades are as follows in Fig. 5: (when discussing the effect of any variable on roll force, the other variables remain consistent.)

The influence of the above factors on roll force is single trend. Roll force increases with the increase of reduction rate, friction coefficient and the roll radius, and decreases with the increase of tensile stress. The figure clearly shows the quantitative relationship of each factor on the roll force for different steel grades. By changing the input parameters of the model, the quantitative relation on roll force of different steel grades and specifications can be studied, just as shown in the above figure for steel A and steel B, providing a theoretical basis for the process optimization of roll force and the new steel grade development in the actual rolling process.

Flatness Control Model

The flatness control model is used to adjust the shape of loading roll gap contour by means of adjusting mechanisms such as roll bending, roll lateral shifting and downward tilt of pressure, in order to control (eliminate or generate) the exit flatness distribution and strain distribution in the width direction of the strip. The function of this model is transferring the data to level 1 and providing the basis for the online adjustment according to the calculated results.

Flatness control model of DCR in current research includes four submodels: work roll flattening model, roll gap model (material flow model), roll bending model and actuators setup model. The flow chart of the flatness model control process is shown

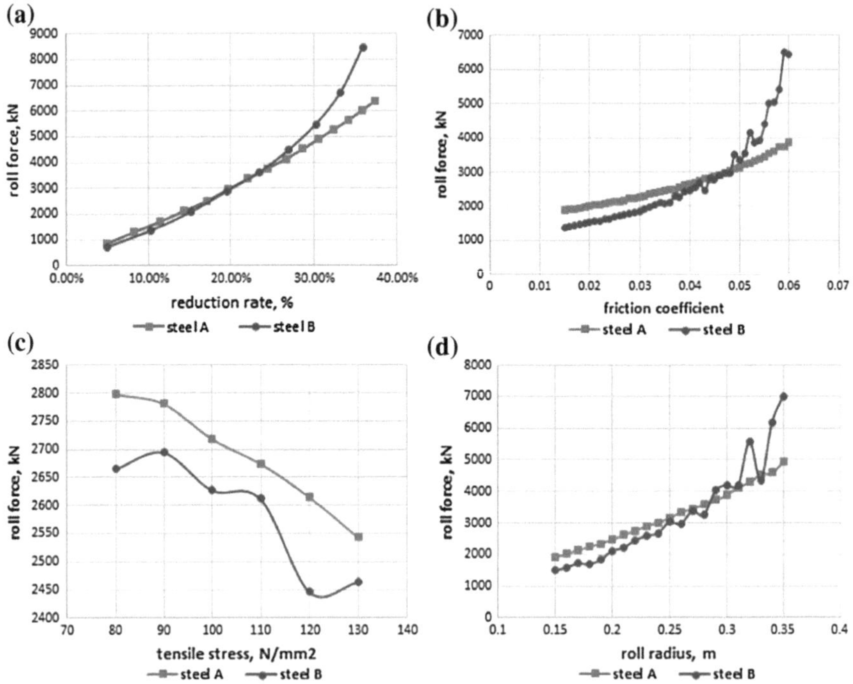

Fig. 5 Effects of variables on yield stress **a** effect of reduction rate on roll force, **b** effect of friction coefficient on roll force, **c** effect of tensile stress on roll force, **d** effect of roll radius on roll force, where steel A is MR DR-8M CA, and its entry thickness is 0.313 mm, exit thickness is 0.257 mm, steel B is MR DR-7M CA, and its entry thickness is 0.189 mm, exit thickness is 0.157 mm

in Fig. 6. According to the target shape curve, considering bending deformation, the distribution of actuators is determined, and then the strip contour, flatness difference, thickness difference and efficiency coefficients (roll force efficiency coefficient, roll bending force efficiency coefficient, and roll shifting efficiency coefficient) are calculated.

Wedge Control Model

DCR of continuous annealing process has the function of unsteady state control. The front and subsequent strips are welded together by welding machine to realize temper rolling. The setup calculation of the front and subsequent strips in the transition section should be considered in this process. The setup accuracy seriously affects the thickness difference and quality of the head/tail strip.

Figure 7 shows that when the strip reaches the weld of the temper rolling mill, it is necessary to control the constant mass flow to keep the front strip A and the

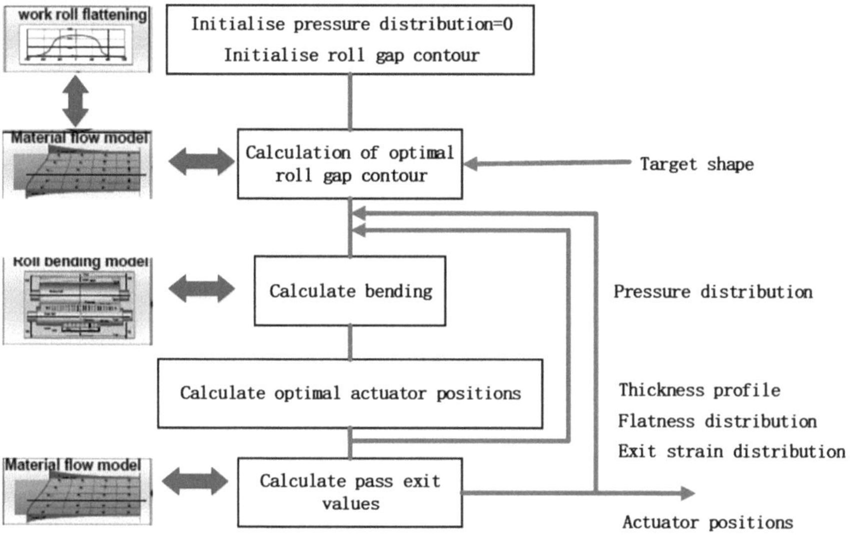

Fig. 6 Calculation process of flatness control model

subsequent strip B rolling steadily according to their own setup values. The strip speed, thickness and tension change from the state of the front strip A to the state of the subsequent strip B. This transformation cannot be done suddenly. Otherwise, it impacts the mill, even causes strip breakage and the failure of full continuous rolling process. The precise setup of weld thickness and length during wedge control process is very important, which guarantees the continuous and steady rolling of the weld to avoid the occurrence of strip breakage. The wedge control model mainly includes wedge position and thickness calculation model, additional tension calculation model, speed calculation model and roll gap setting model at low speed. The specific calculation process is shown in Fig. 8.

Adaptation Model

Although the results of the physical models are fairly accurate, small deviations remain between calculated set points and measured values. To reduce these deviations, this model is modified by both integrating control and neural network of adaptation model. The adaptation model includes the rolling adaptation and flatness adaptation, and there are long inheritance and short inheritance, respectively. The long and short inheritance are distinguished according to four factors: alloy composition code, strip width, entry thickness and exit thickness. If one of the above factors is different or in different range, the strip is considered as long inheritance strip.

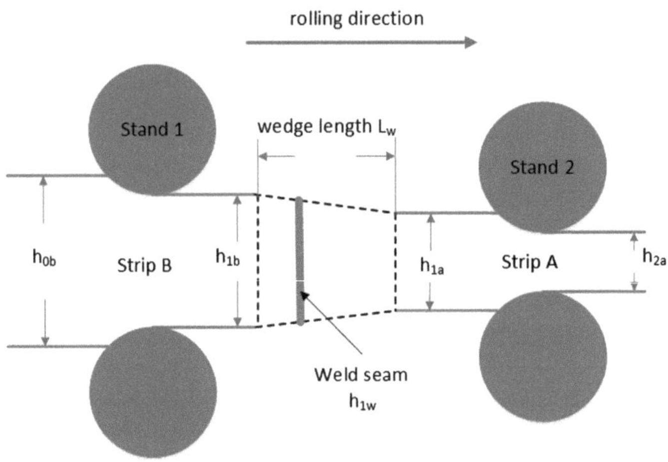

Fig. 7 Dynamic rolling program change and wedge position sketch map

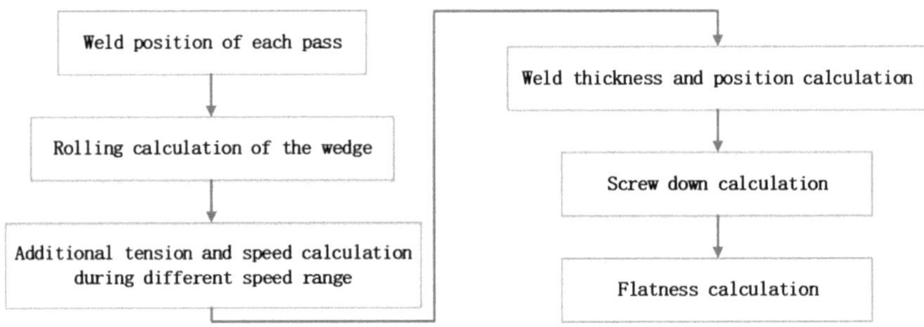

Fig. 8 Flow chart of the main models in wedge control process

In the rolling model, the yield stress model requires neural network inheritance. The deviation of post calculation results of roll force, forward flip and torque without yield stress neural network correction and with yield stress neural network correction is used to determine the stand network correction to train, in order to reduce the deviation from the target value.

The actuators are adjusted according to the flatness error and the roll force variation. According to actual rolling force and actual actuator's position, the actual roll gap contour is calculated based on bending model. Then, the model is modified by neural network inheritance algorithm according to quadratic deviation between the actual roll gap contour and the setup roll gap contour.

The adaptive process of the model is executed section by section. Each strip section carries out its respective rolling adaptation and shape adaptation. The rules of L2 adaptation strategy are shown in Table 1. After two times' rolling adaptations (two sections of strip), the next strip section begins flatness adaptation. The flatness

Table 1 Rules of the L2 adaptation strategy of DCR during continuous annealing process

	1	2	3	4	5	6	7	8	9	10	11	12	13	14	15	16	17	18	19	20	…
Execution	R	R	F	R	R	F	R	R	F+ In	R	R	F	R	R	F	R	R	F+ In	R	R	…
Rolling adaptation times	1	2	2	3	4	4	5	6	6	7	8	8	9	10	10	11	12	12	13	14	…
Rolling inheritance times	0	0	0	0	0	0	0	0	0	0	0	0	0	0	0	0	0	0	0	0	…
Flatness adaptation times	0	0	1	1	1	2	2	2	3→0	0	0	1	1	1	2	2	2	3→0	0	0	…
Flatness inheritance times	0	0	0	0	0	0	0	0	1	1	1	1	1	1	1	1	1	2	2	2	…
Total times of adaptation	0	1	2	3	4	5	6	7	8	9	10	11	12	13	14	15	16	17	18	19	…

Where, R—rolling adaptation, F—flatness adaptation, F+In—flatness adaptation and inheritance neural network inheritance begins at the third time of flatness adaptation. According to the above rule, the whole coil strip completes the adaptation.

Improvement and Application Prospects of DCR L2 Control Model

The requirements for a sophisticated process automation system have become increasingly severe. Due to the growing requirements and complexity of technology, rolling mills need a sophisticated process automation level for process optimization, to make the functions transparent and to enable quick interventions by the operator. However, due to the complexity of L2 model mechanism and control process, the automation degree of DCR process control has not been fully realized at present, which has a broad prospect for further research and application.

According to the research on the L2 control model of DCR, we know that the L2 model can control all the important process parameters in the actual rolling process, especially having a significant impact on the production of ultra-thin, extremely hard, and high-pressure rolled tin plate. In this paper, the mechanism and algorithm of the L2 DCR control model are studied, and the quantitative influence of process variables in the model is explored. It is also found that the model has some inadaptability behaviors especially in the thin strip rolling (please refer to another paper "Research on Level 2 Rolling Model of Tin Plate Double Cold Reduction Process" in TMS annual meeting), and relevant guidance and measures are proposed. This lays a stable theoretical foundation for the realization of L2 process automation in the process of DCR for extremely thin specification tin plate and has great significance and effect on the optimization of online process parameters, the improvement of model algorithm and new steel grade development.

References

1. Wei LQ, Lu DH (2002) BP network based skin rolling force calculation. Iron Steel 37(12):33–35
2. Bai ZH, Feng XZ, Jiang YF (2007) Research on reform program of roll shape in skin rolling process of super thin strip. China Mech Eng 18(23):2887–2889
3. Liu ZL, Li WQ, Wang YJ (2011) Research on shape control model for DSR skin mill. China Mech Eng 22(11):1624–1628
4. Yu M, Zhang QD, Li R et al (2010) Control of surface roughness for R2 grade tin mill black plate in two-stand temper mill rolling. Iron Steel 45(12):44–49
5. Li XJ, Bai ZH, Li LL et al (2009) Study of color aberration combination controlling technique for steel strip in temper rolling processing. Iron Steel 44(11):60–63
6. Wu SM, Chen C, Ge L (2009) Technology research of 2-high UCM temper mill for thin strip. Mech Eng Autom 4:113–114
7. Robert WL translated by Li YH (1973) Approximation theory of temper rolling. Heavy Mach 4: 41–59

8. Domanti SA, Edwards WJ (1994) Application of foil rolling model in thin strip and flat rolling. Paper presented in translation of the 6th international steel rolling conference, vol 3. Beijing, China, pp 180–188
9. Fleck NA, Johnson KL (1992) Cold rolling of foil. Proc Inst Mech Eng 206:119–131
10. Le HR, Sutcliffe MPF (2001) A robust model for rolling of thin strip and foil. Int J Mech Sci 43:1405–1419
11. Johnson KL (1985) Contact mechanics. Cambridge University Press

Research on Level 2 Rolling Model of Tin Plate Double Cold Reduction Process

Hui Wang, Wei Guo, Yanglong Li, Fei Chen, Jie Wen, Meng Yu and Fengqin Wang

Abstract The double cold reduction (DCR) process of tin plate has the characteristics of thin entry thickness and small reduction compared with conventional cold rolling. In this case, the hypothesis of circular roll profile is no longer reasonable, and non-circular arc theory is adopted to ensure the rolling model accuracy. In this paper, the rolling model is carried out using non-circular arc theory, based on the actual process data of tin plate rolling. The effects of thickness, reduction ratio, tensile stress, entry temperature of strip, and work-roll radius on the roll force were studied, which described the various trends and changes of the roll force with strip thickness, tensile stress, roll radius and other technological parameters under different simulation conditions. The research results provide an effective guidance for developing rolling strategy of tin plate DCR process.

Keywords Non-circular arc theory · Double cold reduction of tin plate
Rolling model

Introduction

Double cold reduction (DCR) technology is a common process for the production of thin tin plate, and the thinnest target thickness can even be 0.12 mm [1, 2]. The strip of DCR process has smaller reduction and shorter contact arc than conventional cold rolling strip. Therefore, the deformation mechanism of the strip is quite different from that of conventional cold rolling. The elastic deformation of the strip and the elastic flattening of work roll have great influence on the distribution of roll force. The common assumption of circular roll profile is no longer reasonable. Therefore, Conventional theories of cold rolling such as those developed by von Karmann [3], Orowan [4] and Bland and Ford [5] are known to be unsatisfactory for predicting the roll force for the process of rolling thin hard strip. A major development in modeling

H. Wang (✉) · W. Guo · Y. Li · F. Chen · J. Wen · M. Yu · F. Wang
Shougang Research Institute of Technology, Beijing 100043, China
e-mail: athenaco@163.com

G. Lambotte et al. (eds.), *Materials Processing Fundamentals 2019*, The Minerals, Metals & Materials Series, https://doi.org/10.1007/978-3-030-05728-2_12

thin strip rolling process with the influence function method was achieved by Fleck and Johnson [6]. The model was based on non-circular arc roll theory and widely used to calculate roll force for thin strip rolling process [7]. The deformation zone is separated into several zones, for which the boundaries have to be solved, and the solving process is numerically unstable and time-consuming [8].

In this study, the non-circular arc roll theory was used to simulate DCR process and calculate roll force for tin plate DCR process. Effect of entry thickness, reduction ratio, backward and forward tensile stress, entry temperature and radius of work roll on roll force was studied. The overall trend of roll force calculated by non-circular model is the same as that by Hitchcock model with tensile stress and other variables, but with obvious fluctuations. It generally leads to unreasonable Level 2 roll force setup, and sometimes even causes major technological accidents such as strip breakage. Consequently, the roll force setup of Level 2 on-site is usually based on the operator's experience. Manual operation without Level 2 control system cannot achieve accurate control, which results in shape and surface defects, and large cutting loss. The fluctuations of roll force can be suppressed by improving convergence standard, but the solution time is increased correspondingly. It is not applicable to the actual online continuous production process. The appropriate convergence standard plays an important role in Level 2 setup of the DCR process of tinplate production. Therefore, it is extremely necessary to investigate the changing regularities of calculated roll force with different convergence standard. In addition, the complete control mode of DCR process can be found in another submitted paper «Research on the L2 control model technology of double cold reduction during continuous annealing process» in TMS annual meeting.

Theory and Equations of Thin Strip Rolling Model

Basic Equations of Roll Force Model

Fleck and Johnson have shown that the rolling conditions are divided into three kinds according to the characteristics of the rolling deformation zone, which correspond to different methods of calculating the roll force. Sutcliffe developed a new approach to calculate roll force by Hooke's law in deformation zone which is separated into entry elastic deformation zone, entry plastic deformation zone, neutral zone in which the roll profile is flat, exit plastic deformation zone, and exit elastic deformation zone. According to the homogeneous deformation theory, the equation can be derived as follows from the forces balance on strip [9].

$$h\frac{d\sigma_x}{dx} + (p_x + \sigma_x)\frac{d\sigma_x}{dx} + 2\tau_x = 0 \tag{1}$$

where h is strip gauge, p_x is normal pressure, σ_x is tensile direct stress in strip, τ_x is shear stress.

Hooke's Law holds in the entry and exit elastic deformation zones, and the elastic strain can be represented as

$$\varepsilon_S{}^{ze} = -\frac{1-\nu_S{}^2}{E_S}[p_x + \frac{\nu_S}{1-\nu_S}(\sigma_x - \sigma_i)] \tag{2}$$

where E_S is Young modulus of strip, ν_S is Poisson's ratio of strip, σ_i is the tensile stress at entry or exit side of the deformation zone.

The normal pressure p_x can be solved by Eq. (2).

Based on Mises yield criterion, the relationship among normal pressure p_x, tensile direct stress σ_x and averaged deformation resistance σ_S in plastic zone is described by Eq. (3).

$$p_x + \sigma_x = \frac{2}{\sqrt{3}}\sigma_S \tag{3}$$

where σ_S is the averaged deformation resistance.

The total strain of the roll in rolling direction can be derived from elastic deformation theory applied in an infinite half-plane [10].

$$\varepsilon_R{}^x = -\frac{p_x(1-2\nu_R)(1+\nu_R)}{E_R} - \frac{2(1-\nu_R{}^2)}{\pi E_R}\int_{x_1}^{x_2}\frac{\tau_x}{x-s}ds \tag{4}$$

where E_R is elasticity modulus of roll, ν_R is Poisson's ratio of roll, x_1 and x_2 are the contact width from centre-line to entry and exit location, respectively.

The plane deformation obeys the law of constant volume, which means the sum of plastic strain of strip in rolling direction and in height direction is 0.

$$\varepsilon_S{}^x = \frac{(1-\nu_S^2)}{E_S}\left(\sigma_x + \frac{\nu_S}{1-\nu_S}p_x\right) + \varepsilon_S{}^{xp} \tag{5}$$

$$\varepsilon_S{}^z = -\frac{(1-\nu_S^2)}{E_S}\left(p_x + \frac{\nu_S}{1-\nu_S}\sigma_x\right) + \varepsilon_S{}^{zp} \tag{6}$$

$$\varepsilon_S{}^{xp} + \varepsilon_S{}^{zp} = 0 \tag{7}$$

where $\varepsilon_S{}^x$ and $\varepsilon_S{}^z$ are the total strain in rolling direction and height direction, respectively. $\varepsilon_S{}^{xp}$ and $\varepsilon_S{}^{zp}$ are plastic strain of strip in rolling direction and in height direction, respectively.

There is no relative sliding between the roll and the strip in neutral zone, so the difference between total strain of strip and total strain of roll in rolling direction is constant.

$$\varepsilon_R{}^x - \varepsilon_S{}^x = c \tag{8}$$

where c is a constant.

The distribution of normal pressure p_x, namely the roll force per unit in the plastic zone, can be obtained by solving the above integral equations.

Friction Model

The Coulomb friction law is satisfied between the shear stress and normal stress outside the neutral zone.

$$\tau_x = \pm\mu p_x \tag{9}$$

where μ is the Coulomb friction coefficient.

In this model, the Coulomb friction coefficient in neutral zone is assumed to vary linearly from μ at the start to $-\mu$ at the end.

Results and Discussion

Modeling and off-line simulation were carried out according to the non-circular arc theory, and the effects of different variables on the results of the rolling model were obtained. The variables mainly referred to strip specifications, process parameters and mill parameters, which include strip thickness, backward and forward tensile stress, entry temperature of strip, and roll radius.

Roll Force

The changing regularities of roll force with different variables are studied with control variate method and shown in Fig. 1. When the relationship between rolling force and a variable is studied, the other variables remain unchanged.

In Fig. 1a, each curve corresponds to a given reduction ratio. At thin thickness, roll force decreases sharply and fluctuates as the entry thickness increases. However, when the entry thickness increases to conventional thickness, the fluctuation disappears and the roll force increases linearly as the entry thickness increases, which is in good agreement with the results reported by Xue [11]. Because the change of rolling force is more violent and the fluctuation is more obvious when the thickness is thinner, it is difficult to obtain the target exit thickness and achieve stable rolling in practical production. In Fig. 1b, at the same entry thickness, the roll force increases as the reduction ratio increases, and the increasing rate decreases. Both Fig. 1a and b indicate that the thinner the strip is, or the higher the reduction ratio is, the greater the hardening intensity is, and the more dramatically the roll force increases.

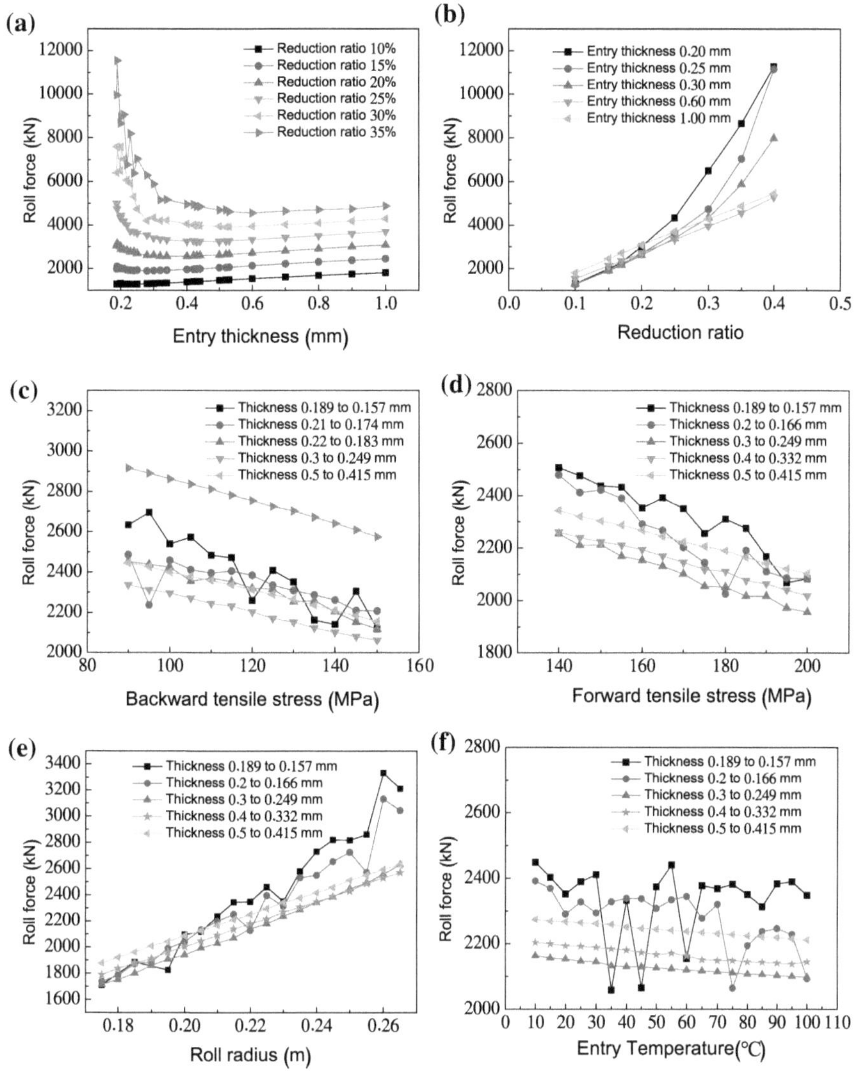

Fig. 1 Relationship of different variables and rolling force; **a** entry thickness, **b** reduction ratio, **c** backward tensile stress, **d** forward tensile stress, **e** roll radius, **f** entry temperature

In Fig. 1c, d, the roll force decreases as the backward or forward tensile stress increases and the calculated results for extremely thin strip fluctuate as well. In Fig. 1e, the roll force increases as roll radius increases, and Fig. 1f shows the opposite trend of roll force changing with entry temperature. The roll force fluctuates with increasing roll radius or entry temperature when the thickness is extremely thin in both of Fig. 1e, f.

Wang et al. [12] found through experiments that the roll speed affects the roll force through influencing the friction coefficient, which has a great impact on the rolling force. The effects of roll speed on friction coefficient and roll force are shown in Fig. 2a and b, respectively. The variation trend of friction coefficient of extremely thin strip with roll speeds is dependent on speed range. As seen in Fig. 2a, when the roll speed is below 10 m/s, friction coefficient decreases rapidly with increasing the roll speed, then remain nearly constant as roll speed increases above 10 m/s, which is in accordance with the findings reported by Wang et al. [12].

As shown in Fig. 2b, the change trend of roll force with roll speed is similar to that of friction coefficient. The roll force decreases as the roll speed increases when it is less than 10 m/s. When the rolling force exceeds 10 m/s, the rolling force is almost constant. On the same reduction ratio, there are also fluctuations in the relationship of roll force with roll speed, and the roll force fluctuates first and then decreases slowly as the roll speed increases in the case of thin specifications. However, for the conventional thickness, the fluctuation disappears and roll force decreases linearly as the roll speed increases (within 10 m/s).

Analysis of Calculated Results Fluctuation

The results in section "Roll Force" show the fluctuation in the case of thin specification. As shown in Fig. 1c, the reduction ratios are all 16.93%. The fluctuation of roll force with tensile stress is obvious when entry thickness is below 0.22 mm, but disappears when thickness is above 0.22 mm. In this paper, 0.22 mm is defined to be critical thickness of sensibility of roll force to thickness when the reduction ratio is 16.93%. It is found that different reduction ratio corresponds to different critical thickness of the sensitivity of roll force to thickness. The relationship between critical thickness and reduction ratio satisfies conic characteristic and is shown in Fig. 3.

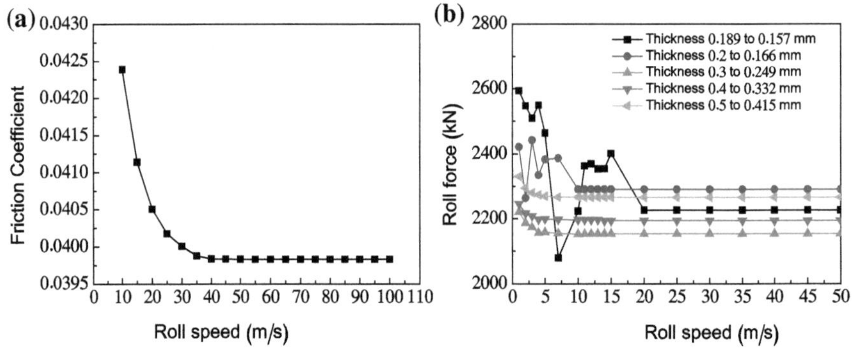

Fig. 2 Effect of stand speed on **a** friction coefficient, **b** roll force

Fig. 3 The critical thickness on different reduction ratios

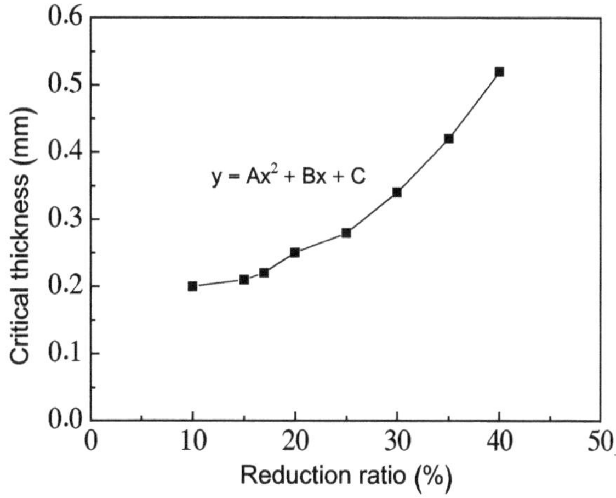

The bigger the reduction ratio is, the thicker the critical thickness is and the sharper the slope of critical thickness with reduction ratio is.

The reason of the calculated roll force fluctuating with different variables of thin strip is that the improper convergence standard leads to wrong results in the calculation, and the force balance for discrete spice of rolling deformation zone is unreasonable when modeling based on non-circular theory. The stress of each spice is calculated by elastic mechanics formula, and boundary condition is the backward tensile stress σ_{in} and forward tensile stress σ_{out}. The neutral zone is assumed to be flat as the contact pressure is approximated by the Hertzian elliptical shape, onto which is added a perturbation caused by the plastic strain in the strip [6]. The start and the end of neutral zone are determined by calculating elastic deformation of roll produced by shear strain. Normally, the neutral zone is from the start to the end, but sometimes the start and the end are the same point when the thickness is very small. The abnormal neutral zone means that the convergence standard is unreasonable and the stress calculation is not accurate. As a result, the roll force fluctuates violently with each variable when the thickness is very small, and even the wrong result is obtained.

In view of the complexity of the non-circular arc theory, the model is difficult to converge. Improving the convergence standard grindingly will lead to a surge in computing time, which is unsuitable for online applications. In order to improve accuracy and restrict computing time at the same time, the basic idea for model optimization is setting different convergence standards for different cases and improving convergence standard only when the thickness is below critical thickness. The calculated results using different convergence standards are shown in Fig. 4. The original convergence standard is that the deviation between roll force of this iteration and the average roll force of the last 10 iterations is no more than 0.01%, and the improved convergence standard is that all of the deviations between roll force of this iteration and the last 2 to N (N is 5 and 9) iteration cannot exceed 0.01%. N is set to be 5 and

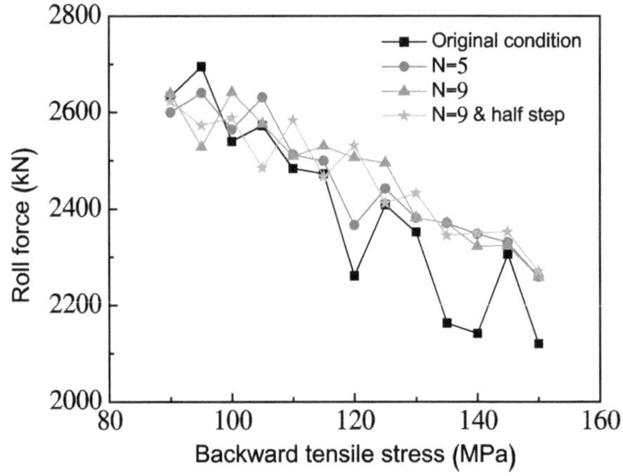

Fig. 4 The relationship of calculated roll force with forward tensile stress on different convergence standards

9, and "N = 9&half step" means N = 9 and the iteration step is reduced to be half simultaneously. Simulation indicated that fluctuation is effectively suppressed when convergence standard is optimized to N = 5, and cannot be further eliminated when optimized to N = 9. The iteration step of searching the lowest point of roll is reduced to half with N = 9. Although the calculated results fluctuate more frequently than the previous results, the amplitude of fluctuation decreases within 100 kN and the fluctuation is more regular. The regular and slight fluctuation is within the margin of error in practical production. Further improving the convergence standard and reducing the iteration step, the fluctuation cannot be further eliminated. On the one hand, it is the normal phenomenon of iterative computation; on the other hand, it is caused by the non-circular arc model. In conclusion, the convergence standard with "N = 5" is the most suitable for industrial application in consideration of both accuracy and compute time.

Conclusions

In this study, the non-circular arc theory is used to model, and the effect of different variables on the roll force of tin plate DCR process in continuous annealing are studied and summarized as follows.

The findings regarding the effect of entry thickness, reduction ration, backward and forward tensile stress, roll radius, entry strip temperature, and roll speed on roll force is consistent with results reported by Xue et al. [13], when the strip is in conventional specification. However, the roll force of thin strip fluctuates with the above variables.

The non-circular model is sensitive to thin thickness of strip such as tin plate. When the thickness is very small, the roll force fluctuates as different variables increase. There is a critical thickness on certain reduction ratio. Fluctuation of calculated roll force is obvious below the critical thickness and disappears above the critical thickness. The relationship between critical thickness and reduction ratio is quadratic polynomial.

The fluctuation of roll force is caused by non-circular model intrinsic deficiency, normal fluctuation from iteration and improper convergence standard. In this paper, the convergence standard is optimized only. After optimization, the fluctuation of calculated roll force is eliminated to some extent. The roll force set in Level 2 control system is more accurate, and cutting loss of strip is greatly reduced, which satisfies the technical requirements of DCR rolling process of tin plating line.

References

1. Wu SM, Chen C, Ge L (2009) Technology research of 2-high UCM temper mill for thin strip. Mech Eng Autom 4:113–114
2. Ji J, Hu H, You L et al (2018) Development and application of skin pass and double cold reduced mill processing for high quality ultra thin uncoated tin plated sheet. Steel Rolling 35(1):49–51
3. von Karmann (1925) Beitrag zür theorie des Walzvorganges. Z angeur Math Mech 5:136–139
4. Orowan E (1943) The calculation of roll pressure in hot and cold flat rolling. Proc Inst Mech Eng 150:140–167
5. Bland DR, Ford H (1948) The calculation of roll force and torque in cold strip rolling with tension. Proc Inst Mech Eng 159:144–163
6. Fleck NA, Johnson KL (1992) Cold rolling of foil. Proc Inst Mech Eng 206:119–131
7. Wang DC, Wang YH (2015) Simplified rolling force model for temper rolling mill based on non circular arc theory. China Mech Eng 26:2677–2681
8. Liu YL, Lee WH (2005) Mathematical model for the thin strip cold rolling and temper rolling process with the influence function method. ISIJ Int 45:1173–1178
9. Le HR, Sutcliffe MPF (2001) A robust model for rolling of thin strip and foil. Int J Mech Sci 43:1405–1419
10. Johnson KL (1985) Contact mechanics. Cambridge University Press
11. Xue T (2014) Tandem rolling process simulation and nonlinear online model research for UCM mills. Yanshan University
12. Wang DC, Peng Y, Liu HM (2008) A high-resolution high-speed rolling force model for cold strip temper rolling mill. J Plast Eng 15:172–177
13. Xue T, Du FS, Sun JN et al (2013) Rolling force prediction of cold strip rolling based on FEM-ANN. J Cent So Univ (Sci Technol) 11:4456–4460

Part III
Alloys Processing and Properties Modeling

Numerical Modelling and Influence of Cu Addition on the Microstructure and Mechanical Properties of Additive Manufactured Ti–Al–Cu/Ti–6Al–4V Composite

E. T. Akinlabi, O. S. Fatoba and S. A. Akinlabi

Abstract Laser metal deposition technique was used for the fabrication of Ti–Al–Cu coating on Ti–6Al–4V Alloy. The microstructure and elemental and phase composition of coatings were studied. The SEM images showed the homogeneous distribution of Cu addition in Ti–10Al–9Cu at scanning speed of 1.0 m/min. Strong metallurgical bond without pores and cracks were observed between the coating and the substrate. Grain refinement was observed within the microstructure as the grains grew in a columnar and dendritic pattern in a counter direction to heat flow. However, the cross-section microstructures of Ti–10Al–6Cu and Ti–10Al–3Cu at 0.8 and 1.0 m/min scanning speed and laser power of 1000 and 1100 W showed minute pores and cracks. The existence of amorphous phase revealed via XRD was also observed in the coatings. The microstructure of these alloys is highly influenced by processes involving plastic deformation and thermal treatments which, in effect, determines the mechanical properties adhering to desired properties. The microhardness testing results indicated that the fabricated coatings had enhanced by 61.9% as compared to the micro-hardness of the Ti–6Al–4V alloy substrate.

Keywords Ti–6Al–4V alloy · Microstructure · Ti–Al–Cu coatings · Hardness Copper · Temperature distribution

E. T. Akinlabi · O. S. Fatoba (✉)
Department of Mechanical Engineering Science, Faculty of Engineering and the Built Environment, University of Johannesburg, Johannesburg, South Africa
e-mail: drfatobasameni@gmail.com; proffatobasameni@gmail.com

E. T. Akinlabi
e-mail: etakinlabi@uj.ac.za

S. A. Akinlabi
Department of Mechanical and Industrial Engineering Technology, Faculty of Engineering and the Built Environment, University of Johannesburg, Johannesburg, South Africa
e-mail: stephenakinlabi@gmail.com

G. Lambotte et al. (eds.), *Materials Processing Fundamentals 2019*, The Minerals, Metals & Materials Series, https://doi.org/10.1007/978-3-030-05728-2_13

Introduction

The development of laser-based advanced coatings are continually implemented in order to meet the desired demands and performances of materials in industries. A way of achieving specified properties is the fabrication of hard and high wear resistant coatings suitable to protect the bulk substrate against any form of surface degradation. Surface modification technique can be a realistic approach in addressing these limitations associated with titanium and its alloy [1]. The enhancement of a metal surface by changing the microstructure and elemental constituent composition is known to achieve reduction in friction coefficient, enhanced wear resistance and increase in the surface hardness of the material without changing the bulk properties of the substrate [2]. Surface modification is an essential approach in mitigating material loss due to erosion and wear. Application of coatings resistant to abrasion can be observed as a practical solution to safeguard the surface of the alloy from wear actions. Enhancement of the surface can also lead to the development of novel microstructures which can support outstanding properties without affecting the bulk properties of the alloy [3].

The vital factor to be considered during direct laser metal deposition process is the simultaneous action of melting and fusion of the coating material to the base metal. The appropriate selection of laser processing parameter will produce desired result and properties [4].

Titanium alloy delivers the best all-round performance for a wide range of weight savings applications, strength requirement, less serviceable effective materials and the demands for advanced engineering materials [5]. Ti–6Al–4V indeed has emerged as a global material to address maximum performance and energy efficiency and high performance for offshore structures, marine environment, chemical and petrochemical applications.

However, the distinct shortcomings such as relatively low hardness and poor wear resistance restrict the utilization of titanium alloys especially in friction conditions. Ti–6Al–4V has a poor sliding characteristic [6, 7]. This leads to its proneness to failure via galling and shows undesirable coefficients of friction [8]. Improving the mechanical properties will require the incorporation of alloying elements, material processing and composite development [9].

In this research, Ti–Cu–Al powder was used as reinforcement to prepare coatings Ti–6Al–4V alloys (Grade 5), using laser deposition technique. The appropriate fraction of Ti and Al was chosen to fabricate the amorphous alloys, with the addition of a small amount of Cu powder to enhance the bonding between the coating and the substrate.

Experimental Details

Materials Specifications and Sample Preparation Method

The substrate material used in the present investigation was Ti–6Al–4V alloy with the chemical composition (wt%) 6.2 Al, 3.9 V, 0.2 Fe, 0.12 C, 0.15 O, 0.005 N, 0.0003 H and Balanced Ti. The substrate was cut and machined into dimensions 72 × 72 × 4 mm^3. Prior to laser treatment, the substrates were sandblasted, washed, rinsed in water, cleaned with acetone and dried at room temperature before exposure the laser beam to minimize reflection of radiation during laser processing and enhance the absorption of the laser beam radiation. Al (99.8% purity), Ti (99.97% purity), Cu (99.9% purity), reinforcement metallic powders were used as alloying powders mixed in Ti–10Al–3Cu (A_1), Ti–10Al–6Cu (A_2), and Ti–10Al–9Cu (A_3) ratios, respectively, in a shaker-mixer (Turbular T2F; Glenn Mills, Inc.) for 18 h at a speed of 72 rpm to obtain homogeneous mixture. The particle shape of the powder used was spherical with 50–105 μm particle sizes. Samples were characterized for SEM and energy dispersive spectroscopy (EDS) analysis. Specimens for SEM (JSM-7600F; JOEL, Ltd.) were prepared by cutting samples in such a way to reveal the transverse section of the coatings. Laser cladding was performed using a 3-kW continuous wave (CW) Ytterbium Laser System (YLS) controlled by a KUKA robot which controls the movement of the nozzle head and emitting a Gaussian beam at 1064 nm. The nozzle was fixed at 2 mm from the steel substrate. The admixed powders were fed coaxially by employing a commercial powder feeder instrument equipped with a flow balance to control the powder feed rate. An argon gas flowing at a rate of 2.5 L/min was used as a shielding gas to prevent oxidation of the sample during laser surface alloying. Overlapping tracks were obtained by overlapping of melt tracks at 70%. To determine the best processing parameters, optimization tests were performed with the laser power of 900–1100 W and scanning speed varied from 0.6 to 1.2 m/min. The final selection criteria during optimization tests were based on a surface having homogeneous layer free of porosity and cracks determined from SEM analysis. The optimum laser parameters used was 1000 and 1100 W power, a beam diameter of 2 mm, gas flow rate of 2.5 L/min, powder flow rate of 2.5 g/min and scanning speeds of 0.8 and 1.0 m/min respectively.

Results and Discussion

Microstructural Analysis and XRD

The XRD spectrum for Ti6Al4V/Ti–10Al–9Cu composites at a laser power of 1100 W and scanning speed of 1.0 m/min is shown in Fig. 1. Due to the effect of dilution, a part of Ti entered into the molten pool from the substrate. The results

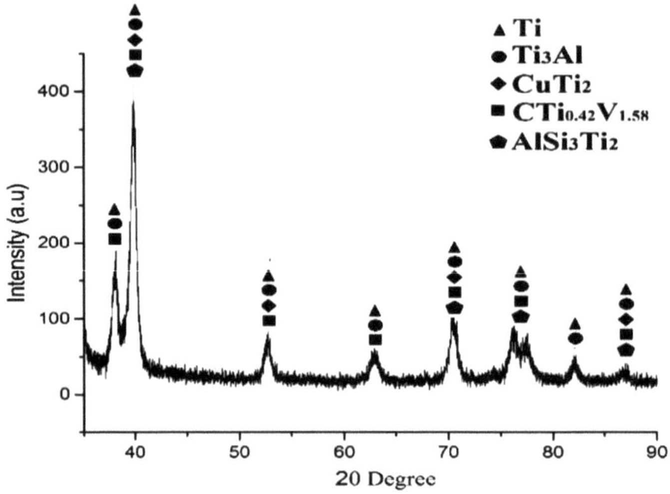

Fig. 1 XRD Spectra of Ti–10Al–9Cu Coatings at 1.0 m/min scanning speed and laser power of 1100 W

indicate that Ti, Al_3Ti, Ti_3Al, $CuTi_2$ is formed through the in situ metallurgical reactions during the laser metal deposition process. Increasing laser power and scanning speed will significantly increase the diffraction peak of Ti and Ti_3Al. It is noted that Al and Ti concentration is highest which leads to the formation of titanium aluminide (Ti_3Al) and presence of α and γ phases. Increase in Cu content increases hardness and tensile strength and yield strength of materials as reported in the literature. Increase in amount of Al and a decrease in the amount of Cu results in a change from FCC structure to a combined FCC and BCC structure. Thus, Al promotes bcc phase formation while Cu promotes FCC phase formation [10–12].

One of the strong β-stabilizing elements and well recognised is Cu. Atomic migration of Cu into Ti lattice results in the β-Ti phase formation during cooling and travels a longer distance in the Ti lattice than other elements which opens more crystallographic structure of the β-matrix. Due to high thermal diffusivity and reflectivity, aluminium and copper are difficult to process but in many situations these metals are required to have a metallurgical bonding. Thus, aluminium and copper powder must be coated with good materials for better photons absorption resulting in better deposit quality [13]. Enhancement in the mechanical properties of Ti–Cu could be explained to be caused by solid–solution strengthening of the α-phases and by precipitation of intermetallic compounds. The presence of gamma titanium aluminides (γ-TiAl) as well as the presence of an alternate layer of shallow grey phase (α-Al).

Figure 2 shows the coating microstructure without cracks and porosities. The microstructure of Fig. 2a and b distribution is similar under various process parameters. Planar crystal and then cellular crystal along the direction of solidification are seen at the bottom of the coating. From the microstructures, dendrites are seen at the interior and top regions in coatings. In addendum, along with the distance from the

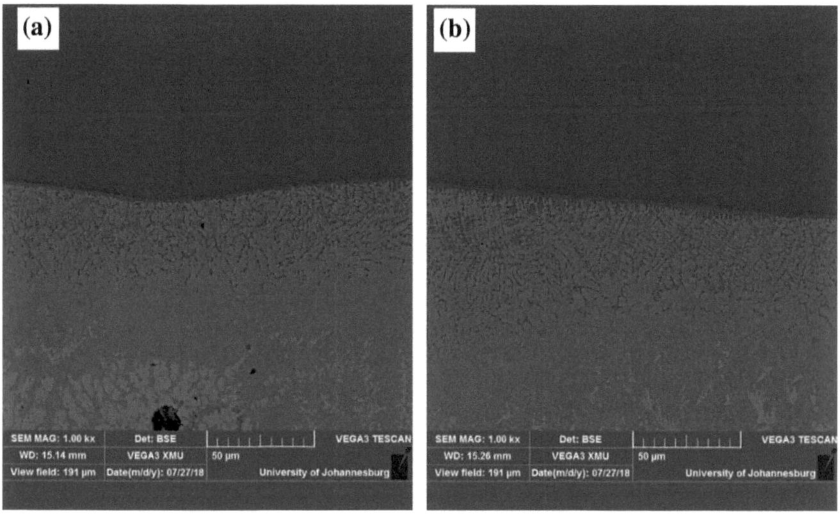

Fig. 2 SEM images cross-section of Ti-10Al–6Cu and Ti–10Al–9Cu coatings at laser power of 1000 and 1100 W and scanning speed of 1.0 m/min

fusion line, grain size decreased which can be explained by the cooling theory. The solidification of the microstructure depends on the temperature gradient and growth rate. The planar grain seen at the interface between coating and substrate is attributed to the high temperature gradient and low growth rate. Grains grow in the form of columnar dendrites are induced in the interior and top region of the coating as a result of high rapid cooling which is parallel to the direction of heat flow. However, the direction of dendrites at the upper part of the coating is more even compared with that in the middle, due to the instability of heat flow in the molten pool [10–17]. Figure 3a and b show slight cracks in the microstructure and this may be due to in-built stress as a result of too much laser power.

It is found that low solidification velocity and high thermal gradient at the bottom of melt pool and above the solid/liquid interface results in the presence of columnar dendrites (Figs. 2b and 3a). An equiaxed microstructure is developed from middle to the top section of the composite and this due to high solidification velocity and low thermal gradient at the upper part of the melt pool. It is evident that an increase in laser power results in an increase in the degree of melting. The columnar grains and dendritic microstructure were also increased by increasing laser power (Figs. 2a, b and 3a, b). Refined microstructure was observed as the laser speed increases while lower scanning speed leads to coarse microstructure. The clad area above the substrate decreases more significantly than the clad area below the substrate which results in an increase in dilution with an increase in the laser scanning speed. Increasing laser

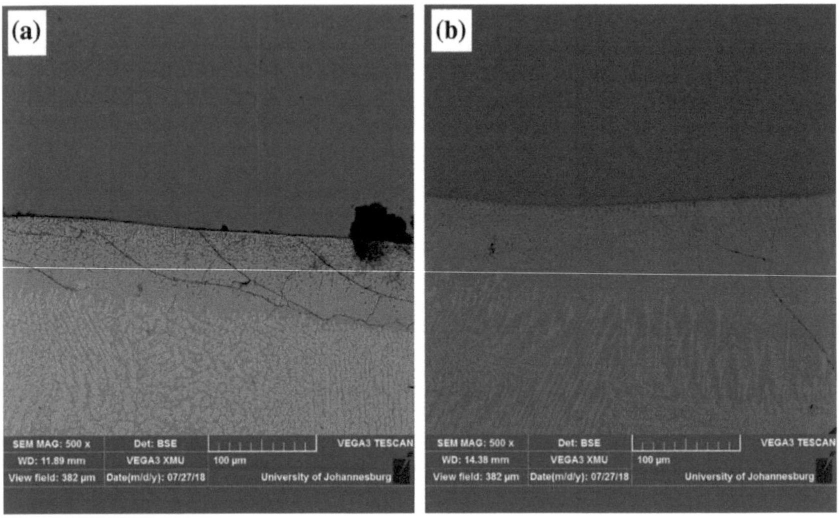

Fig. 3 SEM images cross-section of Ti–10Al–3Cu coatings at scanning speed of 0.8 and 1.0 m/min and laser power of 1000 W

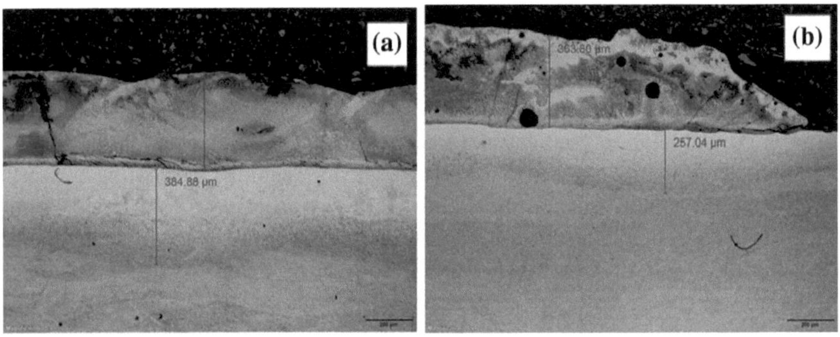

Fig. 4 Optical micrographs cross-section of Ti–10Al–6Cu at 1.0 m/min scanning speed and laser power of 1000 and 1100 W

scanning speed decreases the clad height while it also has a slight effect on the clad width. Hence, the influence of laser scanning speed on the clad height and clad width are slightly the same (Fig. 4). The heat affected zone of Fig. 4b is lower than that of Fig. 4a. While the deposit height in Fig. 4b is lower than that of Fig. 4a due to the influence of scanning speed.

Micro-hardness Property

The micro-hardness values are shown below under different laser powers and scanning speeds. It can be seen that laser scanning speed and the addition of Cu have greatly improved the microhardness of coatings. With increasing laser power, the hardness of the laser metal deposition of Ti–Al–Cu/Ti–6Al–4V composite coatings increased in general terms and the maximum hardness increased by 61.9%. The reason can be attributed to the interaction between the elements Ti, Al, Cu, Fe, C, and other elements in the remelting layer during the rapid solidification process. In addition, according to the Hall–Petch formula [18], the smaller the grain sized, the higher the hardness of the material H.

After laser deposition, the newly formed hard phase interacts with the fine parts of the resulting dense microstructure formed by rapid melting and cooling. The grains were refined and the formation of a hard intermediate phase was promoted in the coating. Therefore, the addition of Cu improved the micro-hardness of the coating. The average hardness obtained for samples Ti–10Al–3Cu, Ti–10Al–6Cu, Ti–10Al–9Cu at scanning speed of 0.8 and 1.0 m/min are 441 $\pm$ 0.71 $HV_{0.1}$ (A_1 – 0.8); 449 $\pm$ 0.71 $HV_{0.1}$ (A_1 – 1.0); 502 $\pm$ 0.71 $HV_{0.1}$ (B_1 – 0.8); 494 $\pm$ 2.12 $HV_{0.1}$ (B_1 – 1.0); 694 $\pm$ 0.07 (C_1 – 0.8); 798 $\pm$ 2.12 HV (C_1 – 1.0) respectively against the hardness of the substrate (304 $\pm$ $HV_{0.1}$).

Mathematical Modelling

The Computational Fluid Dynamics (CFD) simulation was done on hybrid coating (Ti–Al–Cu) as it is deposited on grade five titanium alloy (Ti–6Al–4V). COMSOL Multiphysics was used for the CFD simulation.

A model derived by Jouvard et al. [19] for calculating laser power required to melt the substrate and reinforcement powders is stated as follows:

$$P(substrate) = \sqrt{\pi} K_{stc}(T_{sm} - T_{si})/2\beta\sqrt{\alpha_{std} T_{int}} \tag{1}$$

$$P(powder) = M_{ppm} C_{pt} \left(T_{pmt} - T_{pit}\right)/\gamma, \tag{2}$$

where: T_{sm} is the substrate melting temperature, T_{si} is the substrate initial temperature, K_{stc} is the substrate thermal conductivity, α_{std} is the substrate thermal diffusivity, Tint is the laser beam/substrate interaction time, M_{ppm} is the powder particle mass, C_{pt} is the thermal heat capacity of powder material, T_{pmt} is particle melting temperature, T_{pit} is particle initial temperature. The simulation results show the contour plot temperature distribution at various distances in Fig. 5. The temperature (ht) and isothermal contours (ht) distributions obtained. Figures 5a–d represent contour temperature distributions as the coating was deposited on the Ti–6Al–4V substrate. The temperature was at a maximum at the contact area of the laser and gradually decreased as molten powder comes into contact and was distributed over the substrate.

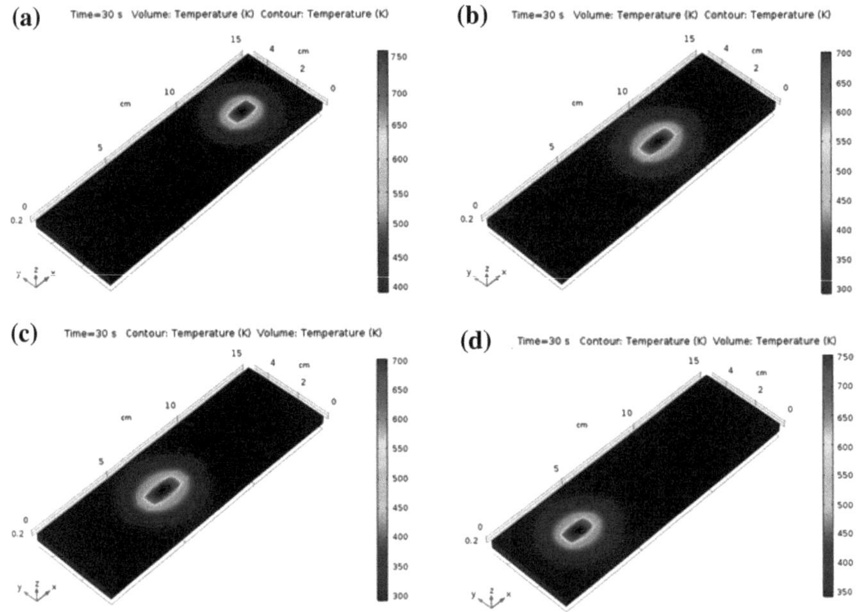

Fig. 5 Contour plot of temperature distribution results at various distances

The initial temperature was constant before the coming of the laser. Temperature rises rapidly by the irradiation of laser on the substrate because of laser power high density. Temperature decreases slowly as the laser moves from one point to another as a result of conduction and convection processes. The material expands thermally due to heating and contracts upon the removal of the heat source. Moreover, non-preheated part of the substrate showed a rapid change in temperature compared to preheated zone and this lead to high thermal gradients. In laser materials processing, thermal strain occurs as a result of temperature gradients caused by laser heating the solid substrate. Thermal stresses (compressive and tensile stresses) induced in preheated zones were less than the non-preheated zones due to the high thermal gradients in non-preheated zones. The coefficient of thermal expansion of the material and the substrate thermal gradients determine the thermal strains [20–23].

Molten pool is generated at the front of the track with a moving beam and the melt solidifies very rapidly as the laser beam moves on. Heat transfer in the molten pool is enhanced by the both the Marangoni and buoyancy forces by driving the molten fluid at the free surface from the center of the melt pool towards the boundaries [24, 25]. The temperature was observed to be below 750 °C (Fig. 5a), and as it moves towards the centre of the laser track, the temperature rises above 700 °C (Fig. 5b). As it moves towards the end of the track, the temperature starts to decrease once again below 750 °C. This is due to the fact that the centre of the track receives higher amount of powder and it requires higher temperature that allows the melt pool to absorb predefined powder material which is then redistributed to the rear section of the melt pool, and thereafter accumulated resulting in the formation of waves. More amount of powder takes more heat from the melt pool, thus decreases the temperature as it reaches to the end of the laser track as shown in Figs. 5c and d.

Conclusion

- The hardness property was enhanced by 61.9% at the composition of Ti–10Al–9Cu at a laser power of 1100 W and scanning speed of 1.0 m/min. This was attributed to newly formed hard phase interacts with the fine parts of the resulting dense microstructure formed by rapid melting and cooling. The grains were refined and the formation of a hard intermediate phase was promoted in the coating.
- It was found that low solidification velocity and high thermal gradient at the bottom of melt pool and above the solid/liquid interface results in the presence of columnar dendrites. Likewise, an equiaxed microstructure was also developed from middle to top section of the composite and this due to high solidification velocity and low thermal gradient at the upper part of the melt pool.
- Increased laser power and scanning speed significantly increased the diffraction peak of Ti and Ti_3Al. It is noted that Al and Ti concentration is highest which leads to the formation of titanium aluminide (Ti_3Al) and presence of and phases.

Acknowledgements The authors which to acknowledge the National Research Foundation (NRF) South Africa for their funding support.

References

1. Chikarakara E, Naher S, Brabazon D (2012) High speed laser surface modification of Ti-6Al-4V. Surf Coat Technol 206(14), 3/15/:3223–3229
2. Dutta Majumdar J, Manna I (2015) 21—Laser surface engineering of titanium and its alloys for improved wear, corrosion and high-temperature oxidation resistance. In: Waugh JLG (ed) Laser surface engineering. Woodhead Publishing, pp 483–521
3. Katta, S, Chaitanya G (2017) Key improvements in machining of Ti6Al4V alloy: a review. In: AIP conference proceedings, vol. 1859, no 1. AIP Publishing, p 020048
4. Weng F, Chen C, Yu H (2014) Research status of laser cladding on titanium and its alloys: a review. Mater Des 58:412–425
5. Mokgalaka MN, Pityana SL, Popoola PAI, Mathebula T (2014) NITI intermetallic surface coatings by laser metal deposition for improving wear properties of Ti–6Al–4V substrates. Adv Mater Sci Eng 2014:8
6. Ganesh B, Ramanaiah N, Rao PC (2012) Effect of surface treatment on tribological behavior of Ti–6Al–4V implant alloy. J Min Mater Charact Eng 11(07):735
7. Revankar GD, Shetty R, Rao SS, Gaitonde VN (2016) Wear resistance enhancement of titanium alloy (Ti–6Al–4V) by ball burnishing process. J Mater Res Technol 1(1):1–20
8. Qin L, Liu C, Yang K, Tang B (2013) Characteristics and wear performance of borided Ti6Al4V alloy prepared by double glow plasma surface alloying. Surf Coat Technol 225:92–96
9. Peters M, Kumpfert J, Ward CH, Leyens C (2003) Titanium alloys for aerospace applications. Adv Eng Mater 5(6):Pp.419-427

10. Gharehbaghi R, Fatoba OS, Akinlabi ET (2018) Influence of scanning speed on the microstructure of deposited Al–Cu–Fe coatings on a titanium alloy substrate by laser metal deposition process. In: Proceedings at the 2018 IEEE 9th international conference on mechanical and intelligent manufacturing technologies (ICMIMT 2018), Cape Town, South Africa, pp 44–49. https://doi.org/10.1109/icmimt.2018.8340418
11. Gharehbaghi R, Akinlabi ET, Fatoba OS (2018) Experimental investigation of laser metal deposited icosahedral Al–Cu–Fe coatings on grade five titanium alloy. In: Proceedings at the 2018 IEEE 9th international conference on mechanical and intelligent manufacturing technologies (ICMIMT 2018), Cape Town, South Africa, pp 31–36. https://doi.org/10.1109/icmimt.2018.8340416
12. Fatoba OS, Akinlabi ET, Akinlabi SA (2018) Effects of Fe addition and process parameters on the wear and corrosion properties of laser deposited Al–Cu–Fe coatings Ti–6Al–4V alloy. In: Proceedings at the 2018 IEEE 9th international conference on mechanical and intelligent manufacturing technologies (ICMIMT 2018), Cape Town, South Africa, pp 74–79. https://doi.org/10.1109/icmimt.2018.8340424
13. Fatoba OS, Akinlabi ET, Akinlabi SA (2018) Numerical investigation of laser deposited Al-based coatings on Ti–6Al–4V alloy. In: Proceedings at the 2018 IEEE 9th international conference on mechanical and intelligent manufacturing technologies (ICMIMT 2018), Cape Town, South Africa, pp 85–90. https://doi.org/10.1109/icmimt.2018.8340426
14. Popoola API, Fatoba OS, Aigbodion VS, Popoola OM (2017) Tribological evaluation of mild steel with ternary alloy of Zn-Al-Sn by laser deposition. Int J Adv Manuf Technol 89(5–8):1443–1449. https://doi.org/10.1007/s00170-016-9170-7
15. Fatoba OS, Popoola API, Aigbodion VS (2018) Electrochemical studies and surface analysis of laser deposited Zn-Al-Sn coatings on AISI 1015 Steel. Int J Surf Sci Eng 12(1):40–59. http://dx.doi.org/10.1504/IJSURFSE.2017.10009146
16. Fatoba OS, Popoola API, Fedotova T, Pityana SL (2015) Electrochemical studies on the corrosion behaviour of laser alloyed Zn-Sn coatings on UNS G10150 steel in 1M HCl Solution. Silicon 7(4):357–369
17. Fatoba OS, Akinlabi ET, Makhatha ME (2017) Effect of process parameters on the microstructure, hardness and wear resistance properties of Zn-Sn-Ti coatings on AISI 1015 steel: laser alloying technique. Int J Surf Sci Eng 11(6):489–511
18. Anderson PM, Li C (1995) Hall-petch relations for multilayered materials. Nanostruct Mater 5:349–362
19. Jouvard JM, Grevey DF, Lemoine F, Vannes AB (1997) Continuous wave Nd:Yag laser cladding modeling: a physical study of track creation during low power processing. J Laser Appl 9(1):43–50
20. Fatoba OS, Popoola API, Aigbodion VS (2016) Experimental study of hardness values and corrosion behaviour of laser alloyed Zn–Sn–Ti coatings of UNS G10150 mild steel. J Alloy Compd 658:248–254
21. Fatoba OS, Adesina OS, Popoola API (2018) Evaluation of microstructure, microhardness, and electrochemical properties of laser-deposited Ti–Co coatings on Ti–6Al–4V alloy. Int J Adv Manufact Technol. http://dx.doi.org/10.1007/S00170-018-2106-7
22. Makhatha ME, Fatoba OS, Akinlabi ET (2018) Effects of rapid solidification on the microstructure and surface analyses of laser-deposited Al–Sn coatings on AISI 1015 steel. Int J Adv Manuf Technol 94(1–4):773–787
23. Fatoba OS, Popoola API, Aigbodion VS (2018) Laser alloying of Al–Sn binary alloy onto mild steel: in-situ formation, hardness and anti-corrosion properties. Lasers Eng 39(3–6):292–312
24. Fatoba OS, Akinlabi SA, Gharehbaghi R, Akinlabi ET (2018) Microstructural analysis, microhardness and wear resistance properties of quasicrystalline Al–Cu–Fe coatings on Ti–6Al–4V alloy. Mater Express Res 5(6):1–14. https://doi.org/10.1088/2053-1591/aaca70
25. Akinlabi SA, Fatoba OS, Akinlabi ET (2018) Investigating resulting Residual stresses during mechanical forming process. IOP Conf Ser: Mater Sci Eng 328:1–7. https://doi.org/10.1088/1757-899X/328/1/012012

High-Cycle Fatigue Behaviour of Ultrafine Grained 5052 Al Alloy Processed Through Cryo-Forging

K. K. Yogesha, Amit Joshi, Raviraj, A. Raja and R. Jayaganthan

Abstract Mechanical properties of ultrafine grained 5052 Al alloy processed through multi-directional forging were investigated in the present work. The as-received 5052 Al alloy was solution-treated (ST) at temperature 540 °C for two hours and subjected to multi axial forging at room temperature as well as liquid N_2 temperature to a cumulative true strain of 4.2. The cryo-forged samples have exhibited a significant improvement in strength (380 MPa) and hardness (130 Hv) with 7.1% ductility, as compared to other conditions. Similarly, the high-cycle fatigue behaviour of the cryo-forged samples is found to be 80 MPa, which is better than other conditions. It was due to the formation of ultrafine grained microstructure with an average grain size of 230 nm in the cryo-forged samples. The formation of nanoshear bands in the cryo-forged samples, which accommodates the applied strain during cyclic loading is also responsible for dislocation accumulation along with broken/deformed impurity phase particles. The microstructure of the samples was characterized by optical microscopy, X-ray diffraction, and TEM to substantiate the mechanisms of grain refinement and its influence on the mechanical properties. Fractography of the tensile, as well as fatigue, tested samples were carried out using a Scanning Electron Microscope (SEM) to reveal the type of fracture.

Keywords Ultrafine grains · Microstructure · Characterization · Fractography

K. K. Yogesha (✉) · A. Joshi · Raviraj · A. Raja · R. Jayaganthan
Department of Metallurgical and Materials Engineering and Centre of Nanotechnology, Indian Institute of Technology Roorkee, Roorkee 247667, India
e-mail: yogeshakk@gmail.com

K. K. Yogesha
Department of Mechanical Engineering, National Institute of Engineering, Mysore 570008, India

A. Joshi
Department of Mechanical Engineering, G. B. Pant Institute of Engineering and Technology Pauri (Garhwal), Pauri (Garhwal), India

R. Jayaganthan
Department of Engineering Design, Indian Institute of Technology Madras, Chennai 600036, India

G. Lambotte et al. (eds.), *Materials Processing Fundamentals 2019*, The Minerals, Metals & Materials Series, https://doi.org/10.1007/978-3-030-05728-2_14

Introduction

5052 Al alloy is a non-heat treatable material used for components that require greater design efficiency, better forming properties and better welding characteristics. It exhibits a better combination of mechanical properties like yield strength (YS), ultimate tensile strength (UTS), fracture toughness and fatigue durability. The fatigue strength of an alloy is strongly governed by its ultimate tensile strength, which in turn depends on the grain size of the materials, as described by Hall–Petch [1, 2]. The severe plastic deformation techniques like Equi-channel angular pressing (ECAP) [3–6], High pressure torsion (HPT) [7–10], Accumulative roll bonding (ARB) [11–14], Repetitive corrugation and straightening (RCS) [15–18], Multiaxial forging (MAF) [19–22] are being used in producing ultrafine grained materials from its bulk alloy from the past two decades.

P. Cavaliere has ECAPed different metals like Al, Cu, Ti, Ni and conducted their tensile and fatigue test under stress control mode with load ratio R = 0.25. He noticed an improvement in their fatigue properties, which was attributed to the grain refinement caused during ECAP (2).

A. Vinogradov et al. have ECAPed four different compositions of Al–Mg–Sc–Zr alloys at various temperatures and performed both high- and low-cycle fatigue test. They observed an improvement in fatigue properties with increasing UTS and a percentage of Mg content in the alloy [23]. Further, in another investigation, the same researcher has conducted the fatigue test of 5056 Al–Mg alloy after ECAP processing. They observed an improvement in fatigue life of the processed samples at low-stress amplitudes [24].

G. Khatibi et al. have compared the fatigue properties of HPT processed and ECAP processed pure copper with its bulk alloy. They observed a better fatigue property in HPT processed samples compared to other conditions. This was attributed to the higher stability of extremely fine microstructure occurred, as because of the existence of impurities that avoids cyclic encouraged recrystallization as well as coarsening of grains [25]. The drawbacks associated with SPD processes like design difficulties, expensive tooling, requirement of a higher strain of the order 5 to 6 had insisted the researchers to identify an alternate potential route cryorolling, which suites for large-scale productions with medium strain requirement of the order 2 to 3 [26–29].

D. Singh et al. have processed 5083 Al alloy through cryorolling and cryorolling followed by warm rolling. Further, they have studied its high-cycle fatigue (HCF) properties, wherein they observed that the cryorolling followed by warm rolling samples exhibited better fatigue strength, which was due to enhancement in YS of the material through various strengthening mechanisms like solid solution strengthening, precipitation strengthening, grain boundary strengthening, and dislocation strengthening [30, 31]. However, the literature on HCF behaviour of AA 5052 is scarce. Therefore, the present work is envisaged on the HCF performance of cryo-forged 5052 Al alloy. The detailed microstructural characterization of the alloy was made through SEM and TEM to substantiate the fatigue properties.

Experimental

Processing

The commercially available 5052 Al alloy with chemical composition 2.3 wt% Mg, 0.26 wt% Cr, 0.2 wt% Si, 0.33 wt% Fe, 0.10 wt% Cu, 0.10 wt% Mn, 0.10 wt% Zn and rest Aluminium, is used in the present work. The received materials were cut into rectangular samples of dimensions 20 × 25 × 40 mm. Further, these samples were homogenized at 540 °C for 2 h followed by water quenching. The Solution Treated (ST) samples were forged in a friction screw press to a true strain of 4.2, both under room temperature as well as liquid nitrogen temperature.

Hardness and Tensile Strength Measurements

Hardness and tensile strength of the samples were conducted at room temperature to analyze the effect of deformation strain on the grain refinement. The hardness value of the samples was measured on Vickers hardness testing machine with a load of 5 kg and 15 s of dwell time. The tensile test was conducted on H25 K-S Tinius Oslen tensile testing machine with a strain rate of 0.6×10^{-3}/s. The samples were prepared as per the ASTM Standard E-8/E8 M-09 sub-size specimen of gauge length 25 mm.

High Cycle Fatigue Test

The stress-life fatigue test of the solution treated and cryo-forged samples were conducted on Instron 8802 universal testing machine as shown in Fig. 1. The fatigue samples were prepared as per ASTM E466-15 standards. The required smooth surface was obtained by polishing the samples with emery papers up to 1500 grit size. These specimens were axially loaded under stress control mode, using stress ratio, R of 0.1 and frequency, f of 20 Hz.

The optical microstructures of the samples were characterized by using Leica DMI 5000 optical microscope. TEM samples of ST, processed samples were prepared as discussed in the reference papers [27–29].

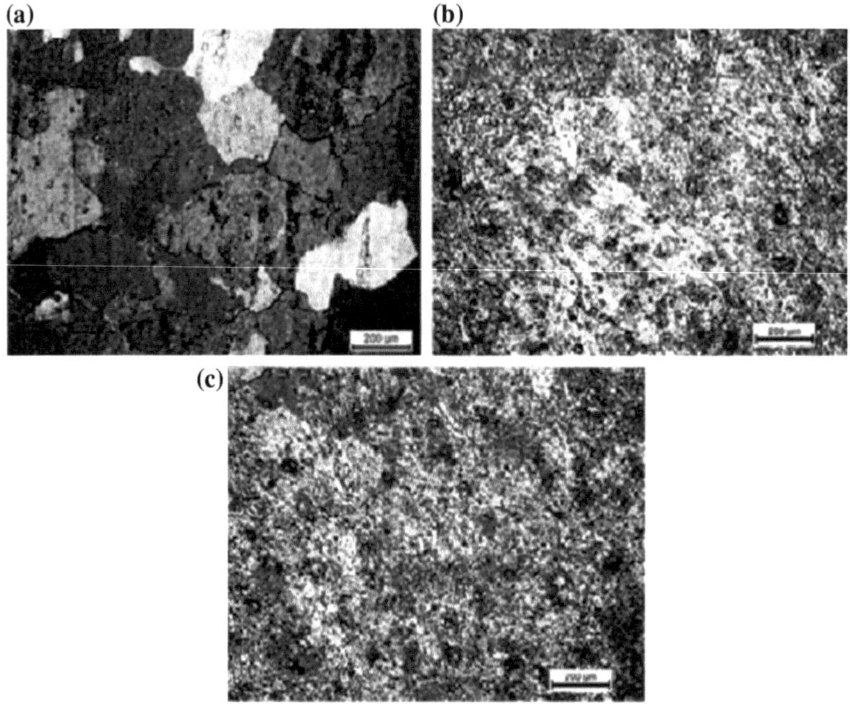

Fig. 1 Optical micrograph; **a** starting material (ST), **b** RF, **c** CF

Results and Discussion

Optical Microstructures

The ST alloy possesses an equiaxed grain structure (grain size 352 μm) Fig. 1a. Figure 1b is the magnified view of ST samples, which shows the presence of impurity phase particles which promotes accumulation of dislocation by obstructing its movement. In the case of Room temperature Forged (RF) samples, grain structures with fragmentation are seen. The cryo-forged (CF) samples exhibit more deformed structure than RF samples Fig. 1c. This is due to the suppression of more dislocations in the cryogenic temperature.

Tensile Properties

Figure 2 shows the tensile results of all condition samples. The ST samples have exhibited an ultimate tensile strength (UTS)(σ_{ut}) of 170 MPa and yield strength

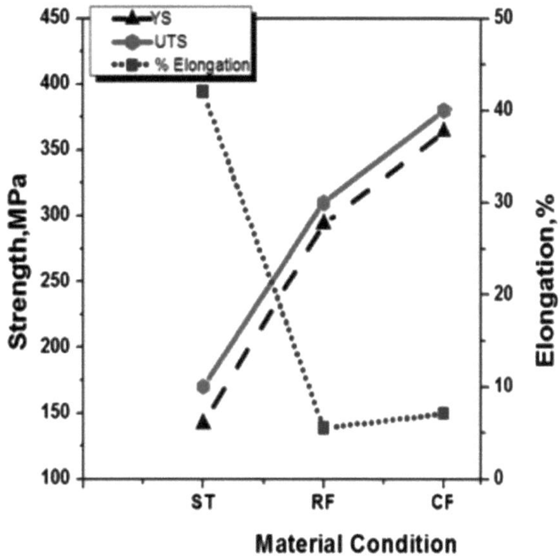

Fig. 2 Variation in UTS, YS and Percentage elongation of ST, RF and CF samples

(YS)(σ_y) of 152 MPa, with percentage elongation to failure 42%. The CF samples have shown better tensile properties than other conditions (UTS of 380 MPa and YS of 354 MPa). This improvement in mechanical properties in CF samples is due to accumulation of dislocation density developed during cryo-forging process [32, 33]. The deformed impurity phase particles are also responsible in effective hindering of dislocation movement.

High-Cycle Fatigue Properties

The S–N plot shown in Fig. 3 describes the relationship between stress amplitude v/s numbers of cycles to failure in ST and CF samples. From this plot fatigue strength of the material has been evaluated based on 10^6 cycles as a measure of high-cycle fatigue life [34–36]. Fatigue strength (σ_f) for ST samples is 40 MPa, whereas it is 80 MPa in CF samples. This fatigue strength is dependent on grain size as well as the strength of the material.

TEM Analysis

Figure 4 shows the TEM images corresponding to ST, RF and CF samples. In ST sample the presence of impurity phase particles is seen in Fig. 4a. These particles are of different shapes (needle, plate and rod shape) and sizes. They do a significant role

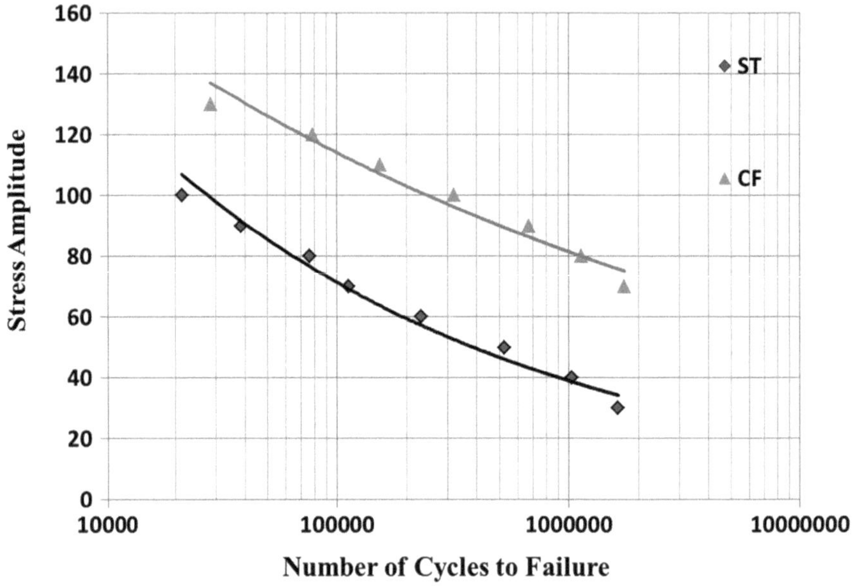

Fig. 3 S-N plot of ST, CF samples

in increasing the strength of the material by hindering the movement of dislocation through effective pinning at grain boundary [37]. Figure 4b shows the TEM image corresponding to RF samples, wherein, the dislocation accumulated and subgrains formed during room temperature forging are seen. The average size of these subgrains is around 300 nm. The TEM image corresponding to CF samples is shown in Fig. 4c, in which very fine subgrains of average grain size 230 nm with more dislocation density are seen. These ultrafine subgrains, which minimize the size of the nucleating flaws and enhances the resistance to the crack propagation in the CF sample is responsible in increasing the UTS and fatigue strength of this condition [28, 30]. The nanoshear bands, which were formed during the cryo-forging are shown in Fig. 4c. They accommodate the applied strain during cyclic loading and also responsible for dislocation accumulation. The Selected Area Electron Diffraction (SAED) pattern corresponding to CF samples showed in Fig. 4c (in inset) exhibit a number of elongated spots than other conditions, which depicts the presence of high angle grain boundaries due to grain refinement. The deformed or broken impurity phase particles are effectively hindering the dislocation movement and thereby delays crack propagation, also responsible in increasing the fatigue strength of CF samples.

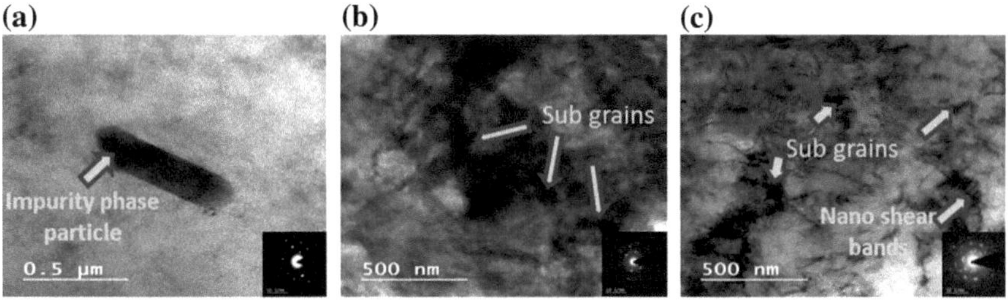

Fig. 4 TEM images AA 5052 at different condition **a** ST, **b** RF and **c** CF samples

Conclusions

Following conclusions are drawn through the HCF strength study of 5052 Al alloy processed by cryo-forging.

- 5052 Al alloy has been cryo-forged to a true strain of 4.2. These deformed samples exhibit an UTS of 380 MPa and fatigue strength of 80 MPa, which is better than its ST condition ($\sigma_{ut} = 170$ MPa and $\sigma_f = 40$ MPa).
- Presence of subgrains of an average grain size 230 nm in CF samples does a significant role in increasing the σ_{ut} and σ_f, as grain refinement is responsible in increasing these properties.
- The deformed or broken impurity phase particles present in the CF samples are effectively hindering the dislocation movement, which in turn improves the resistance of crack propagation, also plays a prominent role in increasing σ_{ut} and σ_f.

References

1. Hansen N (2004) Hall-Petch relation and boundary strengthening. Scripta Mater 51(8):801–806
2. Cavaliere P (2009) Fatigue properties and crack behavior of ultra-fine and nanocrystalline pure metals. Int J Fatigue 31(10):1476–1489
3. Zhao YH, Liao XZ, Jin Z, Valiev RZ, Zhu YT (2004) Microstructures and mechanical properties of ultrafine grained 7075 Al alloy processed by ECAP and their evolutions during annealing. Acta Mater 52(15):4589–4599
4. Chang SY, Lee KS, Choi SH, Shin DH (2003) Effect of ECAP on microstructure and mechanical properties of a commercial 6061 Al alloy produced by powder metallurgy. J Alloy Compd 354(1):216–220
5. Zha M, Li Y, Mathiesen RH, Bjørge R, Roven HJ (2015) Microstructure evolution and mechanical behavior of a binary Al–7Mg alloy processed by equal-channel angular pressing. Acta Mater 84:42–54
6. Dadbakhsh S, Taheri AK, Smith CW (2010) Strengthening study on 6082 Al alloy after combination of aging treatment and ECAP process. Mater Sci Eng, A 527(18):4758–4766
7. Zhilyaev AP, Nurislamova GV, Kim BK, Baró MD, Szpunar JA, Langdon TG (2003) Experimental parameters influencing grain refinement and microstructural evolution during high-pressure torsion. Acta Mater 51(3):753–765

8. Zhilyaev AP, Oh-Ishi K, Langdon TG, McNelley TR (2005) Microstructural evolution in commercial purity aluminum during high-pressure torsion. Mater Sci Eng, A 410:277–280
9. Straumal BB, Baretzky B, Mazilkin AA, Phillipp F, Kogtenkova OA, Volkov MN, Valiev RZ (2004) Formation of nanograined structure and decomposition of supersaturated solid solution during high pressure torsion of Al–Zn and Al–Mg alloys
10. Xu C, Horita Z, Langdon TG (2008) The evolution of homogeneity in an aluminum alloy processed using high-pressure torsion. Acta Mater 56(18):5168–5176
11. Tsuji N, Saito Y, Lee SH, Minamino Y (2003) ARB (accumulative roll-bonding) and other new techniques to produce bulk ultrafine grained materials. Adv Eng Mater 5(5):338–344
12. Lee SH, Saito Y, Tsuji N, Utsunomiya H, Sakai T (2002) Role of shear strain in ultragrain refinement by accumulative roll-bonding (ARB) process. Scripta Mater 46(4):281–285
13. Saito Y, Utsunomiya H, Tsuji N, Sakai T (1999) Novel ultra-high straining process for bulk materials—development of the accumulative roll-bonding (ARB) process. Acta Mater 47(2):579–583
14. Höppel HW, May J, Göken M (2004) Enhanced strength and ductility in ultrafine-grained aluminium produced by accumulative roll bonding. Adv Eng Mater 6(9):781–784
15. Huang JY, Zhu YT, Jiang H, Lowe TC (2001) Microstructures and dislocation configurations in nanostructured Cu processed by repetitive corrugation and straightening. Acta Mater 49(9):1497–1505
16. Rajinikanth V, Arora G, Narasaiah N, Venkateswarlu K (2008) Effect of repetitive corrugation and straightening on Al and Al–0.25 Sc alloy. Mater Lett 62(2):301–304
17. Thangapandian N, Prabu SB, Padmanabhan KA (2016) Effects of die profile on grain refinement in Al–Mg alloy processed by repetitive corrugation and straightening. Mater Sci Eng, A 649:229–238
18. Bhovi PM, Patil DC, Kori SA, Venkateswarlu K, Huang Y, Langdon TG (2016) A comparison of repetitive corrugation and straightening and high-pressure torsion using an Al–Mg–Sc alloy. J Mater Res Technol
19. Rao PN, Singh D, Jayaganthan R (2014) Mechanical properties and microstructural evolution of Al 6061 alloy processed by multidirectional forging at liquid nitrogen temperature. Mater Des 56:97–104
20. Cherukuri B, Srinivasan R (2006) Properties of AA6061 processed by multi-axial compressions/forging (MAC/F). Mater Manuf Processes 21(5):519–525
21. Fuloria D, Kumar N, Goel S, Jayaganthan R, Jha S, Srivastava D (2016) Tensile properties and microstructural evolution of Zircaloy-4 processed through rolling at different temperatures. Mater Des 103:40–51
22. Rao PN, Singh D, Jayaganthan R (2014) Mechanical properties and microstructural evolution of Al 6061 alloy processed by multidirectional forging at liquid nitrogen temperature. Mater Des 56:97–104
23. Vinogradov A, Washikita A, Kitagawa K, Kopylov VI (2003) Fatigue life of fine-grain Al–Mg–Sc alloys produced by equal-channel angular pressing. Mater Sci Eng, A 349(1):318–326
24. Vinogradov A, Nagasaki S, Patlan V, Kitagawa K, Kawazoe M (1999) Fatigue properties of 5056 Al–Mg alloy produced by equal-channel angular pressing. Nanostruct Mater 11(7):925–934
25. Khatibi G, Horky J, Weiss B, Zehetbauer MJ (2010) High cycle fatigue behaviour of copper deformed by high pressure torsion. Int J Fatigue 32(2):269–278
26. Panigrahi SK, Jayaganthan R (2008) A study on the mechanical properties of cryorolled Al–Mg–Si alloy. Mater Sci Eng, A 480(1):299–305
27. Yogesha KK, Joshi A, Kumar N, Jayaganthan R (2016) Effect of cryo groove rolling followed by warm rolling (CGW) on the mechanical properties of 5052 Al alloy. Mater Manuf Processes, 1–9
28. Joshi A, Kumar N, Yogesha KK, Jayaganthan R, Nath SK (2016) Mechanical properties and microstructural evolution in Al 2014 alloy processed through multidirectional cryoforging. J Mater Eng Perform 25(7):3031–3034

29. Joshi A, Yogesha KK, Jayaganthan R (2017) Influence of cryorolling and followed by annealing on high cycle fatigue behavior of ultrafine grained Al 2014 alloy. Mater Charact
30. Singh D, Nageswara Rao P, Jayaganthan R (2014) High cyclic fatigue behaviour of ultrafine grained Al 5083 alloy. Mater Sci Technol 30(14):1835–1842
31. Yogesha KK, Kumar N, Joshi A, Jayaganthan R, Nath SK (2016) A comparative study on tensile and fracture behavior of Al–Mg alloy processed through cryorolling and cryo groove rolling. Metall Microstruct Anal 5(3):251
32. Lee YB, Shin DH, Nam WJ (2005) Effect of annealing temperature on tensile behavior of 5052 Al alloy deformed at cryogenic temperature. J Mater Sci 40(5):1313–1315
33. Panigrahi SK, Jayaganthan R (2008) A study on the mechanical properties of cryorolled Al–Mg–Si alloy. Mater Sci Eng, A 480(1):299–305
34. Srivatsan TS, Anand S, Sriram S, Vasudevan VK (2000) The high-cycle fatigue and fracture behavior of aluminum alloy 7055. Mater Sci Eng, A 281(1):292–304
35. Yogesha KK, Joshi A, Jayaganthan R (2017) Fatigue behavior of ultrafine-grained 5052 Al alloy processed through different rolling methods. J Mater Eng Perform 26:2826–2836
36. Beachem CD (1965) Electron fractographic studies of mechanical fracture processes in metals. J Basic Eng 87(2):299–306
37. Singh D, Rao PN, Jayaganthan R (2013) Effect of deformation temperature on mechanical properties of ultrafine grained Al–Mg alloys processed by rolling. Mater Des 50:646–655

Effect of Heat Treatment on Microstructure of Continuous Unidirectional Solidified Cu–Ni–Sn Alloy

Ji Hui Luo, Qin Li, Yan Hui Chen, Shu Liu, Qiu Yue Wen and Hui Min Ding

Abstract Cu–15%Ni–8%Sn alloy was prepared by continuous unidirectional solidification (CUS) processing and the as-cast CUS Cu–15%Ni–8%Sn alloy was homogenized. The evolution of microstructure and composition distribution of as-cast and annealed CUS Cu–15%Ni–8%Sn alloy was analyzed. The results show that the microstructure of as-cast CUS Cu–15%Ni–8%Sn is mainly composed of coarse columnar grains and there exist segregation phenomenon in the grain boundaries. After annealing at temperature of 850°C for 30 min, it was found that the composition of the alloy began to become uniform and the solutes gathered at the grain boundaries diffused into the columnar grains.

Keywords Cu–Ni–Sn alloy · Continuous unidirectional solidification
Microstructures · Annealing

Introduction

Cu–Ni–Sn alloys are widely used in space, electronics, transportation and other fields due to their high strength, good wear resistance and corrosion resistance [1–4]. However, the plasticity of the alloys is low [5], so wide application of the alloys is limited. Under certain conditions of the alloy composition, changing microstructure of the alloy is the main way to improve the mechanical properties of the alloy. Columnar grain, obtained by continuous unidirectional solidification (CUS) processing [6–8], can effectively improve the plastic properties of the alloy [9, 10].

On the other hand, the composition of the as-cast microstructure of Cu–Ni–Sn alloy is quite uneven [11]. Segregation is easy to occur. In order to eliminate the component segregation of the alloy, it must be annealed. Accordingly, the aim of homogenization is achieved through diffusion.

J. H. Luo (✉) · Q. Li · Y. H. Chen · S. Liu · Q. Y. Wen · H. M. Ding
College of Mechanical and Electrical Engineering, Yangtze Normal University, Chongqing 408100, People's Republic of China
e-mail: luo_jihui@163.com

G. Lambotte et al. (eds.), *Materials Processing Fundamentals 2019*, The Minerals, Metals & Materials Series, https://doi.org/10.1007/978-3-030-05728-2_15

In this work, Cu–15%Ni–8%Sn alloy was prepared by CUS processing. The microstructure and composition distribution of as-cast and annealed CUS Cu–15%Ni–8%Sn alloy were studied.

Experimental

CUS Cu–15%Ni–8%Sn Experiment

Currently, the CUS experiments were performed using the CUS technology. The method and technology for alloy fabrication were described in elsewhere [12]. The main process parameters are as follows: a mold temperature of 1100°C, melt temperature of 1200°C, and continuous casting speed of 10 mm/min. An alloy sheet with a size of 20 × 5 mm was continuously pulled out by traction wheels.

Annealing Experiment

The prepared alloy sheet was subjected to an annealing test. The alloy was heated to 850°C under a hydrogen atmosphere and held for 30 min, and was slowly cooled to room temperature.

Microscopic Observation and Composition Distribution

The as-cast and annealed samples were cut, sanded and polished, and the surfaces were etched. The surface after etching was observed using a field emission scanning electron microscope (SEM). The energy dispersive spectrometry (EDS) was used to analyze the composition of the alloy.

Results and Discussion

Figure 1a, b show the SEM images of CUS Cu–15%Ni–8%Sn alloy. From Fig. 1a, it can be seen that the microstructure of the alloy is mainly coarse columnar grain, and the average diameter of each columnar grain is about 75 μm. The composition of the columnar grain is uniform. However, there exists another kind of phase between the columnar grain boundaries. Figure 1b is a further amplification of area A in Fig. 1a, which indicated by arrow A. It can be seen that the phase in the grain boundary mainly shows lamellar structure, which has regular distribution and has a certain orientation

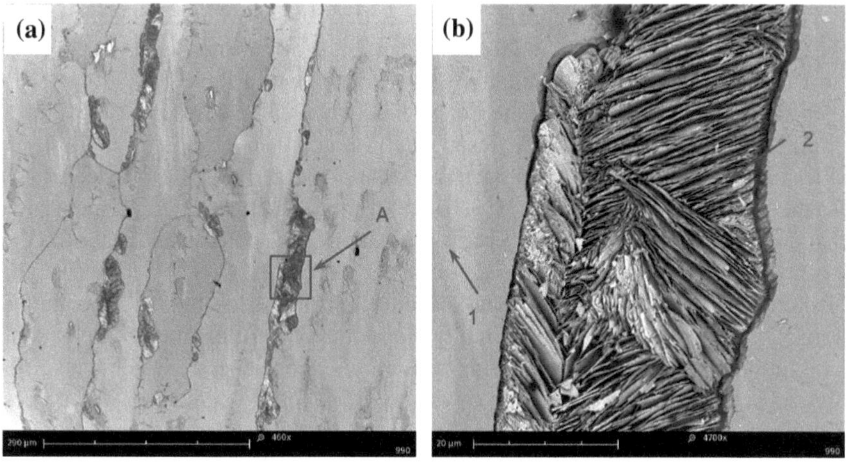

Fig. 1 SEM of as-cast CUS Cu–15%Ni–8%Sn alloy; **a** microstructure, **b** a magnified structure of point A

Table 1 Composition of as-cast CUS Cu–15%Ni–8%Sn

Position	Element		
	Cu(wt%)	Ni(wt%)	Sn(wt%)
1	80.29	15.66	4.05
2	57.67	22.76	19.57

relationship. The composition analysis of columnar grain and lamellar structure was carried out, and the test positions are shown in Fig. 1b. Arrow 1 indicates the test position in columnar grain, and arrow 2 indicates the lamellar structure. The results are shown in Table 1.

It can be seen from Table 1 that the Cu, Ni and Sn content in the columnar grain is 80.29, 15.66 and 4.05%, respectively, which is lower than that of nominal composition of Sn. The reason may be that the alloy will precipitate Sn solute in the liquid phase during solidification, resulting in the difference between the liquid and the solid. Thus leads to less solute in the final solidified alloy.

Table 1 also shows that the Cu, Ni and Sn content in the lamellar structure is 57.67, 22.76 and 19.57%, respectively, which is much higher than that of nominal composition of Ni and Sn. The above results show that most of the precipitated solute are concentrated in columnar grain boundaries, forming a serious segregation.

In order to obtain homogeneous microstructures, as-cast CUS Cu–15%Ni–8%Sn alloy was annealed at temperatures of 850°C for 30 min followed by cooling with furnace. Figure 2 depicts the microstructure of CUS Cu–15%Ni–8%Sn alloy after the annealing treatment. It can be seen that the lamellar structure of the alloy begins to disappear at the grain boundary. Only a large number of pits and holes were left in the area where the lamellar structure appeared.

The composition of the annealed alloy was analyzed by using EDS technique. In Fig. 2, arrow 1 is the test position for columnar grain, and arrow 2 is the test position for the lamellar structure. Quantitative results are shown in Table 2.

In Table 2, it can be seen that the Cu, Ni and Sn content in position 1 (in columnar grain) is 77.16, 14.52 and 8.32%, respectively. Compared with the as-cast alloy, in which the Sn content tends to approach the nominal composition. It can be concluded that the composition of CUS Cu–15%Ni–8%Sn alloy is homogenized after annealing treatment. The Cu, Ni and Sn content in position 2, where the lamellar structure appears in as-cast alloy, is 73.06, 16.74 and 10.2%, respectively. It is observed that the composition of alloy show a small difference from the location of the columnar grain. The Sn and Ni content are slightly higher than that of nominal composition.

Solute redistribution occurs during the solidification process, which causes the alloy to precipitate solute at the front of the solid–liquid interface. In the process of upward growth of columnar grains, these precipitated solutes are concentrated in front of the solid–liquid interface between the columnar grains and the liquid phase and thus resulted in decreasing of Sn and Ni content in columnar grains. The Sn and Ni content in front of the solid–liquid interface begin to increase. When two grains grow up and start to contact, a grain boundary is formed. These solute enriched

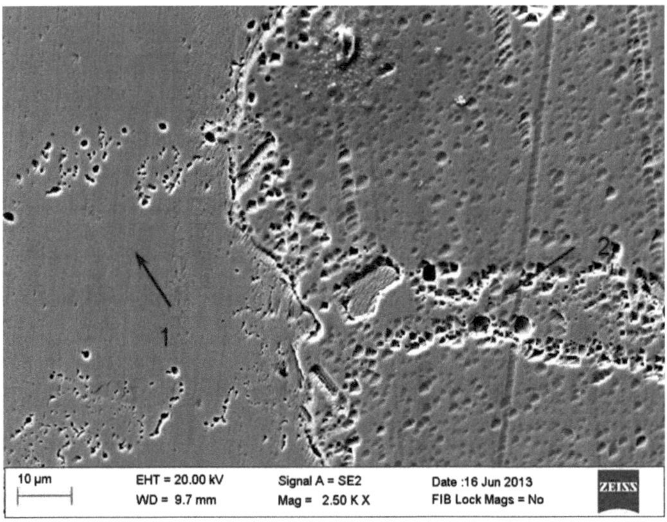

Fig. 2 Microstructure of annealed CUS Cu–15%Ni–8%Sn alloy

Table 2 Composition of annealed CUS Cu–15%Ni–8%Sn

Position	Element		
	Cu(wt%)	Ni(wt%)	Sn(wt%)
1	77.16	14.52	8.32
2	73.06	16.74	10.2

in front of the solid–liquid interface begin to accumulate and solidify at the grain boundaries of the columnar grains, eventually forming the morphology shown in Fig. 1a.

While the CUS Cu–15%Ni–8%Sn alloy is annealed under high temperature, the solute Sn and Ni begin to diffuse from the higher concentration in grain boundaries to the lower concentration in columnar grains, thus causing the composition of the CUS Cu–15%Ni–8%Sn alloy to become uniform. After the Sn and Ni at the grain boundaries diffusing into the columnar grains, some pits and holes are left, as shown in Fig. 2.

Conclusions

Cu–15%Ni–8%Sn alloy was prepared by CUS process. The composition distribution of the as-cast and annealed alloy was studied. The main results are as follows:

(1) The CUS Cu–15%Ni–8%Sn alloy is mainly composed of coarse columnar grains. There exists segregation of Ni and Sn at the grain boundary.
(2) After annealing at 850 °C for 30 min, the segregation at the grain boundary begin to disappear and the composition of CUS Cu–15%Ni–8%Sn alloy become uniform, which meets the requirements of nominal composition.

Acknowledgements This work was supported by the project of Yangtze Normal University (2017KYQD130).

References

1. Deyong L, Elboujdaïni M, Tremblay R, Ghali E (1990) Electrochemical behaviour of rapidly solidified and conventionally cast Cu–Ni–Sn alloys. J Appl Electrochem 20:756–762
2. Alili B, Bradai D, Zieba P (2008) On the discontinuous precipitation reaction and solute redistribution in a Cu–15%Ni–8%Sn alloy. Mater Charact 59:1526–1530
3. Wang Y, Wang M, Hong B (2005) Microstructures of spinodal phase in Cu–15Ni–8Sn alloy. J Univ Sci Technol Beijing 12(3):243–247
4. Zhao JC, Notis MR (1998) Spinodal Decomposition, ordering transformation, and discontinuous precipitation in a Cu–15Ni–8Sn alloy. Acta Mater 46(12):4203–4218
5. Ouyang Y, Gan X, Zhang S, Li Z, Zhou K, Jiang Y, Zhang X (2017) Age-hardening behavior and microstructure of Cu−15Ni−8Sn−0.3Nb alloy prepared by powder metallurgy and hot extrusion. Trans Nonferrous Met Soc China 27:1947–1955
6. Motoyasu G, Soda H, McLean A, Shimizu T (1997) Al–CuAl2 eutectic structure in unidirectionally solidified rods by the Ohno continuous casting process. J Mater Sci Lett 16:566–568
7. Ozawa S, Motegi T, Kuribayashi K (2004) Unidirectional solidification of aluminum–indium monotectic alloys by Ohno continuous casting. Mater Trans 45(2):353–356
8. Okayasu M, Takeuchi S (2014) Mechanical strength and failure characteristics of cast Mg–9%Al–1%Zn alloys produced by a heated-mold continuous casting process: Fatigue properties. Mater Sci Eng, A 600:211–220

9. Okayasu M, Yoshie S (2010) Mechanical properties of Al–Si_{13}–$Ni_{1.4}$–$Mg_{1.4}$–Cu_1 alloys produced by the Ohno continuous casting process. Mater Sci Eng, A 527:3120–3126
10. Wang Y, Huang HY, Xie JX (2011) Enhanced roofm-temperature tensile ductility of columnar-grained polycrystalline Cu–12 wt. %Al alloy through texture control by Ohno continuous casting process. Mater Lett 65(7):1123–1126
11. Virtanen P, Tiainen T, Lepisto T (1998) Precipitation at faceting grain boundaries of Cu–Ni–Sn alloys. Mater Sci Eng, A 251:269–275
12. Luo JH (2018) Formation mechanism of surface segregation in heated mold continuous casting Al–Cu alloy. Light Metals 2018:435–439

Part IV
Multiphysics—Process and Properties Modeling

Modeling of Fluid Flow Effects on Experiments Using Electromagnetic Levitation in Reduced Gravity

Gwendolyn Bracker, Xiao Xiao, Jonghyun Lee, Marcus Reinartz, Stefan Burggraf, Dieter Herlach, Markus Rettenmayr, Douglas Matson and Robert Hyers

Abstract Electromagnetic levitation experiments provide a powerful tool that allows for the study of homogeneous nucleation, solidification and growth in a containerless processing environment. However, in these experiments it is important to understand the magnetohydrodynamic flow within the sample and the effects that this fluid flow has on the experiment. A recent solidification study found that

---

G. Bracker (✉) · R. Hyers
University of Massachusetts, Amherst, USA
e-mail: gbracker@umass.edu

R. Hyers
e-mail: hyers@umass.edu

X. Xiao · D. Matson
Tufts University, Medford, USA
e-mail: Xiao.Xiao@tufts.edu

D. Matson
e-mail: Douglas.matson@tufts.edu

J. Lee
Iowa State University, Ames, USA
e-mail: jolee@iastate.edu

M. Reinartz · M. Rettenmayr
Otto-Schott-Institut für Materialforschung, Friedrich-Schiller-Universität, Jena, Germany
e-mail: Marcu.reinartz@uni-jena.de

M. Rettenmayr
e-mail: M.Rettenmayr@uni-jena.de

S. Burggraf
Deutsches Zentrum für Luft- und Raumfahrt, Institut für Materialphysik im Weltraum, Cologne, Germany
e-mail: Stefan.burggraf@dlr.de

D. Herlach
Deutsches Zentrum für Luft, Institut für Experimentalphysik IV, Institut für Materialphysik im Weltraum, Ruhr-Universität Bochum, Bochum, Germany
e-mail: Dieter.Herlach@dlr.de

G. Lambotte et al. (eds.), *Materials Processing Fundamentals 2019*, The Minerals, Metals & Materials Series, https://doi.org/10.1007/978-3-030-05728-2_16

aluminum-nickel alloy samples have an unusual growth response to the degree of undercooling. These aluminum-nickel alloys experienced a decrease in the growth velocity as the initial undercooling deepened instead of the expected increase in solidification velocity with deepening undercoolings. Current work is exploring several different theories to explain this phenomenon. Distinguishing between several of these theories requires a comprehensive understanding of the behavior of the internal fluid flow. USTIP has done flow modeling to support this and multiple other collaborators on ISS-EML. The fluid flow models presented provide critical insights into the nature of the flow within the aluminum-nickel alloy experiments conducted in the ISS-EML facility. These models have found that for this sample the RNG k-ε model should be used with this sample at temperatures greater than 1800 K and the laminar flow model should be used at temperatures lower than 1600 K.

Keywords Aluminum-Nickel · Electromagnetic levitation
Containerless processing · Solidification · Fluid flow simulation · ISS-EML

Introduction

According to classical solidification theory, the solidification velocity is a function of the undercooling of a melt [1, 2]. As a melt experiences deeper undercooling, the difference of the Gibbs free energies between the solid and liquid phases scales with undercooling, resulting in an increased driving force for solidification and an increased growth velocity. However, recent work on aluminum-nickel alloys processed in reduced gravity has shown the growth velocity to decrease with increased undercooling [3]. The phenomenon was observed across a range of different aluminum-nickel sample compositions tested terrestrially in an electromagnetic levitation field. The results of these terrestrial experiments are shown in Fig. 1 along with two data points testing $Al_{68.5}Ni_{31.5}$, shown as filled triangles, obtained from a reduced gravity test during the TEXUS 44 flight mission [3].

Current work is exploring several different theories to explain this anomaly including additional experiments in reduced gravity in the Material Science Laboratory-Electromagnetic Levitator (MSL-EML) on board the international space station (ISS). To differentiate between solidification theories, it is important to have a comprehensive understanding of the fluid flow in the sample during cooling to account for the effects of stirring and convection in the liquid [3, 4].

Sample Properties

The sample tested using the oscillating drop method on the ISS was an $Al_{75}Ni_{25}$ sample that had a mass of 527.24 mg and was 6.5 mm in diameter [5]. Based on prior work, this alloy is known to have a liquidus temperature at 1398 K and a solidus

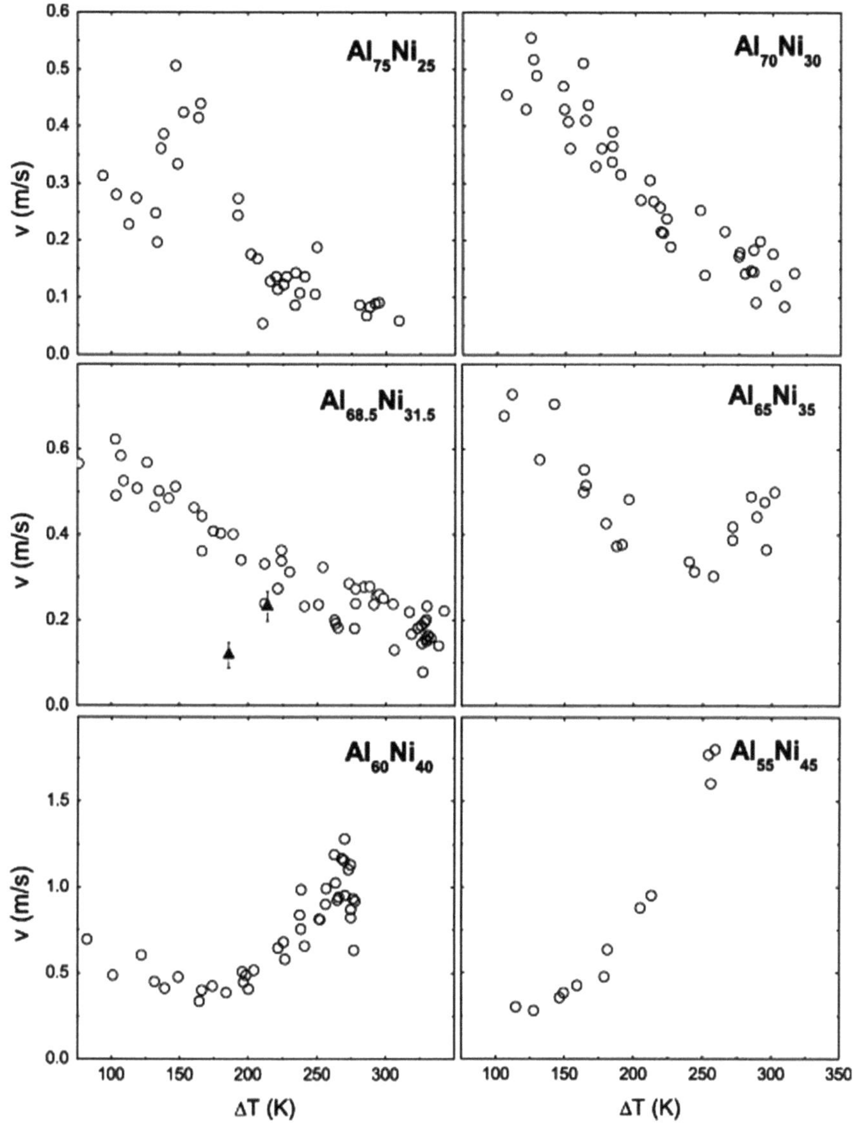

Fig. 1 Dendrite growth velocity as a function of undercooling measured under terrestrial conditions (open circles) and reduced gravity samples (filled triangles) on electromagnetically levitated Al-Ni samples of various compositions [3]

temperature at 1132 K [6]. The same prior work by I. Egry also provided other property-temperature relationships that are necessary to simulate the flow within the drop under different conditions [6]. The electrical conductivity of the melt at a given temperature is necessary to relate the electromagnetic field to the force that is applied to the liquid sample and drives the flow. The electrical conductivity of this Al-Ni alloy is given in Eq. 1 [6].

$$\sigma(T) = 10165 + 0.59(T - T_l) \quad \left[\Omega^{-1}\mathrm{m}^{-1}\right] \tag{1}$$

In addition, the density and viscosity of the melt are critical to determining the flow within the drop. In the same prior work, the density was measured in contactless processing and image analysis to measure the volume of the sample, assume axially symmetry. The density as a function of temperature was then fit to the linear relation given in Eq. 2. The viscosity was also measured using an oscillating cup viscometer and the linear fit given in Eq. 3 was fit to the viscosity data for temperature (T) in K [6].

$$\rho(T) = 3.59 - 4.2x10^{-4}(T - T_l) \quad \left[\mathrm{g/cm^3}\right] \tag{2}$$

$$\eta(T) = 7.94 - 0.0034 * T \quad [\mathrm{mPa\,s}] \tag{3}$$

Model Set-up

The models were run in ANSYS Fluent using the pressure-based solver and an axisymmetric geometry in 2D space to simulate the fluid flow within the drop. There were several different assumptions that define the boundary conditions for the models. The flow cannot cross the free surface of the drop or the symmetry axis. Additionally, the free surface of the drop is free of traction. Finally, the derivatives must be zero at the axis of symmetry to maintain the symmetry boundary.

Using these assumptions with prior work, a model was built in ANSYS Fluent to correlate the voltage applied to the levitation coils and the resulting electromagnetic field with the resulting flow velocity in the droplet. The model was then validated against experimental results in a copper-cobalt sample and found to have an excellent agreement between the directly observed fluid velocity and the velocity predicted by the model [7]. In this copper-cobalt sample, oxide particles on the surface of the drop made it possible to directly observe the flow velocity on the surface of the sample by tracking the movement of the particles [7]. The code used in this prior work was used to derive the code for the simulations presented here. Further validation was done against prior models [8] and found further support for the accuracy of the present model.

Using the validated model, a range of different processing conditions was simulated according to the experimental cycles through which the sample was tested. Details of the electromagnetic levitation facility are presented in by Lohöfer and Piller [9] and the experimental process during which the sample is levitated, melted, and tested using the oscillating drop method is described by Hyers et al. [8]. The cycles for the experiment had two different phases—one for heating and one for cooling each with different currents applied through the heater and positioner coils. During the heating phase, the control voltage applied on the heater coil was 7.8 V

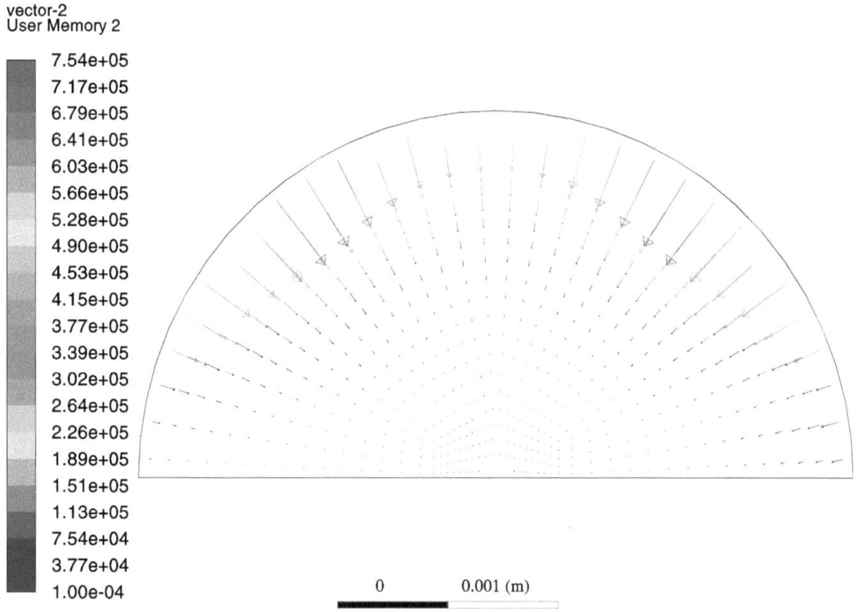

Fig. 2 Electromagnetic field applied by the combine heater and positioner coils during the heating phase of the cycle [10]

and the control voltage applied on the positioner coil was 9.7 V. While the sample was allowed to cool, the control voltage on the heating coil was reduced to zero and the control voltage for the positioner coil was reduced to 5.7 V. During the heating phase of the cycle, the heating coils dominate the forces applied to the droplet and are the primary drivers of flow within the sample. The resultant forces of the combined electromagnetic field are shown in Fig. 2. During the cooling phase of the cycle, the positioner coil dominated the applied electromagnetic field, shown in Fig. 3.

While there were several cycles of the experiment on this alloy, Cycle 3 and Cycle 5 were of particular interest and was the focus of the modeling. These cycles covered conditions expected to be relevant to the solidification theories being explored. Cycle 3 experienced a maximum temperature of 2050 K as the peak temperature during heating. The sample was then allowed to cool until recalescence at 1077 K. Cycle 5 reached a maximum temperature of 1785 K and experienced a recalescence at 1150 K.

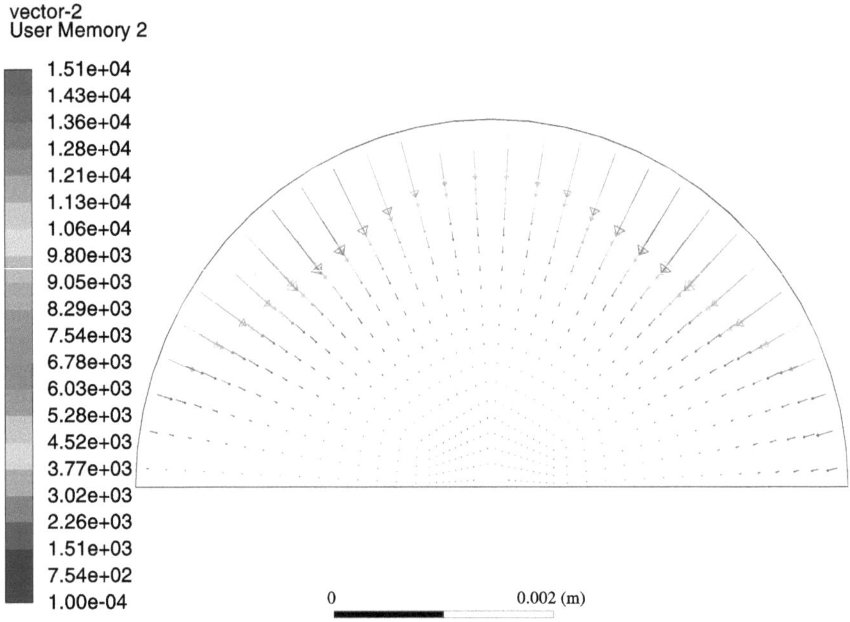

Fig. 3 Electromagnetic field applied by the combine heater and positioner coils during the cooling phase of the cycle [10]

Results and Discussion

Both cycles were modeled at their high temperature condition during heating to establish the maximum flow the drop could experience. This model was then repeated along the cooling process using both laminar and turbulent flow models.

Under the high temperature conditions both Cycle 3,2050 K, and Cycle 5,1785 K, displayed a similar flow pattern between both the laminar and the turbulent model with differences in the flow velocity. This heater dominated flow pattern is shown in Fig. 4 with variations in the magnitude of the flow vectors distinguishing cases. Under these conditions, both models have high Reynolds numbers indicating highly turbulent flow. The maximum flow velocity and calculated Reynolds numbers are given in Tables 1 and 2 where it can be seen that the flow is expected to be turbulent based on the predicted velocity and calculated Reynolds numbers.

When the samples were allowed to cool, the flow was driven by the positioner. This flow pattern is shown in Fig. 5 with variations in the velocity of the flow differing between flow models and temperature conditions. The resultant maximum flow velocity and Reynolds numbers are given for each model and each cycle in Tables 3 and 4 where it can be seen that immediately before recalescence in both cycles, both models predict laminar flow.

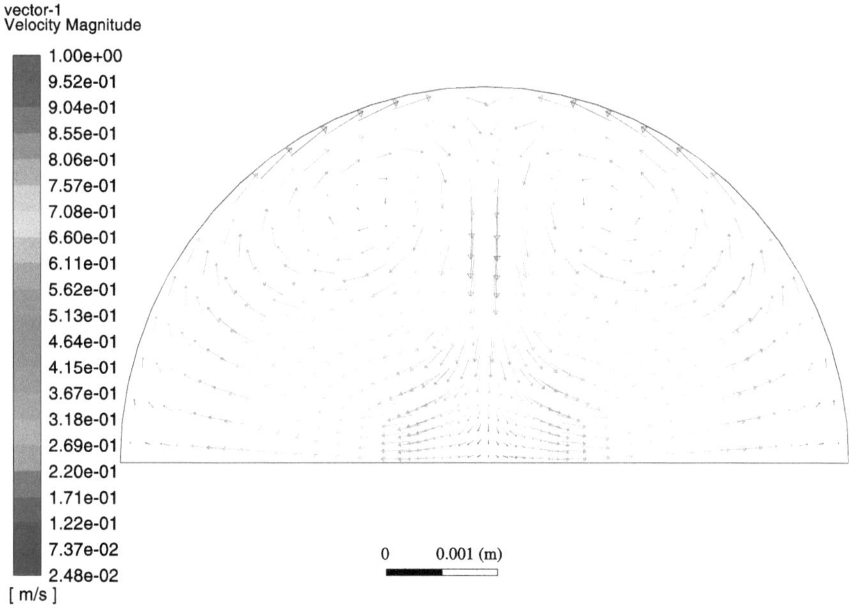

Fig. 4 Fluid flow pattern for the heater dominated flow that occurs during the heating phase of the cycle. Shown is the flow model calculated using the laminar flow model, heater dominated EML field, and liquid assumed to be at 2050 K

Table 1 Maximum flow velocity and calculated Reynolds numbers for both the laminar and turbulent flow models for the maximum temperature conditions achieved during Cycle 3

	Maximum flow velocity (m/s)	Reynolds number
Laminar model	1.00	22,200
RNG k-ε turbulence model	0.449	9990

Table 2 Maximum flow velocity and calculated Reynolds numbers for both the laminar and turbulent flow models for the maximum temperature conditions achieved during Cycle 5

	Maximum flow velocity (m/s)	Reynolds number
Laminar model	0.739	8670
RNG k-ε turbulence model	0.432	5070

In addition to the cases discussed, laminar and turbulent flow models were evaluated at intervals over the cooling range. From these models, the change in the flow velocity and the corresponding changes in the Reynolds number can be clearly observed as a function of temperature, plotted in Fig. 6. Combining these results with prior work where the character of the flow was directly visible allow prediction of the temperature at which the laminar-turbulent transition occurs.

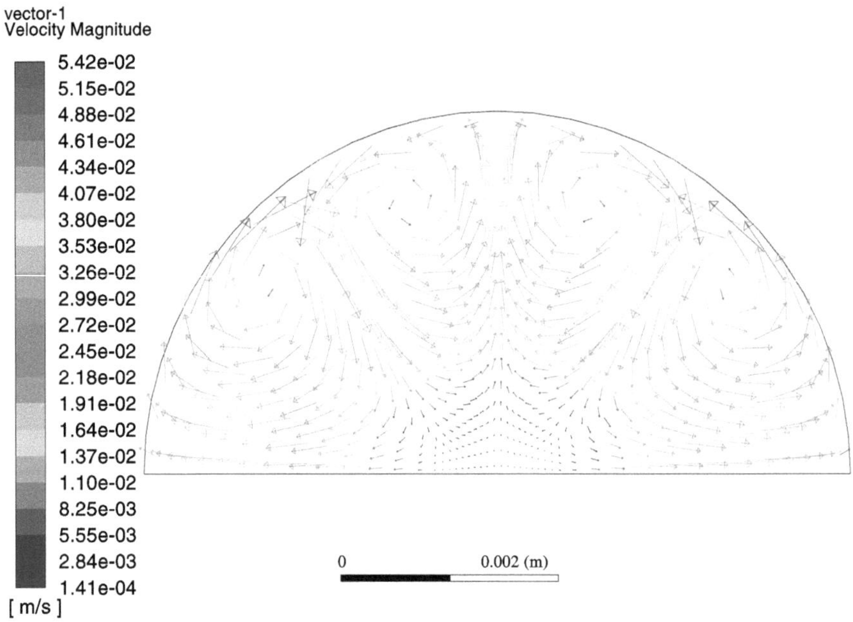

Fig. 5 Fluid flow pattern for the positioner dominated flow that occurs during the cooling phase of the cycle. Shown is the flow model calculated using the laminar flow model, positioner dominated EML field, and liquid assumed to be at 1077 K

Table 3 Maximum flow velocity and Reynolds numbers for both laminar and turbulent flow models as simulated based on the temperature of recalescence in cycle 3

	Maximum flow velocity (m/s)	Reynolds number
Laminar model	0.0530	300
RNG k-ε turbulence model	0.039	220

Table 4 Maximum flow velocity and Reynolds numbers for both laminar and turbulent flow models as simulated based on the recalescence temperature in cycle 5

	Maximum flow velocity (m/s)	Reynolds number
Laminar model	0.0545	325
RNG k-ε turbulence model	0.0395	236

Prior work in EML drops has found that the flow became turbulent at a Reynolds number of about 600 [11]. The laminar model assumes that there is no additional redistribution of momentum by turbulent eddies in the flow. These models are good predictors for the behavior of the flow until turbulence is introduced into the system as the Reynolds number increased above 600. The laminar models predicted this to occur at temperatures above 1600 K. To model the behavior of turbulence, the RNG k-ε turbulence model was used to allow for the redistribution of momentum within

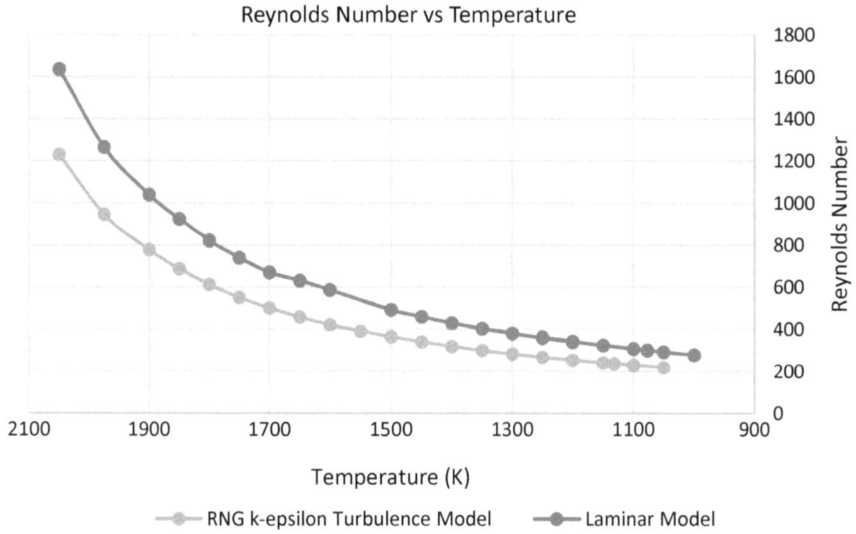

Fig. 6 Reynolds numbers vs temperature of the molten sample during cooling. Above Reynolds numbers of about 600, shown with a dotted line [11], the flow will be turbulent and the curve marked by triangles is applicable. Below this value, the flow will be laminar and the curve marked by squares applies

the drop due. This redistribution of momentum reduces the maximum velocity of the flow in the model. These models predict the laminar-turbulent transition to have occurred when the drop was between 1600 and 1800 K with laminar flow occurring at temperatures lower than 1600 K and turbulent flow occurring above 1800 K. Since the video recording of these experiments does not demonstrate clear evidence of the behavior of the flow in the aluminum-nickel drop, these models provide the only insight into the nature of the flow in the drop during cooling.

Conclusions

Though the fluid flow models run on the aluminum-nickel alloy sample, an improved understanding of the flow during cooling and solidification was gained. In both Cycle 3 and Cycle 5, the highest temperatures of the cycles were modeled to display clear turbulent behavior. However, it is also clear that at the time of recalescence the flow had transitioned to laminar behavior. Based on previous work, the laminar-turbulent transition has been shown to occur near a Reynolds number of 600. Lacking clear video evidence of the behavior of the flow in this sample, these simulations provide the only insight into the nature of the flow.

The RNG k-ε model should be used when the Reynolds number is greater than 600, which corresponds to conditions at temperatures greater than 1800 K during cooling. Laminar flow models should be used during cooling when the Reynolds numbers for the flow is less than 600, which corresponds to temperatures less than 1600 K.

Acknowledgements The authors thank Stephan Schneider for his technical assistance with the ISS experimental data archives. The experiment simulated was run in the ISS-EML facility, formerly MSL-EML. Support for this project was provided through NASA grant NNX16B40G.

References

1. Porter DA, Easterling KE (1992) Phase transformations in metals and alloys. Springer, Boston
2. Herlach DM, Cochraine RF, Egry I, Fecht HJ, Greer AL (1993) Containerless processing in the study of metallic melts and their solidifciation. Int Mater Rev 38(6):273–347
3. Lengsdorf R, Holland-Moritz D, Herlach DM (2010) Anomalous dendrite growth in undercooled melts of Al–Ni alloys in relation to results obtained in reduced gravity. Scr Mater 62(6):365–367
4. Reinartz M (2018) AlNi solidification velocity, 06 Aug 2018
5. Voss D (2014) SCI-ESA-HSO-ESR-EML2, no 1, p 142 15 Apr 2014
6. Egry I et al (2010) Thermophysical properties of liquid Al-Ni alloys. High Temp-High Press 38(4):343–351
7. Lee J, Matson DM, Binder S, Kolbe M, Herlach D, Hyers RW (2014) Magnetohydrodynamic modeling and experimental validation of convection inside electromagnetically levitated Co-Cu droplets. Metall Mater Trans B 45(3):1018–1023
8. Hyers RW, Matson DM, Kelton KF, Rogers JR (2004) Convection in containerless processing. Ann N Y Acad Sci 1027(1):474–494
9. Lohoefer G, Piller J (2002) The new ISS electromagnetic levitation facility—'MSL-EML'". In: 40th AIAA aerospace sciences meeting & exhibit. American Institute of Aeronautics and Astronautics, 2002
10. Bracker GP, Hyers RW (2018) fluid flow results modeling molten aluminum-nickel. University of Massachusetts, Amherst
11. Hyers RW, Trapaga G, Abedian B (2003) Laminar-turbulent transition in an electromagnetically levitated droplet. Metall Mater Trans B 34(1):29–36

Optimal Stator Design for Oxide Films Shearing Found by Physical Modelling

Agnieszka Dybalska, Dmitry G. Eskin and Jayesh B. Patel

Abstract A new technology suggests breaking oxide films into small fragments or particles to play the role of a grain refiner. A high-shear mixer (HSM) with a rotor-stator impeller can produce mechanical breakage. Physical modelling with powders demonstrates the defragmentation potency of HSM. Optimisation methods are considered and a new design of HSM is proposed according to the experimental findings. This design improves the uniformity of mixing in the pseudo-cavern volume and exhibits the dispersion efficiency better than the design previously used. The understanding and development of high shear technology for processing of liquid metals is of great interest to the industry.

Keywords Liquid metal · High shear · Rotor-stator · Pseudo-cavern
Defragmentation · Stator design

Introduction

The necessity of the research focused on the oxides in the liquid metals, especially light ones as aluminium and magnesium, is caused by their harmful role of factor facilitative for the porosity and cracking of the cast material. The large oxide films and clusters in aluminium are usually distributed non-uniformly in the melt and have poor wettability [1–4]. Then again, wetted and dispersed oxides may act as good nucleating substrates for aluminium and magnesium [5–8]. Thus, mechanical breakage of clusters [9, 10] and dispersion thereof in the liquid metal can change the situation. The intensive melt shearing by special devices, for example, a rotor-stator impeller is believed to be a reason for the films breaking into small fragments or particles. The underlying mechanisms, which explain the observed reduction in grain size after melt shearing, were reported elsewhere [11–14].

A. Dybalska (✉) · D. G. Eskin · J. B. Patel
BCAST, Brunel University London, Uxbridge UB8 3PH, UK
e-mail: dybalska.ag@gmail.com

G. Lambotte et al. (eds.), *Materials Processing Fundamentals 2019*, The Minerals, Metals & Materials Series, https://doi.org/10.1007/978-3-030-05728-2_17

Materials and Methods

The mixer used was a Silverson L5M Laboratory Mixer. The motor has a power of 750 W, the impeller used has a diameter of about 25 mm. For the stators that were used, the rotational speed (N) of the HSM could be chosen for up to 9000 rpm. The rotor was inside the impeller, surrounded by the stator, which could be changed. The specification of the ordered stators is shown in the inset in Fig. 4. The first stator (RH, Fig. 4b), with round holes arranged in rows is designated to repeat a design that up to now had been used in liquid metal processing (prototype in Fig. 4a). The second stator (SqH) has a similar design, except for the shape of the holes, which are square (Fig. 4c). The last stator has other round holes arrangement (in cross lines) than used before (crossH, Fig. 4d). The holes have a specific size, 3 mm (diameter or height). More exact technical details could be found in [15] in the technical drawings.

In this choice of stator shape, we were able to easily compare the influence of the shape of the stator holes. The flow patterns have been observed in the water-model system. Water can mimic aluminium flow behaviour and is often used as a modelling material (i.e., Refs. 16, 17, 18, 19) because the essential properties of water at room temperature are similar to those of liquid aluminium and because both are Newtonian fluids. The investigation in the xz-plane was done by alumina powder pattern studies. PIV (particle image velocimetry) photographs were taken in the xy-plane to complete and confirm previous research results. On each occasion, the HSM was placed in a slightly off-centre position in the tank (sized 260 mm × 260 mm × 800 mm and a water level of 200 mm from the bottom) to avoid unnecessary surface vortexing in an unbaffled vessel [20].

Still photos and movies of the water were taken using a Fuji digital camera and a PIV system [21]. PIV relies on sequential images of visible features within the flow, which is not invasive for flow behaviour. To improve visibility fluid flow was seeded with particles that follow the fluid motion. Spherical particles, usually coated to reflect laser light were used to avoid alteration. In this research, we used the 8–12 μm hollow glass spheres [22]. A laser sheet in the region of interest illuminates the particles and the scattered light produces a tracer field for image capture [21]. To avoid the scattering of the light from the metal impeller, a laser light sheet was placed just in front of the head of the HSM. A schematic diagram of the system is shown in Fig. 1.

Intervals of 100 μs for recording the photographic sequences were chosen experimentally to achieve a reasonable quality of the pictures. For each experiment, those photographic sequences were taken at least 50 times when the laser beam was turned on using the high-speed camera. Using the Insight-6 software, the vector and velocity magnitude plots were drawn with the expected maximum error of around 0.1 pixels [23]. To confirm the results done by observing the powder pattern, we have to keep the height around 30 mm above the bottom [15] and that clearance results in a scattered flow from the bottom, but not from the walls.

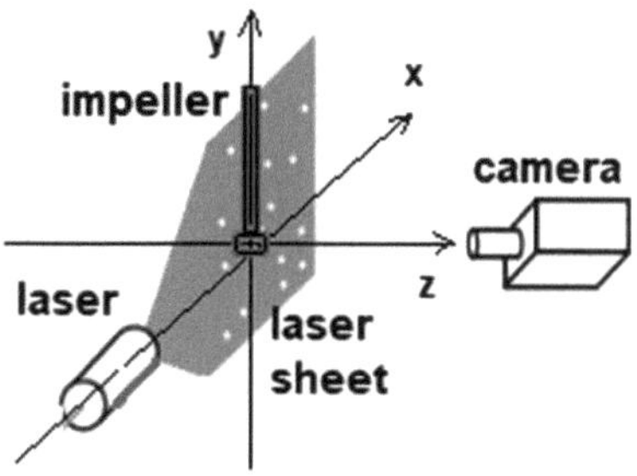

Fig. 1 Schematic diagram of the PIV system. The laser sheet illuminates the particles moving inside a glass tank (not shown in the diagram)

Results and Discussion

To improve the processing conditions we aim to find the optimal stator design and that is what this paper discusses. The main issue is the uniform dispersion of the sheared clusters as well as their effective fragmentation. Our consideration focused on the uniformity of the mixing inside the well-mixed region compared (the so-called pseudo-cavern [24, 25]), for each stator design. The different stator designs were checked by experimental shearing of alumina powder, which allows us to compare the fragmentation potency.

Commercial dispersion stators used for dispersion purposes have square holes [26]. The rotor-stator head used to improve the liquid metal quality is made from ceramic-like materials different from commercially used steel and each significant change will increase the costs. Because ceramic-like materials have high elastic modulus and hardness, preparing the square-shaped holes is much more difficult and expensive than for round-shaped holes. This is why a prototype head has simple rows of circular holes. The main question was how to improve this design to achieve the best results without the unnecessary increase in costs. To find an answer we have to compare the mixing efficiency and potency in the defragmentation of all the stators.

Mixing Uniformity

The first simple observations were made with alumina powder placed in the tank filled with water. During powder processing, the trace on the bottom reveals some uniformity different for different stators used (see Fig. 2). Analytical consideration lets us suppose that the strong jets localised in one direction can be the reason for that non-uniformity. To improve the uniformity of the mixing, the crossH design has been proposed. The idea is explained in Fig. 3.

In the case of the RH, the jets emerging from the stator openings are not distributed randomly around the head and there is a possibility that stagnation areas occur between them. To avoid this problem the proposed head has a cross-line arrangement of holes. Secondly, this change costs little as we only changed the placement of the

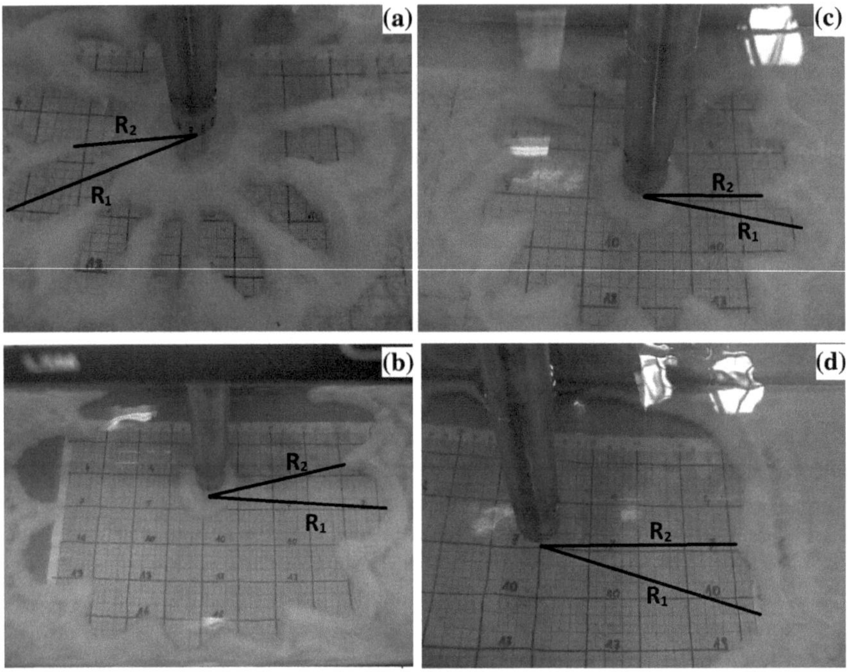

Fig. 2 The non-uniformity of mixing observed for RH and crossH stator. **a** RH, 3000 rpm, **b** RH, 5000 rpm, **c** crossH, 3000 rpm, **d** crossH, 5000 rpm

holes on the stator. To check how the proposed changes influence the uniformity of the mixed region we took the photographs of the created pattern after shearing the powder by both heads. The difference between the radii of the mixed region measured along the jets (R_1) and along the more stagnant zones (R_2) between them is taken as the non-uniformity coefficient $\Delta R = R_1 - R_2$ (see Fig. 2).

The measurements presented in Table 1 suggest that non-uniformity of the mixed region is smaller with the crossH design and it is comparable to the SqH head. The non-uniformity coefficient found for the RH stator is about 2–5 times larger than the one measured for the crossH. The low uniformity level observed has been seen not only for powder patterns, but it is also indicated by the PIV velocities pattern for this stator—the agitated region observed in the cross section is much smaller than for other stators, what can be associated with the presence of stagnant regions around this head. PIV photographs of the flow that will be discussed further (see Fig. 5).

An additional analysis, confirmed by the PIV observations, gives more specific information about that well-mixed region volume. In Fig. 6, the volume dependence on N for all is compared for each stator. The volume has been approximated as an ellipsoid and the size of the well-mixed region along x and y directions was established by physical modelling in water. The detailed procedure of volume calculations can be found elsewhere [15].

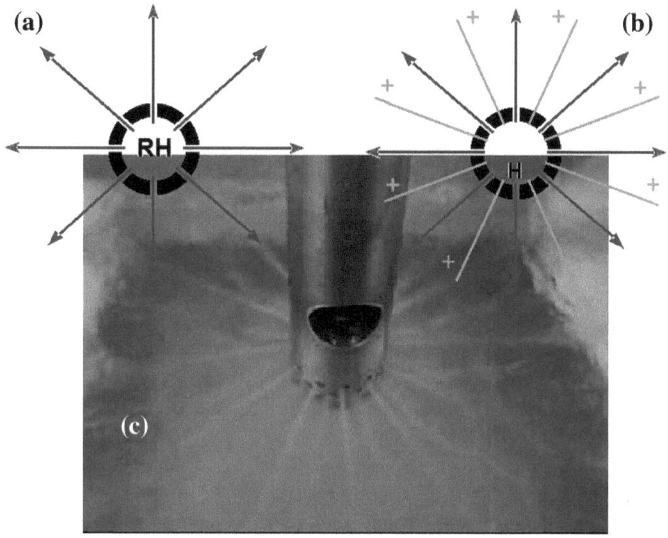

Fig. 3 **a** Schematic illustration of the jets' directions around the RH stator, **b** Schematic illustration of the jets' directions around the crossH stator, additional (+) positions of jets are the effect of the holes' rearrangement on the stator that improves uniformity. **c** Jets observed in 3D for crossH. To observe them only the bottom of the head was immersed in water. The sucking force brought a liquid to the head and spread it out as the jets travelled in the air

Table 1 The average non-uniformity coefficient ΔR measured experimentally from the traces of the well-mixed region on the bottom of the tank in water and alumina powder system

N	ΔR for RH (mm)	ΔR for crossH (mm)	ΔR for SqH (mm)
2000	10 ± 2	5 ± 1	10 ± 2
3000	10 ± 2	5 ± 1	5 ± 1
4000	10 ± 2	5 ± 1	0 ± 1
5000	10 ± 2	2 ± 1	0 ± 1
6000	5 ± 1	0 ± 1	0 ± 1

At least 10 coefficients for each stator design and specific N were measured

A predicted volume of the well-mixed region calculated for crossH is in the similar range as the one calculated for SqH and slightly smaller than the one observed for RH (Fig. 4). The prediction is based on jet length, and since the jets come from the holes arranged in the rows, they join causing a strong movement of the fluid which travels farther. It does not necessarily mean the full agitation of the fluid. In between these conjoined jets, there are stagnant zones. The strong jets have a longer travel distance, and calculations of the volume are based on maximal jet length. Therefore, the non-uniformity is not taken into account in Fig. 4. The mixed volume is also smaller when the holes are square-shaped than when they are round. The reason for this can be explained by the jets' impairment on the hole boundary. Round holes

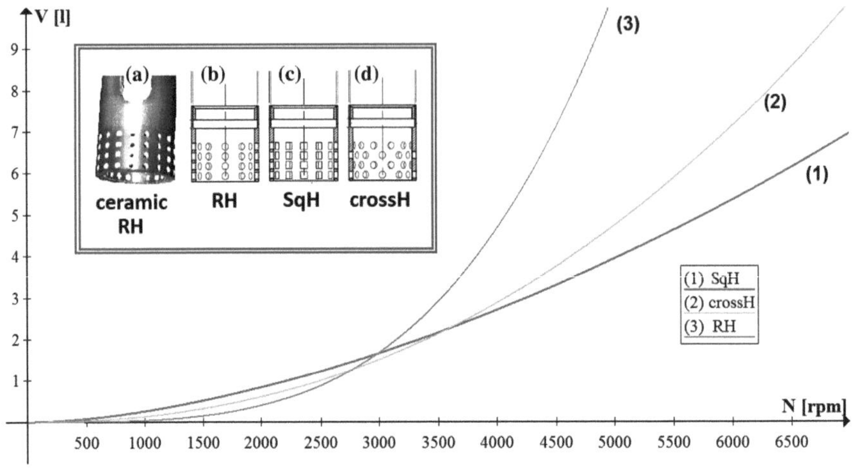

Fig. 4 Predicted volumes of the pseudo-cavern observed with different stators for a wide range of the N. In the inset: different shapes of stators, **a** prototype ceramic stator used to shear molten, **b** model of the ceramic stator, RH stator with round holes in the rows, **c** SqH stator with square-shaped holes, **d** crossH stator with proposed change of holes arrangement

do not disturb jets and allow them to move more freely than the square ones. The free jets tend to acquire a cylindrical shape, which is an effect of the surface tension holding droplets together. If no other forces are present, the free jet maintains its shape for a specific length and will be broken in droplets in the air [27, 28; after 29] or will change from the concentrated form of the jet into the turbulence of the fluid [30].

Before this happens, just after passing the hole, the contraction can be observed and the jet diameter will become smaller than the hole aperture [31]. Thus, the jet shape is always close to a cylinder. The square hole will change this natural shape by mechanical reflections of the fluid, which means that the "square jet" becomes less stable. This effect was illustrated and predicted by computer simulations. Isosurfaces of the velocity gradient tensor for circular and square jets show that the turbulence at the end of the jet occurs earlier in the case of the square jet [30]. Less stability of the jets is a simple reason for the observation presented in Fig. 4 and it explains why the mixed volume is bigger for RH since the expected jets are longer for the RH stator.

Since RH and SqH compared in Fig. 4 have the same design, except for the shape of the holes, in the chosen N, the moving rotor gives the same amount of energy into a fluid. As the mixed region seems to be smaller for the SqH, it means that part of the energy is used for a process other than mixing, or the agitation inside the pseudo-cavern is more intensive. To find an explanation for this difference in volumes we need to compare the flow around all the stators. The flow pattern observed with a PIV system is shown in Fig. 5.

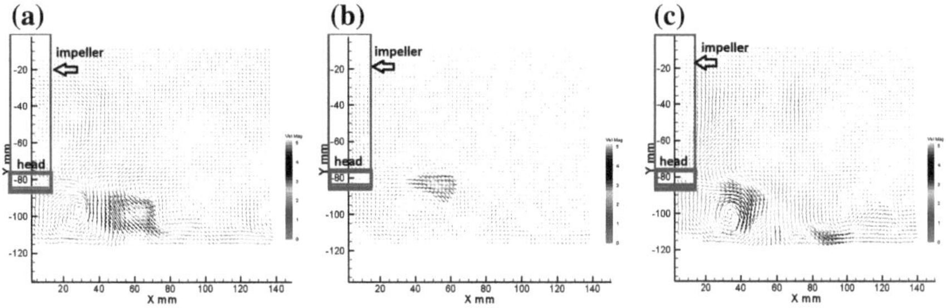

Fig. 5 **a** Flow around the SqH stator; **b** flow around the RH stator, **c** flow around the crossH stator. The head position is shown by the red rectangular, N = 7000 rpm. Velocities on the scale bars are given in m/s

The flow observed in the xy-plane was mainly recorded to confirm the size of the well-mixed region but recorded photographs offer complex information. In the pictures (Fig. 5), we can observe proof of the non-uniform agitation in (b) part of the figure. The flow is much weaker than that recorded with other stators. Surprisingly the jet does not start near the stator. Obviously, the laser beam plane reveals the stagnation zone in this area and part of the jet above point on the x-axis over 30 mm. For SqH and crossH, any of the recorded PIV photographs (taken in the range 2000–9000 rpm with a step of 1000 rpm) have not shown a similar flow pattern. In each case recorded for those stators, the flow pattern was more uniform as in cases (a) and (c). The RH recorded for the same range gives a "weak" pattern twice. Thus, it is a strong indication that sometimes the beam crosses the stagnation area around the RH head, which is consistent with the previous measurements of the non-uniformity coefficient. Similar observations can be found in the Mortensen's et al. research [32], around the used slotted head investigated by the PIV system. The slotted stator is slightly similar to the RH, but instead of the row of round holes, there is one slot.

PIV photographs [32] of the flow reveal a strong jet surrounded by a reverse fluid with much smaller velocities. Since their observation plane was perpendicular to what is shown in Fig. 4, we can complement our knowledge with these findings. According to the results published by Mortensen [32], we can expect strong jets to be accompanied by regions of stagnation or weaker reversed flow, which is the result of fluid inertia. It is not surprising that the RH stator with rows of round holes causes the same effect, since the jets from each hole can join to create a similar flow pattern as for the slotted head.

It should be reminded here that the round holes are the beginning of strong and concentrated jets, while the square-holes stator provides a more uniform mixing since the jets are not so well concentrated. The jets will not travel as far with round holes but the velocity pattern is more uniform because the energy is transferred by the turbulence on the end of the jet into the fluid surrounding the jet. The better agitation can be seen in Fig. 4 as the flow observed with an SqH stator is more agitated and the calculated mean velocities differ by as much as 28% for the case in Fig. 4. When the N increases, the difference increases too [15].

After PIV analyses, we can conclude that even if the volume calculated for RH is slightly bigger than for SqH, it is accompanied by the decrease in the uniformity and intensity of the agitation inside the well-mixed volume. The intensity of the agitation was previously checked for few Silverson stators by the computer simulation [33] and it was found that in the case of the square-hole head from the Silverson provider about 28% of the energy is dissipated in the jet region and in the rest of the tank. For disintegrating the head in a similar region, the amount of energy dissipated equals 43%. Thus, jet energy and tank agitation are on a higher level with wide round holes (8 mm diameter). In the case of narrower (1.6 mm wide slots with the length of 11 mm) and square-shaped holes (2.5 mm), about 18% more energy is dissipated in the rotor swept region and in the holes region than in the case of the disintegrating head. That means that more energy is used for the shearing process that occurs inside the head and in the holes region than in the case of the disintegrating head. The RH (Fig. 4b) has some similarities to the disintegrating head. The holes of the RH are smaller but the same round shape will be promoting the free jets emanating from the stator holes. Therefore, the jets will transfer more energy outside the head. If we compare only the well-mixed region (jets region at [33]), it can be seen that the agitation is slightly less for the disintegrating head. The non-uniform agitation of this region can give an explanation. Similar results are observed in our research for RH as the averaged velocities are smaller in this region than with the SqH stator [15].

The observations presented here are in good agreement with results of computer simulation done previously [33], since square holes, according to PIV results, are accompanied by a more uniform and concentrated flow inside the pseudo-cavern. Author of [33] also described that the energy dissipated around the square-holed head is more intensive inside the jet region than observed with other stators. Additionally, for square holes, only 1% of the energy is transferred to the rest of the tank, which indicates the intensification of the well-mixed volume (especially when compared to 20% of energy transferred by the disintegrating head).

The mixing efficiency of the crossH stator is better than that of the RH design. Using square holes will be even beneficial, but if that is non-economical, the RH stator should be replaced by the crossH design to improve mixing uniformity until the defragmentation potency is comparable, which is discussed in the next section.

Fragmentation Potency

The last presented set of experiments was prepared to check the potency of the defragmentation of all stators. To compare results, the powders were observed and the cross-sectional area of particles was measured. As an effect, we present the frequency distribution of particles area (FDPA) of alumina sheared by each stator. Figure 5 shows the results of the alumina shearing by the SqH compared to the results of shearing by the RH and proposed the crossH stator. Examples of the pictures taken by microscope for the epoxy-mounted powders are given in the inset.

The sheared powder was alumina, bonded with ion bonds into bigger granules. In the molten metal, oxides are present as films and clusters joined by van der Waals attractive forces. The van der Waals bond is the weakest of all bond types; the strength of the van der Waals force is between 0.01 and 0.1 eV per bond [34, 35] while aluminium oxide molecules are bonded by ionic bonds with a typical energy of about 8–10 eV [36], which means that in the experiments described here we are breaking bonds about 100–1000 times stronger than those expected to bond oxide films and agglomerates in the melt.

The agglomeration process is often described as a stochastic process [37] because it depends on local and temporary conditions; for example, the number of particles in agglomerates. May be we cannot predict the exact bonding forces, but we can refer to real shearing effects, which are documented by a grain size refinement [11–14]. This is indirect evidence that the shear applied by using a "round" head is enough to break oxide agglomerates and films. Men et al. [37] investigated the mechanisms of grain refinement by intensive melt shearing and found that it can effectively disperse MgO films into more individual particles using the RH head. The MgO particle density was three orders of magnitude higher than without shearing, as found by analysing the size distribution of the particles found by the pressurised filtration [37]. Thus, if the RH is proven to cause the oxide agglomerate defragmentation, other considered designs can be compared with this base-one. The alumina size reduction is stronger for powder processed by SqH and crossH stators in comparison to RH (see Fig. 6, inset) and both stators should be considered for liquid metal processing.

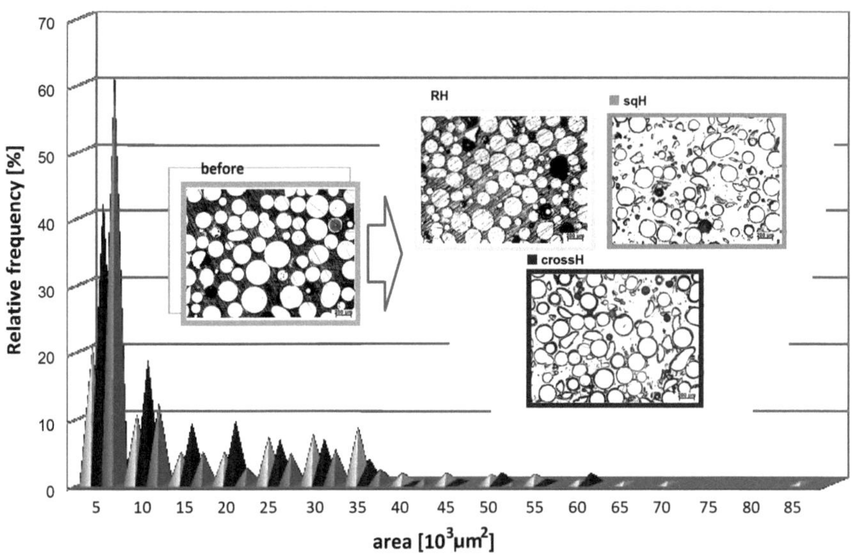

Fig. 6 The histogram of the FDPA for sheared alumina particles with different stators calculated from the powder pictures (an example is given in the insets)

The number of holes, time and mixing speed were the same for both experiments. Thus, we observe the only effect is the change in the shape of the holes. If we compare the percentage results (see Fig. 6) for SqH and RH, we can state, that for a square-holed head over 70% of the particles are in the range of 0–10,000 μm^2. For the round-holed head only 30% of the particles are in this range. It shows that square holes are really effective to shear used hard particles. However, if the costs of the square holes preparation are too high we can improve the effects of shearing by making other changes. The proposed change in holes arrangement (crossH, see Fig. 3) results in the change in local pressure inside the head. In the case of the RH, the pressure exerted on the stator walls by the fluid is stronger close to areas between the rows, where holes are not present. Inside the new head, we expect smaller differences between the local pressures as the area between the holes has a more uniform size. To check how this local change in pressure influences the shearing potency we can check the size of the processed alumina. If we take one more look at Fig. 6, we notice the particles broken by the crossH have mostly cross-sectional areas below or equal to 10,000 μm^2 and about 60% of them are in this range, 30% more than for RH. Obviously, the uniformity of the mixing inside the rotor-stator assembly does not decrease the fragmentation potency and even improves the shear process.

Concluding Remarks

The physical modelling proved that the HS processing of liquid metals is a potent method to achieve the defragmentation of agglomerates, as the defragmentation of the alumina occurs in all the experiments presented. The defragmentation efficiency is strongly influenced by the shape of the stator holes and the best results were obtained with square holes. However, the defragmentation potency of round holes for oxides present in the melt was proven experimentally.

The uniformity of the mixing inside the well-mixed region was checked for different mixing conditions and, according to the results, a new stator design was proposed for treating liquid aluminium. This design improves the uniformity of the mixing in the well-mixed volume and has higher dispersion efficiency than the design used up to now.

Acknowledgements Allocation of the equipment in the BCAST (Brunel University London) is highly appreciated. The first author is grateful for Ph.D. study funding from the Institute of Materials and Manufacturing, Brunel University London. The authors would also like to acknowledge Prof. Z. Fan, who initiated this research. The PIV measuring system was provided by the EPSRC Engineering Instrument Pool.

References

1. Green NR, Campbell J (1993) Statistical distributions of fracture strengths of cast A1-7Si-Mg alloy. Mater Sci Eng A 173:261–266
2. Mi J, Harding RA, Campbell J (2004) Effects of the entrained surface film on the reliability of castings. Metall Mater Trans A 35(9):2893–2902
3. Nyahumwa C et al (1998) Effect of mold-filling turbulence on fatigue properties of cast aluminum alloys. Paper presented at the 102nd Casting Congress, Atlanta, Georgia, 10–13 May 1998
4. Wang QG, Apelian D, Lados DA (2001) Fatigue behavior of A356-T6 aluminum cast alloys. Part I. Effect of casting defects. J Light Met 1(1):73–84
5. Fan Z (2011) Epitaxial nucleation and grain refinement. Paper presented at the John Hunt International Symposium, Uxbridge, London, 12–14 Dec 2011
6. Men H, Fan Z (2011) Effects of lattice mismatch on interfacial structures of liquid and solidified Al in contact with hetero-phase substrates: MD simulations. Paper presented at the ICASP–3, Rolduc Abbey, Aachen, The Netherlands, 7–10 June 2011
7. Tzamtzis S, Zhang H, Xia M, Babu NH, Fan Z (2011) Recycling of high grade die casting AM series magnesium scrap with the melt conditioned high pressure die casting (MC-HPDC) process. Mater Sci Eng A 528(6):2664–2669
8. Li HT et al (2011) Harnessing oxides in liquid metals and alloys. Paper presented at the John Hunt International Symposium, Uxbridge, London, 12–14 December 2011
9. Scamans G et al (2012) Melt conditioned casting of aluminum alloys. Paper presented at ICAA13, Pittsburgh, Pennsylvania, 3–7 June 2012
10. Li HT, Scamans G, Fan Z (2013) Refinement of the microstructure of an Al-Mg_2Si hypereutectic alloy by intensive melt shearing. Mater Sci Forum 765:97–101
11. Patel J et al (2013) Liquid metal engineering by application of intensive melt shearing. Paper presented at LMPC 2013, Austin, Texas, 22–25 Sept 2013
12. Zuo YB et al (2011) Grain refinement of DC cast magnesium alloys with intensive melt shearing. Paper presented at the ICASP–3, Rolduc Abbey, Aachen, The Netherlands, 7–10 June 2011
13. Li HT, Wang Y, Fan Z (2012) Mechanisms of enhanced heterogeneous nucleation during solidification in binary Al–Mg alloys. Acta Mater 60:1528–1537
14. Li HT, Xia M, Jarry P, Scamans G, Fan Z (2011) Grain refinement in a AlZnMgCuTi alloy by intensive melt shearing: A multi-step nucleation mechanism. J Cryst Growth 314(1):285–292
15. Dybalska A (2016) Understanding and development of high shear technology for liquid metals processing. PhD thesis, Brunel University
16. Gupta D, Lahiri AK (1996) A water model study of the flow asymmetry inside a continuous slab casting mold. Metall Mater Trans B 27(5):757–764
17. Xu D, Jones W, Kinzy W, Evans JW (1998) The use of particle image velocimetry in the physical modeling of flow in electromagnetic or direct-chill casting of aluminum: Part I. Development of the physical model. Metall MaterTrans B 29:1281–1288
18. Zhang L, Yang S, Cai K, Li J, Wan X, Thomas BG (2007) Investigation of fluid flow and steel cleanliness in the continuous casting strand. Metall Mater Trans B 38(1):63–83
19. Karcz J, Szoplik J (2004) An effect of the eccentric position of the propeller agitator on the mixing time. Chem Pap-Slovak Acad Sci 58(1):9–14
20. Tzanakis I, Lebon GSB, Eskin DG, Pericleous KA (2017) Characterizing the cavitation development and acoustic spectrum in various liquids. Ultrason Sonochem 34:651–662
21. Raffel M, Willert C, Wereley S, Kompenhans K (2007) Particle imaging velocimetry—a practical guide, 2nd edn. Springer, Berlin
22. Adrian RJ, Westerweel J (2011) Particle image velocimetry. Cambridge University Press, Cambridge
23. Huang H, Dabiri D, Gharib M (1997) On errors of digital particle image velocimetry. MST 8(12):1427–1440
24. Doucet L, Ascanio G, Tanguy PA (2005) Hydrodynamics characterization of rotor-stator mixer with viscous fluids. Chem Eng Res Des 83(10):1186–1195

25. Barailler F, Heniche M, Tanguy PA (2006) CFD analysis of a rotor-stator mixer with viscous fluids. Chem Eng Sci 61(9):2888–2894
26. Apparatus and method for high-shear mixing. US. Patent Application, US20160271575A1. 22 Sept 2016
27. Rayleigh JWS (1891) Some applications of photography. Nature 44:249–254
28. Nollet, JA (1749) Recherches sur les causes particulieres des phénoménes électriques, et sur les effets nuisibles ou avantageux qu'on peut en attendre. A Paris: chez les Freres Guerin, Paris
29. Eggers J, Villermaux E (2008) Physics of liquid jets. Rep Prog Phys 71(3):036601
30. Gohil TB, Saha AK, Muralidhar K (2010) Control of flow in forced jets: a comparison of round and square cross sections. J Vis 13:141–149
31. Brodkey RS, Hershey HC (2003) Transport phenomena: a unified approach. Brodkey Publishing, Columbus, Ohio
32. Mortensen HH, Calabrese RV, Innings F, Rosendahl L (2011) Characteristics of batch rotor–stator mixer performance elucidated by shaft torque and angle resolved PIV measurements. Can J Chem Eng 89(5):1076–1095
33. Utomo AD (2009) Flow patterns and energy dissipation rates in batch rotor-stator mixers. PhD thesis, University of Birmingham
34. Roesler J, Harders H, Baeker M (2007) Mechanical behaviour of engineering materials. Springer, Berlin
35. Smirnov BM (1992) Cluster Ions and van der Waals Molecules. CRC Press, Boca Raton
36. Il'inskii YA, Keldysh LV (2013) Electromagnetic response of material media. Springer, Berlin
37. Men H, Jiang B, Fan Z (2010) Mechanisms of grain refinement by intensive shearing of AZ91 alloy melt. Acta Mater 58(18):6526–6534

An Investigation on Electrodeposition of Titanium in Molten LiCl-KCl

Chenyao Li, Jianxun Song, Shaolong Li, Xuepeng Li, Yongchun Shu and Jilin He

Abstract A molten LiCl-KCl (40.8:59.2, mol%) eutectic salt was used as an electrolyte due to its relatively low melting point. The molar ratio of fluoride ions and titanium ($[F^-]/[Ti^{n+}]$ ratio) was employed as a parameter to illustrate the influence of fluoride anions on the electrochemical behaviors of Ti(III), and KF was used as a source of fluoride ions. A study on the electrochemical properties of Ti(III) was carried out in molten LiCl-KCl-KF. Results suggest that there are two steps for reducing Ti(III) in molten LiCl-KCl: Ti(III) → Ti(II) and Ti(II) → Ti. Ti(III) can be reduced directly to Ti in one step when KF was added in the melt in increasing amount. Metallic titanium was produced when $[F^-]/[Ti^{n+}]$ equals to 10, and it is dendrites with a layer structure. The oxygen contents in the titanium crystal are 1200 ppm.

Keywords LiCl-KCl · Fluoride ions · SWV · Dendrites

Introduction

Although high pure titanium has many excellent properties, the application in the industry was limited by its high cost. Many methods have been investigated for producing high pure titanium at low cost [1–3]. Electrolysis, involving the extraction and preparation of high pure metal from ore using an electrolytic process, is considered as a promising method for titanium metallurgy. It was discussed in the previous paper that the main obstacles for the successful development of an electrochemical route for titanium production are associated with the existence of various valences of dissolved titanium species [4–14]. Thus, it has the same problem for electrorefining

C. Li · J. Song (✉) · S. Li · Y. Shu · J. He
Henan Province Industrial Technology Research Institute of Resources and Materials, Zhengzhou University, Zhengzhou 450001, China
e-mail: jianxun.song@zzu.edu.cn

X. Li
National Engineering Laboratory for Vacuum Metallurgy, Kunming University of Science and Technology, Kunming 650093, China

G. Lambotte et al. (eds.), *Materials Processing Fundamentals 2019*, The Minerals, Metals & Materials Series, https://doi.org/10.1007/978-3-030-05728-2_18

of titanium. The different valences of titanium ions in different electrolytes undergo reoxidation and disproportionation reactions, which result in a low current efficiency. It was investigated by the previous paper that both cation and anion of the electrolyte have a significant influence on the behaviors of titanium ions in molten salt and the electrodeposition process of titanium [14–20]. It was also reported that it is a two-reduction step for reducing Ti(III) in molten chloride system except for CsCl-LiCl (59.5:40.5, mol%) at 793 K, and the polarization power was used to interpret the mechanism [21]. However, it is more complicated in fluoride–chloride mixtures. The principle of electrowining of high pure titanium in fluoride–chloride mixture shows in Fig. 1. Complexes, $[TiCl_xF_y]^{n-}$, was formed with the presenting of metallic titanium.

Ti(III) is supposed to be reduced directly to metallic titanium when the concentration of fluoride is high enough, and an intermediate reduction step will be involved under a lower fluoride concentration. The investigations on the influence of fluoride have been studied for years [22–25]. However, the quantitative analyses and the mechanism were unambiguous before a finding was reported [15]. One significance of the paper is that, a way was found to make the Ti(III) stable so that it can be reduced into metallic titanium in one step, and it was achieved in the way of

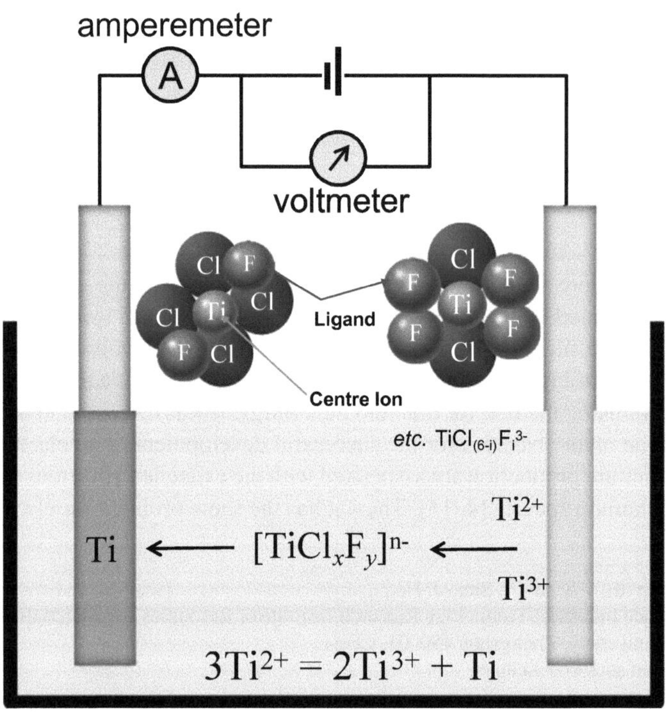

Fig. 1 Principle of electrodeposition of titanium in fluoride-chloride mixture and possible complexes formed in the mixture

detecting an optimal concentration of fluoride or $[F^-]/[Ti^{n+}]$ ratio in alkali chloride melt. Afterwards, the titanium can be produced with a lower energy consumption in fluoride–chloride mixtures. As an electrolyte, LiCl-KCl has a low melting point and saturated vapor pressure. Therefore, it is significant for studying the electrochemical behaviors of Ti(III) in molten LiCl-KCl under various concentrations of fluoride.

Experimental

The majority of experimental methods are identical to previously delivered papers [11–21, 26], in which they are discussed in detail. Thus, here only a brief account of key aspects is given.

A molten LiCl-KCl (Sinopharm Chemical Reagent Co., Ltd. Analytical grade 99.99%) eutectic salt was taken as the electrolyte. The molar ratio equals to 59.2:40.8 for LiCl and KCl in this paper. As can be seen from Fig. 2, the melting point of the eutectic is approximately 626 K. Then, the eutectic was melted in an alumina crucible after being placed in the furnace. It was pre-treated in an inert argon atmosphere previously by heating in a vacuum. Additionally, high-purity hydrogen chloride gas was bubbled into the salt to remove the O^{2-}. Details have been published elsewhere [14–19].

The titanium species used in experiments is $TiCl_3$ which was prepared by the reaction of $TiCl_4$ ($TiCl_4$, Sinopharm Chemical Reagent Co., Ltd., analytical grade 99.0%) and titanium metal. The procedures were reported in Ref. [19].

The $TiCl_3$ and LiCl-KCl were thoroughly mixed in a crucible and sealed in the cell. All experiments were performed in LiCl-KCl maintained at 723 K. It is an

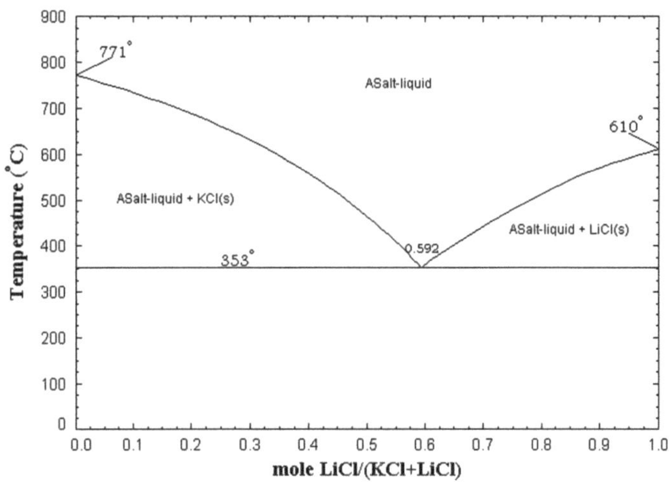

Fig. 2 LiCl-KCl phase diagram [27]

in situ process for adding of KF into the melt, and more details can be found in Ref. [11]. The analytical electrochemical techniques used require a three-electrode setup. The working compartment contained a molybdenum rod (0.102 cm^2) for analytical measurements. A spectral graphite rod with a diameter of 6 mm was used as a counter electrode, and the reference electrode was an Ag^+/Ag [28, 29]. The potentials in this work were all calibrated to Cl_2/Cl^-. Transient electrochemical techniques, square wave voltammetry (SWV), was performed using an electrochemical workstation (Solartron 1287, AMETEK Advanced Measurement Technology). Chronopotentiometry was carried out on a potentiostat (TDK-Lambda, Z10-20) in the melt containing a certain content of fluoride ions while the current density was 0.5 $mAcm^{-2}$. All measurements were carried out under the atmosphere of dried argon.

Results and Discussion

(A) Influence of fluoride on the electrochemical properties of Ti(III)

The $[F^-]/[Ti^{n+}]$ ratio was used to reveal the influence of fluoride anions on the electrochemical properties of titanium ions, and it was defined as the molar ratio of F^- and Ti^{3+}. Figure 3 shows the square wave voltammograms (SWV) curves of titanium ions in molten LiCl-KCl under various $[F^-]/[Ti^{n+}]$ ratios, and they were observed at a potential range of −1.98 to −1.20 V versus Cl_2/Cl^-.

Figure 3a suggests that there are two peaks for reducing of Ti(III) in molten LiCl-KCl when no fluoride was added in the salt. Using the half peak ($W_{1/2}$) of the Gaussian wave, the number of exchanged electrons can be calculated. The relationship between the half peak of the Gaussian wave and the number of exchanged electrons is shown in Eq. (1).

$$W_{1/2} = 3.52\frac{RT}{n\mathrm{F}} \tag{1}$$

where n is the number of electrons involved in the reaction, F is the Faraday constant, R is the gas constant (8.314 J mol^{-1} K^{-1}), T is the temperature in Kelvin. The $W_{1/2}$ could be obtained after these two peaks were fitted. The fitted curve is shown in Fig. 3a. For reduction of A, the number of electrons transferred is 0.93, and it is 1.88 for peak B. It suggests that the cathodic reaction of Ti(III) precedes two steps when no fluoride was added, and they are (2) and (3) as follows:

$$\mathrm{Ti(III)} + e^- = \mathrm{Ti(II)} \tag{2}$$

$$\mathrm{Ti(II)} + 2e^- = \mathrm{Ti} \tag{3}$$

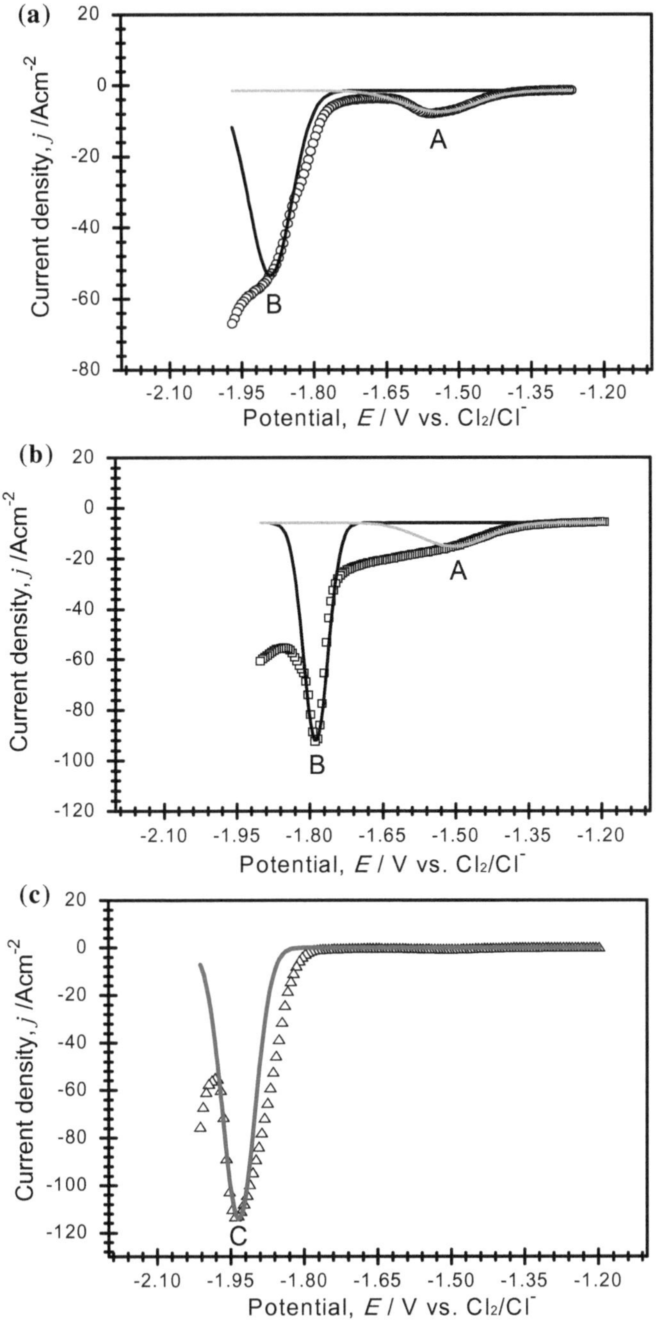

Fig. 3 SWV curves obtained on a Mo working electrode under various of $[F^-]/[Ti^{n+}]$ ratios in molten LiCl-KCl at 723 K, **a** $\left[F^-\right]/\left[Ti^{n+}\right] = 0$; **b** $\left[F^-\right]/\left[Ti^{n+}\right] = 5$; **c** $\left[F^-\right]/\left[Ti^{n+}\right] = 10$, reference: Cl_2/Cl^-. Pulse height: 25 mV, potential step: 3 mV, frequency: 20 Hz

Figure 3b shows the square wave voltammogram of Ti(III) when $[F^-]/[Ti^{n+}]$ ratio equals to 5.0. It is obvious that there are two reduction peaks. However, the potential transferred to a positive value which from −1.52 to −1.49 V versus Cl_2/Cl^- for peak A and from −1.89 to −1.78 V versus Cl_2/Cl^- for peak B, respectively. The transferred electrons number was calculated after the curve was fitted. They are 1.11 and 1.95 for peaks A and B, respectively.

There is only one peak in the square wave voltammogram shown in Fig. 3c when $[F^-]/[Ti^{n+}]$ ratio was 10. The transferred electron number that was determined by Eq. (1) equals to 2.77. It should be a process with three electrons transferred for the calculated number is quite close to 3.0. The reason why it does not equal to 3.0 is that the process may not reversible. Therefore, it is a single-step process via reaction (4) for the Ti(III) reducing when $[F^-]/[Ti^{n+}]$ ratio was 10.

$$\mathrm{Ti(III)} + 3e^- = \mathrm{Ti} \tag{4}$$

Hence, the conclusion can be made from square wave voltammograms that it is a two-step process for reducing Ti(III) when the $[F^-]/[Ti^{n+}]$ is less than or equals to 5.0. However, it is a single step when $[F^-]/[Ti^{n+}]$ is more than 10. Moreover, the reduction potential moved to a more positive value when the fluoride content was increased.

As reported in our previous work, there are two disproportionations existing in the melt in the presence of metallic titanium which can be described as below [30–34]

$$3\mathrm{Ti(II)} = 2\mathrm{Ti(III)} + \mathrm{Ti} \tag{5}$$

$$4\mathrm{Ti(III)} = 3\mathrm{Ti(IV)} + \mathrm{Ti} \tag{6}$$

The Ti(III) ion has a competition with Ti(II) in reaction (5) and it also has competition with Ti(IV) in reaction (6) in the mixture. The equilibrium constant for these reactions can be expressed as follows:

$$K_{c1} = \frac{a^2_{\mathrm{Ti}^{3+}} a_{\mathrm{Ti}}}{a^3_{\mathrm{Ti}^{2+}}} = \frac{a^2_{\mathrm{Ti}^{3+}}}{a^3_{\mathrm{Ti}^{2+}}} \tag{7}$$

$$K_{c1} = \frac{a^3_{\mathrm{Ti}^{4+}} a_{\mathrm{Ti}}}{a^4_{\mathrm{Ti}^{3+}}} = \frac{a^3_{\mathrm{Ti}^{4+}}}{a^4_{\mathrm{Ti}^{3+}}} \tag{8}$$

where K_{ci} is equilibrium constant and a_i is the activity of a species i.

The equilibrium will transfer to the right direction for above two reactions when fluoride was added in the melt. That is due to the relatively higher stability of the complex ions of higher valance cations with the fluoride ion, compared to lower valance cations (Ti^{2+}, Ti^{3+}). It was investigated by previous work that the titanium ions in equilibrium with metallic titanium are Ti^{2+} and Ti^{3+} in the chloride melt with a low concentration of fluoride [15]. Ti^{4+} was detected in the melts with high fluoride concentration, however, Ti^{2+} was undetectable for this situation. It is because that almost all Ti^{3+} and Ti^{4+} forms the complexes with 6 coordination number, $TiF_6{}^{3-}$ and $TiF_6{}^{2-}$ when fluoride ions were added in the melt.

Thus, it is a meaningful investigation for the certain ratio of $[F^-]/[Ti^{n+}]$ was found to make the Ti(III) stable so that it can be reduced into metallic titanium in one step, and it was achieved in the way of detecting an optimal concentration of fluoride or $[F^-]/[Ti^{n+}]$ ratio in molten LiCl-KCl. Afterwards, the titanium can be produced with a lower energy consumption in molten LiCl-KCl-KF.

(B) Electrodeposition of metallic titanium in molten LiCl-KCl-KF

In order to observe the reduction product under a high concentration of fluoride, chronopotentiometry was carried out in the LiCl-KCl-KF melt. The ratio of $[F^-]/[Ti^{n+}]$ equals to 10, and the current density that was applied in this experimental equals to 0.5 $mAcm^{-2}$.

The image of the product collection shows in Fig. 4a. It can be seen clearly that it is one kind of dendrite with a bright and beautiful metallic lustre. It is pure metallic titanium after defined by XRD analysis as shown in Fig. 4b. The oxygen content in titanium crystal was analyzed by Chemiluminescent Nitrogen/Oxygen Analyzer (TCH-600), and the average contents equal to 1200 ppm.

The titanium crystal was also analyzed by an optical microscope (HQ-U300), and the results under different scales show in Fig. 5. It is a crystal with a layer structure, and the thickness for each layer is around 5–20 μm.

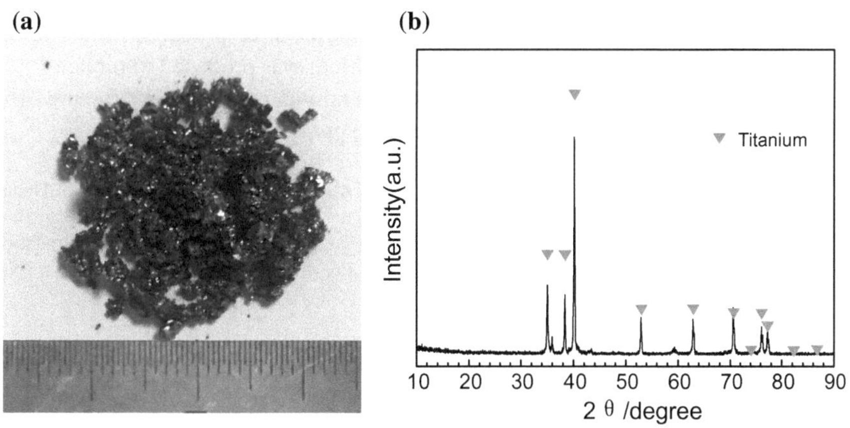

Fig. 4 **a** Image of titanium product and **b** the XRD pattern of the product

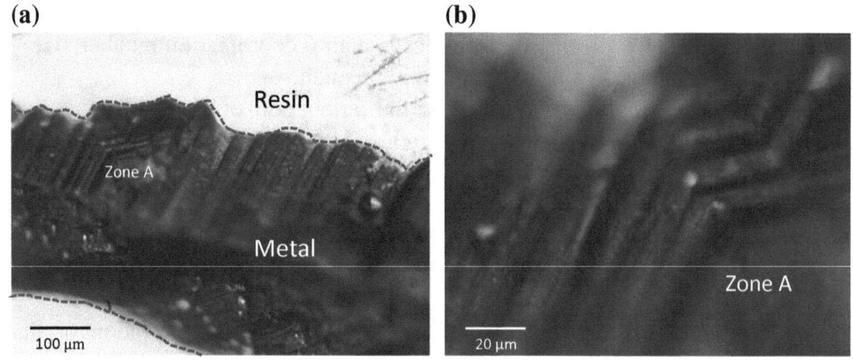

Fig. 5 **a** Optical image of metallic titanium and **b** enlarged area of the optical image

Conclusions

Square wave voltammetry and chronopotentiometry were carried out to investigate the influence of fluoride ions on titanium behaviors in molten LiCl-KCl. The results from square wave voltammograms can be read that it is two-step processes for reducing Ti(III) when the $[F^-]/[Ti^{n+}]$ is less than or equals to 5.0. However, it is a single step when $[F^-]/[Ti^{n+}]$ is more than 10. Moreover, the reduction potential moved to a more positive value when the fluoride content increased in the salt. It provides a way that makes the Ti(III) stable so that it can be reduced into metallic titanium in one step in molten LiCl-KCl. Then, the titanium can be produced with a lower energy consumption. Furthermore, metallic titanium was produced in molten LiCl-KCl when $[F^-]/[Ti^{n+}]$ equals to 10, and it is dendrites with a layer structure. The oxygen contents in the titanium crystal are 1200 ppm.

Acknowledgements The authors thank support from Collaborative Innovation Center of Henan Resources and Materials Industry, Zhengzhou University and Startup Research Fund of Zhengzhou University (No. 32210804). The authors are grateful to the National Natural Science Foundation of China (No. 51804277) and Key Projects of Henan Province Department of Education (No. 19B450004).

References

1. Chen GZ, Fray DJ, Farthing TW (2000) Direct electrochemical reduction of titanium dioxide to titanium in molten calcium chloride. Nature 407:361
2. Suzuki RO (2007) Direct reduction processes for titanium oxide in molten salt. JOM 1:67
3. Zheng H, Ito H, Okabe TH (2007) TH Production of titanium powder by the calciothermic reduction of titanium concentrates or ore using the preform reduction process. Mater Trans 48(8):2244
4. Ferry DM, Picard GS, Tremillon BL (1988) Pulse and AC impedance studies of the electrochemical systems of titanium in LiCl-KCl eutectic melt at 743 K. J Electrochem Soc 135(6):1443

5. Popov BN, Kimble MC, White RE (1991) Electrochemical behaviour of titanium(II) and titanium(III) compounds in molten lithium chloride potassium chloride eutectic melts. J Appl Electrochem 21:351
6. Wei D, Okido M, Oki T (1994) Characteristics of titanium deposits by electrolysis in molten chloride-fluoride mixture. J Appl Electrochem 24:923
7. Stafford GR, Moffat TP (1995) Electrochemistry of titanium in molten 2AlCl3-NaCl. J Electrochem Soc 142(10):3288
8. Ene N, Zuca S (1995) Role of free F- anions in the electrorefining of titanium in molten alkali halide mixtures. J Appl Electrochem 25:671
9. Tsuda T, Hussey CL, Stafford GR, Bonevich JE (2003) Electrochemistry of titanium and the electrodeposition of Al-Ti alloys in the Lewis acidic aluminum chloride-1-Ethyl-3-methylimidazolium chloride melt. J Electrochem Soc 150(4):C234
10. Robin A (2005) Influence of temperature on the reduction mechanism of Ti(III) ions on iron in the LiF-NaF-KF eutectic melt and on the electrochemical behavior of the resultant titanium coatings. Mater Chem Phys 89:438
11. Song J, Huang X, Wu J, Zhang X (2017) Electrochemical behaviors of Ti(III) in molten NaCl-KCl under various contents of fluoride. Electrochim Acta 256:252
12. Wang Q, Song J, Hu G, Zhu X, Hou JG, Jiao S, Zhu H (2013) The equilibrium between titanium ions and titanium metal in NaCl-KCl equimolar molten salt. Metall Mater Trans B 44B:906
13. Zhu X, Wang Q, Song J, Hou JG, Jiao S, Zhu H (2014) The equilibrium between metallic titanium and titanium ions in LiCl-KCl melts. J Alloys Compd 587:349
14. Song J, Wang Q, Kang M, Jiao S, Zhu H (2014) The equilibrium between titanium ions and metallic titanium in the molten binary mixtures of LiCl. Electrochemistry 82:1047
15. Song J, Wang Q, Wu J, Jiao S, Zhu H (2016) The influence of fluoride ions on the equilibrium between titanium ions and titanium metal in fused alkali chloride melts. Faraday Discuss 190:421
16. Song J, Mukherjee A (2016) Influence of F^- on electrochemical properties of titanium ions and Al-Ti alloy electrodeposition in molten $AlCl_3$-NaCl. RSC Adv 6:82049
17. Song J, Wang Q, Hu G, Zhu X, Jiao S, Zhu H (2014) Equilibrium between titanium ions and high-purity titanium electrorefining in a NaCl-KCl melt. Int J Min. Metall Mater 21(7):660
18. Song J, Wang Q, Zhu X, Hou J, Jiao S, Zhu H (2014) The influence of fluoride anion on the equilibrium between titanium ions and electrodeposition of titanium in molten fluoride chloride salt. Mater Trans 55:1299
19. Song J, Wang Q, Kang M, Jiao S (2015) Novel synthesis of high pure titanium trichloride in molten $CaCl_2$. Int J Electrochem Sci 10:919
20. Kang M, Song J, Zhu H, Jiao S (2014) Electrochemical behavior of titanium(II) ion in a purified calcium chloride melt. Metall Mater Trans B 46(1):162
21. Song J, Xiao J, Zhu H (2017) Electrochemical behavior of titanium ions in various molten alkali chlorides. J Electrochem Soc 164(12):E321
22. Polyakova LP, Stangrit PT, Polyakov EG (1986) Electrochemical study of titanium in chloride-fluoride melts. Electrochim Acta 31(2):159
23. Chen G, Okido M, Oki T (1987) Electrochemical studies of titanium ions (Ti^{4+}) in equimolar KCl-NaCl molten salts with 1 wt % K^2TiF_6. Electrochim Acta 32(11):1637
24. Wurm JG, Gravel L, Potvin RJA (1957) The mechanism of titanium production by electrolysis of fused halide baths containing titanium salts. J Electrochem Soc 104:301
25. Lantelme F, Salmi A (1995) Electrochemistry of titanium in NaCl-KCl mixtures and influence of dissolved fluoride ions. J Electrochem Soc 142:3451
26. Song J, Zhang X, Mukherjee A (2016) Electrochemical behaviors of Ce(III) in molten $AlCl_3$-NaCl under various contents of fluoride. J Electrochem Soc 163(14):D757
27. Sridharan K (2012) Thermal properties of LiCl-KCl molten salt for nuclear waste separation. Doctoral thesis. University of Wisconsin, Madison
28. Gao P, Jin X, Wang D, Hu X, Chen GZ (2005) A quartz sealed Ag/AgCl reference electrode for $CaCl_2$ based molten salts. J Electroanal Chem 579(2):321

29. Yasuda K, Nohira T, Ogata YH, Ito Y (2005) Electrochemical window of molten LiCl-KCl-$CaCl_2$ and the Ag^+/Ag reference electrode. Electrochim Acta 51:561
30. Sekimoto H, Nose Y, Uda T, Sugimura H (2010) Quantitative analysis of titanium ions in the equilibrium with metallic titanium in NaCl-KCl equimolar molten salt. Mater Trans 51(11):2121
31. Kado Y, Kishimoto A, Uda T (2013) Electrolysis of TiO_2 or $TiCl_2$ using Bi liquid cathode in molten $CaCl_2$. J Electrochem Soc 160(10):E139
32. Lantelme F, Kuroda K, Barhoun A (1998) A Electrochemical and thermodynamic properties of titanium chloride solutions in various alkali chloride mixtures. Electrochim Acta 44:421
33. Clayton FR, Mamantov G (1973) Electrochemical studies of titanium in molten fluorides. J Electrochem Soc 120(9):1199
34. Wendt H, Reuhl K, Schwartz V (1992) Cathodic deposition of refractory intermetallic compounds from FLiNaK melts I. voltammetric investigation of Ti, Zr, B, TiB_2 and ZrB_2. Electrochim Acta 37:237

Part V
Extractive Process and Thermodynamic Modeling

Effect of Ultrasound on the Extraction of Silicon and Aluminum from the Metallurgical Slag of Laterite Nickel Ore

Pengju Zhang, Jilai Xue, Xuan Liu and Donggen Fang

Abstract Metallurgical slag of laterite nickel ore contains valuable Al and Si that can be recycled by alkali roasting and water leaching process. The leaching solution was further treated to make zeolite materials as a high-value-added by-product. In this work, the contents of silicon and aluminum in the filtered liquor after water leaching was determined by ICP, and the solid residue was characterized by the laser particle-size distribution analyzer and SEM. The results showed that the contents of Al and Si in the filtered liquor increased by 44 and 65%, respectively, by using ultrasound; and the leaching residue was even more fine and uniform, and its grain size was reduced from 100 to 10 μm. This demonstrates that ultrasound can enhance the leaching efficiency of Al and Si for better recycling of the valuable metals and improving the quality of zeolite materials.

Keywords Laterite nickel ore · Ultrasound · Alkali roasting · Water leaching

Introduction

Large volume of metallurgical slag is generated in nickel smelting process as a result of using laterite nickel ore with low content of nickel [1]. It is estimated that production of 1 ton nickel–iron can produce 7–10 tons slag, which large size of land and causes serious environmental pollution [2]. However, the laterite nickel slag still contains valuable metals in the form of magnesium silicate minerals, hematite, etc. Recycling and reusing of silicon, magnesium, iron and other metals from the slag can meet the requirements for cleaner production and reduce the impacts on the environment [3, 4].

Molecular sieves, as a function material, are widely used in the industry as chemical product for deep desiccant, ion exchange, gas adsorption, etc. [5, 6]. Preparation

P. Zhang · J. Xue (✉) · X. Liu · D. Fang
School of Metallurgical and Ecological Engineering, University of Science and Technology Beijing, Xueyuan Road 30, Beijing 100083, China
e-mail: jx@ustb.edu.cn

G. Lambotte et al. (eds.), *Materials Processing Fundamentals 2019*, The Minerals, Metals & Materials Series, https://doi.org/10.1007/978-3-030-05728-2_19

of such molecular sieves usually involves utilization of Si and Al contained in the raw materials. For this purpose, the laterite nickel slag has the potential to serve as an alternative Si- and Al-containing raw material for producing molecular sieves. Such potential technical approach may not only improve the utilization of laterite nickel slag but also reduce the emission of solid waste. Liu et al. [7, 8] carried out a study on the extraction of silicon from the direct treatment of laterite nickel ore by alkali roasting and water leaching. Furthermore, Fang et al. [9] applied the optimized alkali roasting process to extract aluminum and silicon from laterite nickel slag and then, produced 4A and 13X zeolite molecular sieves in laboratory scale. Nevertheless, the effective use of silicon and aluminum recycled from laterite nickel ore remains as a problem in these processes under study. For instance, the relative lower contents of Si and Al in the leaching solution, the large particles of filtered residue and the lack of appropriate ratio of Al to Si are undesirable for producing molecular sieves.

In recent years, power ultrasound has applied to various chemical and physical processes in the medium of liquid [10] in order to produce unique cavitation and agitation [11]. These effects can increase the mass transfer and diffusion capacity of the solution and increase the diffusion rate of the material in aqueous solutions. It can improve the renewal rate of the surface of the crystalline particles and make the concentration of the solution more uniform [12].

The aim of this work is to explore the possibility of using ultrasound to enhance the water leaching process after alkali roasting of laterite nickel slag. The effects of ultrasound on the extraction ratio of Si and Al and the characterization of leaching residue were investigated, in comparison with the common processes without ultrasound.

Experiment

Materials and Chemicals

Laterite nickel slag was provided by Beihai Chengde Nickel Company, China. The main chemical analysis of laterite nickel slag were given as SiO_2 (55 wt%), MgO (25 wt%), Fe_2O_3 (6.7 wt%) and Al_2O_3 (5.3 wt%), respectively. Chemicals used were NaOH (96 wt%), Na_2CO_3 (99.8 wt%), $NaAlO_2$ (CP), and $Al(OH)_3$ (AR).

Experiment Processes

Roasting

The mixture of a certain amount of NaOH and laterite ore slag with the desired size was placed in nickel crucible. It was heated to a given temperature under air atmosphere in the chamber of electric furnace, and then, the mixture roasted was taken out and cooled down to room temperature.

Water Leaching

Table 1 shows various water leaching processes under different operating conditions with hygienically stirring or ultrasound. The roasted sample was crushed and mixed with $Al(OH)_3$, which was put into a heated water bath. As the $NaAlO_2$ was more expensive than $Al(OH)_3$, the latter was purposely added into the solution to replace part of $NaAlO_2$. After this process, the solution and filtered solid residue were separated by water washing and filtrating. The molecular sieves were prepared using silicon and aluminum in the leaching solution, while Mg, Fe, Ti and others metals remained in the leaching residues.

Preparation of Zeolite Material

According to the ICP results on the amount of Si and Al in the filter liquor, certain amount of $NaAlO_2$ was needed to add into the filtered liquor for zeolite materials composition. The main processes include appropriate proportions, rubber mixing, magnetic stirring, crystallization, water leaching, filtration, freeze-drying, etc. Table 2 is experimental conditions for preparing zeolite material.

Table 1 Water leaching processes under various operating conditions

Number	Water leaching process
1	Ultrasound (10% power, alternatively on 3 min and off 1.5 min) for 45 min
2	Magnetic stirring (500 r/min) for 45 min
3	No stirring for 45 min

Table 2 Experimental conditions for preparing zeolite material

Number	Product	Crystallization temperature (°C)	Crystallization time (h)	SiO_2/Al_2O_3	Na_2O/SiO_2	H_2O/Na_2O
1	X zeolite material	100	8	2	1.75	48
2	X zeolite material	100	16	2	1.75	48

Analysis and Characterization

Si and Al contents in the leaching liquor were analyzed by ICP (OPTIMA 7000DV, America). Morphology and particle size analysis of the solid leaching residues were performed by the field-emission scanning electron microscope (JSM6480LV, LEOL, Japan) and laser particle-size distribution apparatus (LMS-30, Japan). Phase components in the roasted mixture and leaching residues were identified using XRD (EDX8000, German).

Results and Discussion

Effect on the Contents of Si and Al in the Filtrate

During alkali-roasting process, Si and Al originally contained in the laterite nickel slag can be transformed to the water-soluble sodium silicate and sodium aluminate. The chemical reactions may occur as follows:

$$SiO_2 + 2NaOH = Na_2SiO_3 + H_2O(g) \tag{1}$$

$$Al_2O_3 + 2NaOH = 2NaAlO_2 + H_2O(g) \tag{2}$$

In the following water leaching process, the products of the Reactions (1) and (2) can dissolve into the water solution, which was further filtered to separate the undissolved solid residue. The filtered liquor is used as starting material to produce molecular sieves in this work.

Si and Al contents in the filtered liquor, as shown in Table 3, can vary with varying operating conditions. It is obvious that the mechanical stirring can increase the contents of Si (0.694 mg/l) and Al content (0.154 mg/l), compared with the values of Si (0.545 mg/l) and Al (0.151 mg/l) without stirring. In general, the water leaching process of the roasted slag mixture could be enhanced kinetically by increased interfacial reaction between the roasted solid particles and the water. Such enhancement might be relatively weak on the Al content, as a side reaction in water leaching can take place.

Table 3 Si and Al contents (ICP-MS analysis) in the filtered liquor obtained from water leaching of the roasted alkali–slag mixture

Number	Water leaching process	Si (mg/L)	Al (mg/L)
1	Ultrasound (10% power, alternatively on 3 min and off 1.5 min) for 45 min	0.785	0.250
2	Magnetic stirring (500 r/min) for 45 min	0.694	0.154
3	No stirring for 45 min	0.545	0.151

$$nNa_2SiO_3 + 2NaAlO_2 + nH_2O = Na_2O{\cdot}Al_2O_3{\cdot}nSiO_2 + 2nNaOH \quad (3)$$

Due to such reason, part of the dissolved Al may come into the "$Na_2O{\cdot}Al_2O_3{\cdot}nSiO_2$" that is considered unsolvable in water, thus influencing the increase in Al and Si content in the leaching solution and the filtered liquor as well. And probably for the same reason, the results of using conventional mechanical stirring in this investigation has demonstrated a limited effect on increasing the Si and Al contents in the filtered liquor.

It has been found that the contents of Al (0.250 mg/l) and Si (0.785 mg/l) in the filtered liquor become higher than those without ultrasound (Table 3) in water leaching process. This is because ultrasound can make cavitation, strong shock wave and microjet in the water medium. These actions can enhance the collisions and the friction among the solid particles of roasted slag, increase the mass transfer in the water medium and reduce the thickness of the boundary layer between solid and liquid reactants surface. After water leaching, the content of silicon and aluminum also depends on the precipitation degree of "$Na_2O{\cdot}Al_2O_3{\cdot}nSiO_2$". Under the action of ultrasound, the slag may dissolve quickly and the precipitation of "$Na_2O{\cdot}Al_2O_3{\cdot}nSiO_2$" may be restrained.

Effect on the Leaching Residue

Table 4 lists the compositions of the roasted alkaline slag and different leaching residues. It is found that the AL and Si contents remained in the leaching residues with and without magnetic stirring are all higher than those with ultrasound. These are in line with the increased Al and Si contents in the filtered liquor, which are beneficial to reduce the solid wastes while improving the quality of the filtered liquor for making molecular sieves.

Table 4 Composition of the roasted alkaline slag and different leaching residues in wt%

Compositions of different slags	SiO_2	Na_2O	MgO	Al_2O_3	Fe_2O_3	CaO	Cr_2O_3	else
Roasted alkaline slag	4.923	89.15	0.738	0.417	2.465	1.417	0.380	0.506
Leaching residues with ultrasound	29.91	20.28	21.91	12.93	7.907	4.797	0.636	1.638
Leaching residues with magnetic stirring	34.68	18.38	19.25	15.54	6.256	4.289	0.641	0.967

Morphology of Leaching Residues

Figure 1 shows the XRD spectra of roasted alkaline slag and different leaching residues, indicating more complex phase composition in the leaching residue. There is relative simple phase composition in the residue of the water leaching process with ultrasound. Figure 2a, b show that the leaching residue with magnetic stirring (500 r/min, 45 min) has coarser and more uniform particles than the one with ultrasound (10%, 3, 1.5, 45 min), and again, the photos in Fig. 2c, d illustrate the same phenomenon. This means that ultrasound can deliver more drastic power than magnetic stirring into the water medium, resulting in stronger reaction between the roasted alkali slag and water solution. Such better completed dissolution of the roasted alkali slag could generate smaller size of residue particles as observed.

Figure 3 is particle-size distribution of leaching residue after different water leaching processes. It is obvious that the particle size of the leaching residue of ultrasound is mainly concentrated at about 10 μm, and the particle size of the leaching residue of magnetic stirring is mainly about 100 μm. The recorded time for the filtration process shows that the time of filtration after ultrasonic assisted water washing was 21 min, while that of mechanical stirring assisted washing was 5 min. This is due to the fine particles of the residue with ultrasound assisted water leaching could be more difficult to pass through the filter paper.

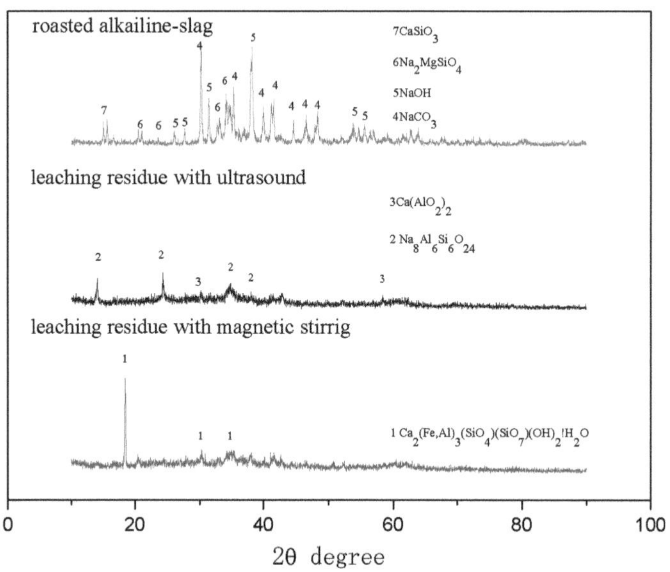

Fig. 1 XRD spectra of roasted alkaline slag and different leaching residues

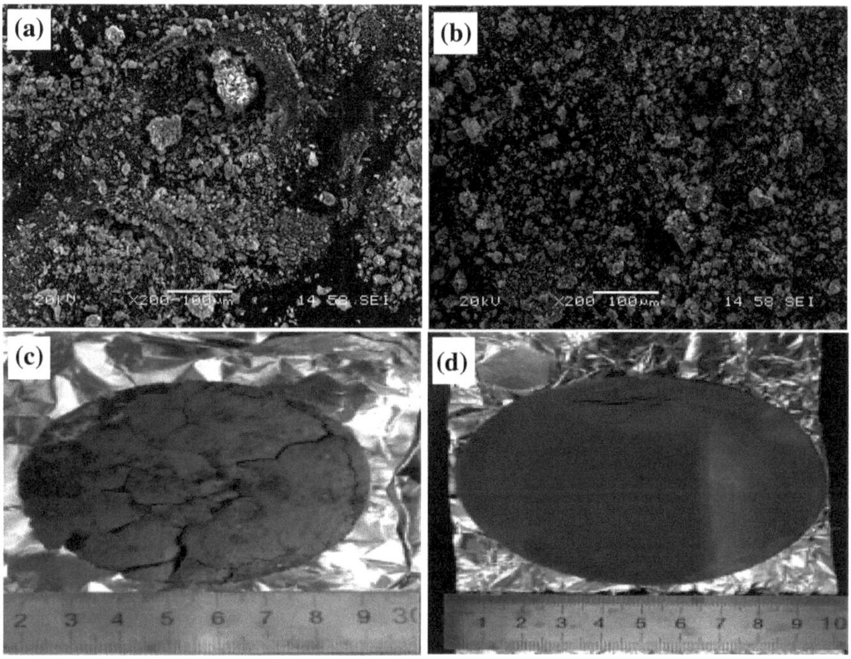

Fig. 2 Images and photos of the leaching residues: **a** SEM images of the leaching residue with magnetic stirring and **b** with ultrasound; **c** Photo of leaching residues with magnetic stirring and **d** with ultrasound

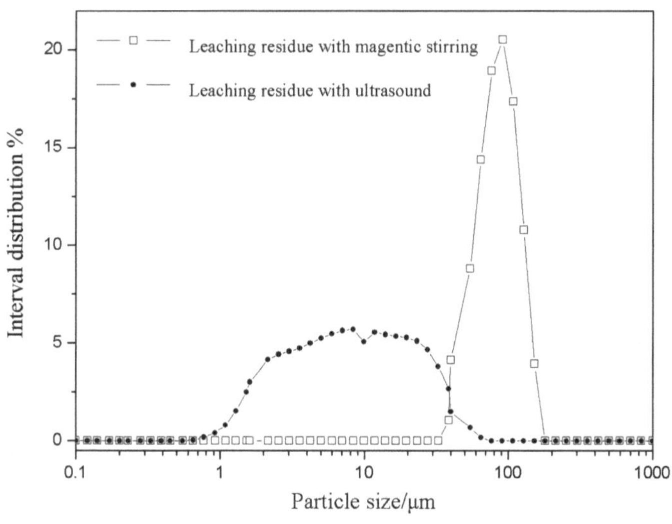

Fig. 3 Particle size analysis of leaching residues with different water leaching processes

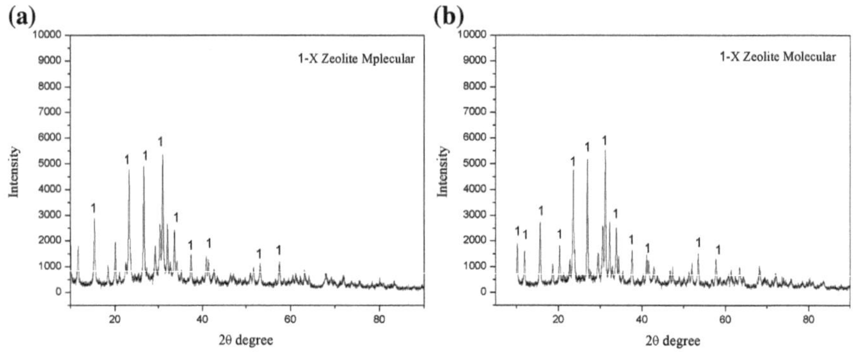

Fig. 4 XRD spectra of X zeolite products prepared from the filtrates with different crystallization time, **a** 8 h and **b** 16 h

Characterization of Zeolite Materials

Figure 4a, b show the characterization patterns of X zeolite materials under the condition of crystallization time 8 h and crystallization time 16 h, respectively, for the samples obtained from the filtrate, as mentioned above, through the process of alkali roasting and water leaching of the laterite nickel slag. The preparation conditions of zeolite were $SiO_2/Al_2O_3 = 2$, $Na_2O/SiO_2 = 1.75$, $H_2O/Na_2O = 100$, $H_2O/Na_2O = 68$, crystallization temperature 100 °C, crystallization time 8h and 16 h, respectively.

The results showed by the XRD spectra is clear enough to characterize the solid samples as results of the recycling process. Under the condition of other experimental parameters were unchanged, the longer the crystallization time is, the stronger and clearer the diffraction peak is. It is indicated that the higher crystallinity and longer crystallization time are beneficial to crystallization.

Conclusions

1. The silicon and aluminium contents in filtered liquor obtained from the water leaching with ultrasound are higher than those with mechanical stirring under identical testing conditions, showing the silicon content increased by 44% and the aluminium content increased by 65%, respectively.
2. Residues particles obtained from water leaching process with ultrasound are more uniform and fine than those with mechanical stirring, and this can be due to completed dissolution of the roasted alkali–slag mixture.
3. X zeolite material with satisfactory quality can be prepared from the filtered liquor as starting material obtained from the ultrasound-assisted water leaching of roasted alkali slag.

Acknowledgements Financial support from National Science and Technology Support Program of China (2015BAB03B00) is gratefully acknowledged.

References

1. Li TP, Xue JL, Luo WB (2015) Effects of power ultrasound on precipitation process of sodium silicate solutions. TMS annual meeting. In: EPD congress 2015, pp 109–116
2. Ettler V et al (2015) Leaching behaviour of slag and fly ash from laterite nickel ore smelting. Appl Geochem 64:118–127
3. Díaz Sandra C, Garcés Adriana, Restrepo Oscar J (2015) Thermodynamic analysis of the reduction process of Colombian lateritic nickel ore. Revista de Metalurgia. 51(4):369–378
4. Mcdonald RG, Whittington BI (2008) Atmospheric acid leaching of nickel laterites review: part I. Sulphuric acid technologies. Hydrometallurgy 91(1–4):35–55
5. Gordina NE et al (2017) Effect of ultrasound on the thermal behavior of the mixtures for the LTA zeolite synthesis based on metakaolin. J Therm Anal Calorim 129:1415–1427
6. Reinoso D, Adrover M, Pedernera M (2018) Green synthesis of nanocrystalline faujasite zeolite. Ultrason Sonochem 42:303–309
7. Liu Y, Tian YW, Zhai YC (2012) A comprehensive utilization of laterite nickel ore. Adv Mat Res 415–417:934–937
8. Shen XY, Shao HW, Zhai YC (2013) Preparation of ammonium jarosite from clinker digestion solution of nickel oxide ore roasted using $(NH_4)_2SO_4$. Trans Trans Nonferrous Metals Soc China 23:3434–3439
9. Fang DG, Xue JL (2018) Recycling SiO_2 and Al_2O_3 from the laterite nickel slag in molten sodium hydroxides. In: 9th international symposium on high-temperature metallurgical processing, pp 245–257
10. Li HY, Li SW, Peng JH (2018) Ultrasound augmented leaching of nickel sulfate in sulfuric acid and hydrogen peroxide media. Ultrason Sonochem 40:1021–1030
11. Li CC, Xie FC, Ma Y (2010) Multiple heavy metals extraction and recovery from hazardous electroplating sludge waste via ultrasonically enhanced two-stage acid leaching. J Hazard Mater 178:823–833
12. Mikhailov I, Komarov S, Levina V (2017) Nanosized zero-valent iron as Fenton-like reagent for ultrasonic-assisted leaching of zinc from blast furnace sludge. J Hazard Mater 321:557–565

Thermal Stability and Thermodynamics of the Ag_2ZnGeS_4 Compound

Mykola Moroz, Fiseha Tesfaye, Pavlo Demchenko, Myroslava Prokhorenko, Daniel Lindberg, Oleksandr Reshetnyak and Leena Hupa

Abstract Phase equilibria in the ZnS–Ag_2GeS_3–Ge–GeS_2 part of the Ag–Zn–Ge–S system were investigated using differential thermal analysis, X-ray diffraction, and EMF methods. The data was used to model Ag_2GeS_3–ZnS polythermal section. Further, the mechanism of formation and thermal stability of the Ag_2ZnGeS_4 compound were established. The results suggest the presence of another quaternary phase $Ag_4ZnGe_2S_7$ in the temperature range of 695–853 K. The determined phase relations were used to express the chemical reactions. Based on the electromotive force versus temperature measurements, experimental thermodynamic data of the Ag_2ZnGeS_4 quaternary phase were derived for the first time. The calculated Gibbs energy, enthalpy and entropy values of the Ag_2ZnGeS_4 compound in both phase regions are consistent, which indicates that Ag_2ZnGeS_4 has stoichiometric composition.

Keywords Chalcogenide semiconductors · Phase equilibria
Thermodynamic properties · EMF method · Gibbs energy

M. Moroz (✉) · F. Tesfaye · L. Hupa
Laboratory of Inorganic Chemistry, Johan Gadolin Process Chemistry Centre, Åbo Akademi University, 20500 Turku, Finland
e-mail: riv018@i.ua

P. Demchenko
Department of Inorganic Chemistry, Ivan Franko National University of Lviv, Lviv 79005, Ukraine

M. Prokhorenko
Department of Cartography and Geospatial Modeling, Lviv Polytechnic National University, Lviv 79013, Ukraine

D. Lindberg
School of Chemical Engineering, Aalto University, 02150 Espoo, Finland

O. Reshetnyak
Department of Physical and Colloid Chemistry, Ivan Franko National University of Lviv, Lviv 79005, Ukraine

G. Lambotte et al. (eds.), *Materials Processing Fundamentals 2019*, The Minerals, Metals & Materials Series, https://doi.org/10.1007/978-3-030-05728-2_20

Introduction

The stannite-type quaternary compounds of the general composition Ag_2BCX_4 (B = Zn, Cd, Hg, Pb, Fe, Mn; C = Si, Ge, Sn; X = S, Se, Te) exist in the Ag–B–C–X systems [1, 2]. These compounds crystallize in tetrahedral structures of the mineral stannite or in the ordered wurtzite structure [1, 3]. These crystallographic forms are close to each other with the only difference in the distribution of the cations in the tetrahedral sites [4]. Each metal cation is tetrahedrally coordinated by four sulfur anions [1]. The stannite-type compounds have been studied as materials for low-cost solar cells, high-efficiency light-emitting diodes [1, 4–6], and thermoelectric applications [7]. In addition, the Ag_2BCX_4 semiconducting compounds are promising materials for photocatalytic hydrogen production [8]. Some of these compounds have acentric crystalline structure and can be used as materials in nonlinear optics, spin- and optoelectronics applications [9]. Such interesting applications are stimulating further research efforts towards to expand the areas of use of these semiconductors.

Four-element compounds with B = Zn, Cd, Hg and X = S are located in the Ag_2X–BX–CX_2 (I) systems [2, 10, 11]. In the Ag_2S–ZnS–GeS_2 (II) system, only one quaternary compound Ag_2ZnGeS_4 (III) is found [2]. It crystallizes in the tetragonal structure (space group $I\bar{4}2m$, a = 0.574996(9) nm, c = 1.03434(3) nm, c/a = 1.799). The phase equilibria at $T \leq 670$ K and glass formation region of the Ag_2S–GeS_2–ZnS system have been investigated by Parasyuk et al. [2]. The energy gap of the Ag_2ZnGeS_4 compound is equal to 2.5 eV [1]. The phase equilibria (II) are represented by the cross sections that connect (III) with binary compounds (II), ternary compounds Ag_8GeS_6, $Ag_{10}Ge_3S_{11}$, Ag_2GeS_3, and the Ag_8GeS_8–ZnS tie line. The phase diagram of the Ag_8GeS_8–ZnS system was investigated by Piskach et al. [12]. It is of the eutectic type, with the invariant point at T = 1201 K and 42 mol% ZnS. The solubility of ZnS in Ag_8GeS_6 at T = 499 K is ~15 mol% ZnS [2]. The mechanism of formation and the thermal stability of the Ag_2ZnGeS_4 compound have not been established.

Two quaternary compounds Ag_2CdGeS_4 and $Ag_4CdGe_2S_7$ exist in the Ag_2S–CdS–GeS_2 system [10]. Four-element phase region of the Ag_2S–HgS–GeS_2 system is characterized by the presence of the ~$Ag_4HgGe_2S_7$, Ag_2HgGeS_4, $Ag_2Hg_3GeS_6$, and $Ag_6Hg_{0.82}GeS_{5.82}$ compounds [11]. The homogeneity range of the $Ag_6Hg_{0.82}GeS_{5.82}$ phase along the Ag_8GeS_6–Hg_4GeS_6 section at T = 670 K is equivalent to 22–31 mol% Hg_4GeS_6 [13].

The presence of a wide glass formation region near Ag_2S–GeS_2 system causes the appearance of kinetic obstacles to the equilibrium crystallization of phases from melts, the polymorphic phase transitions, and change of the regions of stabilities of some phases [2]. In this case, the choice of optimal conditions for the synthesis of compounds to achieve thermodynamic equilibrium is more complicated. These experimental difficulties in the study of phase relations in (I) can partially be solved through theoretical calculations of the T–x phase diagrams using, for example, the CALPHAD methods [14, 15]. These methods are based on thermodynamic data of the pure phases together with experimentally constructed T–x diagrams of separate sections.

In this work, we present the results of the phase equilibria in the vicinity of the Ag_2ZnGeS_4 compound of the Ag–Zn–Ge–S system. Furthermore, we determined the thermodynamic properties of the quaternary phase in this system quantitatively.

Experimental Section

Synthesis

The starting materials for synthesis were high-purity elements (99.99 wt% Ag, 99.999 wt% Ge, 99.99 wt% S, and 99.99 wt% ZnS). The synthesis and annealing were performed in thin-walled evacuated quartz glass ampoules with a residual pressure $\leq$1 Pa. Compounds GeS, GeS_2, and Ag_2GeS_3 were synthesized by cooling appropriately mixed melts of elements from $T = 1170$ K. Compounds Ag_2ZnGeS_4 and $Ag_4ZnGe_2S_7$ were obtained through solid-state synthesis from finely dispersed mixtures of Ag_2GeS_3 and ZnS at $T = 750$ K for 300 h. Differential thermal analysis (DTA) and X-ray diffraction (XRD) methods were used to characterize the composition of the synthesized compounds. The Ag_2GeS_3 glass [16, 17] was obtained by melt quenching of the corresponding elements from $T = 1200$ K in ice water.

Thermal Analysis Methods

DTA curves of the samples were recorded using a Paulik–Paulik–Erdey derivatograph fitted with chromel-alumel thermocouples and an H307-1 XY recorder. The heating and cooling rates in the DTA measurements of the samples were in the range of 6–8 K min^{-1} [18]. The thermocouples were calibrated by the melting temperatures of In (T = 429 K), Sn (T = 505 K), Cd (T = 594 K), Te (T = 723 K), Sb (T = 904 K), NaCl (T = 1074 K), Ge (T = 1209 K), Ag (T = 1236 K), and Cu (T = 1357 K) [19]. Errors in the temperature measurements were below $\Delta T = \pm 3$ K.

Differential scanning calorimetry (DSC) and thermogravimetric (TG) analysis of the compounds were done by using a NETZSCH STA 449 F1 Jupiter® equipment. The calorimeter was calibrated with the melting temperatures and enthalpies of fusion for high-purity chemical elements Sn, In, Bi, Zn, Al, and Au [19]. The average measurement accuracies of temperatures and enthalpies of fusion were determined to be $\Delta T = \pm 1$ K and $\Delta H = \pm 1.14\%$, respectively. To remove traces of reactive gases such as O_2(g), the chamber was evacuated and then backfilled with pure Ar(g) three times before each run. The purging gas Ar(g) at pressure 1×10^5 Pa was also used as a protective gas. The flow rate of the protective Ar(g) was 50 ml min^{-1} in all runs. The heating and cooling rates in the DSC–TG measurements of the samples were 10 K min^{-1}.

XRD patterns were collected on an STOE STADI P diffractometer equipped with a linear position–sensitive detector PSD, in a modified Guinier geometry (transmission mode, Cu$K\alpha_1$ radiation, a bent Ge (111) monochromator, $2\theta/\omega$ scan mode). Preliminary data processing and X-ray phase analyses were performed using STOE WinXPOW 3.03 [20] and PowderCell 2.4 PC programs [21], using data on crystal structures for the phases taken from the database [22].

Electromotive Force Measurements

For electromotive force (EMF) measurements [23, 24], the following electrochemical cells (ECCs) were assembled:

$$(-)C\ |Ag\ |Ag_2GeS_3\ \text{glass}\big|Ag_2ZnGeS_4, GeS, GeS_2, ZnS\ \big|C\ (+) \quad \text{(i)}$$

$$(-)C|Ag\ |Ag_2GeS_3\ \text{glass}\big|Ag_2ZnGeS_4, GeS, Ge, ZnS\big|C(+) \quad \text{(ii)}$$

where C is graphite and Ag_2GeS_3 glass is the fast purely Ag^+ ions conducting electrolyte [16]. Ag_2GeS_3 glass has similar ionic properties with superionic materials Ag_3GeS_3I and Ag_3GeS_3Br [25, 26]. The vertical lines in ECCs (i) and (ii) indicate phase boundaries or contacts between cell components. Commas between compounds mean mechanical mixtures of the phases. The cell polarities and half-cell reactions in the ECCs were established according to rules described in [23, 27].

As positive (right) electrodes of the ECCs, we used the equilibrium samples. The positive electrodes in the ECCs (i) and (ii) were prepared by solid-state synthesis of (Ag_2ZnGeS_4: GeS: GeS_2: ZnS) in the molar ratio 1: 1: 2: 1 and (Ag_2ZnGeS_4: GeS: Ge: ZnS) in the molar ratio 1: 3: 2: 1, respectively. The equilibrium state of the four-phases sample was achieved by vacuum annealing at 750 K for 240 h. Components of the ECCs in powder form were pressed at 10^8 Pa through a 2 mm diameter hole arranged in the fluoroplast matrix up to density $\rho = (0.93 \pm 0.02) \cdot \rho_0$, where ρ_0 is the experimentally determined density of cast samples. Fivefold thermal cycling of ECCs in the range of 400–550 K was performed to eliminate possible defects due to plastic deformation during sample pressing. The heating and cooling rates were of 2 K min^{-1}. Experiments were performed in a horizontal resistance furnace, similar to that described in [28]. As protective atmosphere, we used a continuously flowing highly purified (0.9999 volume fraction) Ar(g) at $P = 1.2 \times 10^5$ Pa, with a flow rate of 2×10^{-3} m^3 h^{-1} from the right to left electrode of the ECCs. The temperature was maintained with an accuracy of ±0.5 K. The EMF of the cells were measured by high-resistance (input impedance of $>10^{12}$ Ω) universal U7-9 digital voltmeter. The equilibrium in ECCs at each temperature was achieved within 2 h. After equilibrium has been attained, the EMF values were constant or their variation did not exceed ±0.2 mV. The dependences of the EMF of the cells on temperature $E(T)$ were analyzed by the method described in [24].

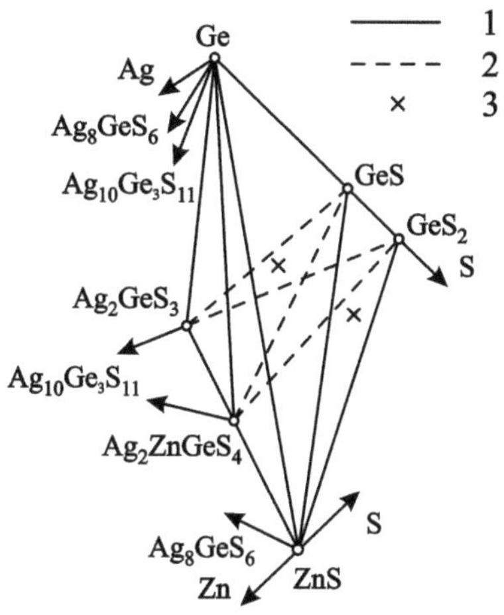

Fig. 1 The phase equilibria of the Ag–Zn–Ge–S system in the $ZnS–Ag_2GeS_3–Ge–GeS_2$ part, below $T = 650$ K. 1 and 2 are the lines of two-phase equilibria, and 3 is compositions of positive electrodes of ECCs

Results and Discussion

Phase Equilibria

Phase equilibria in the $ZnS–Ag_2GeS_3–Ge–GeS_2$ (IV) part of the Ag–Zn–Ge–S system are shown in Fig. 1.

The presence of four subsystems has been established: $Ag_2GeS_3–Ge–GeS–Ag_2ZnGeS_4$, $Ag_2GeS_3–GeS_2–GeS–Ag_2ZnGeS_4$, $Ag_2ZnGeS_4–GeS–GeS_2–ZnS$ (V), and $Ag_2ZnGeS_4–GeS–Ge–ZnS$ (VI). Some studies concerning the phase composition of compounds (IV) have been reported earlier in [2]. Our experimental results are presented in Figs. 2, 3 and 4.

Two-phase state of the Ag_2ZnGeS_4–Ge section without the formation of intermediate phases is confirmed by the diffraction patterns presented in Fig. 2. The diffraction peaks of the Ag_2ZnGeS_4 and Ge matching with JCPDS cards No. 00-059-0249 and 00-0040545, respectively [22]. According to obtained XRD results, the Ag_2ZnGeS_4 crystallized in the structure type of Cu_2FeSnS_4 compound (space group $I\bar{4}2m$, $a = 0.57459(7)$ nm, $c = 1.0332(1)$ nm, $V = 0.34198$ nm^3). The obtained crystallographic data for Ag_2ZnGeS_4 compound are in good agreement with the results reported by Parasyuk et al. [2]. Two-phase state of the ZnS–GeS and Ag_2ZnGeS_4–GeS cross sections at $T < 600$ K have been established by the EMF method [23, 24]. The EMF values of the galvanic cells with the positive electrodes from the phase regions (V) and (VI) at $T =$ const do not depend on molar ratio of phases.

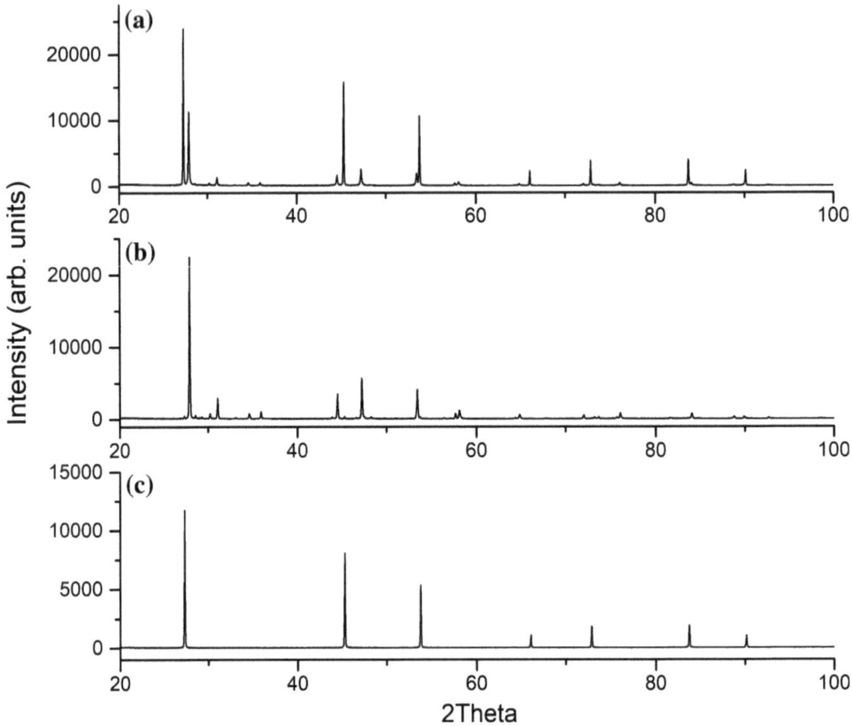

Fig. 2 XRD patterns for different compositions in the Ag_2ZnGeS_4–Ge system: **a** 50 mol% Ag_2ZnGeS_4–50 mol% Ge, **b** Ag_2ZnGeS_4, **c** Ge

The phase diagram of the Ag_2GeS_3–ZnS system is shown in Fig. 3. Some boundary lines in the concentration range of 65–100 mol% Ag_2GeS_3 are marked by dashed lines due to a large viscosity of the melts.

This phase diagram is of the eutectic type with the formation of the Ag_2ZnGeS_4 and $Ag_4ZnGe_2S_7$ quaternary compounds. The eutectic point lies at 833 K and 20 mol% ZnS. According to our experimental results, the Ag_2ZnGeS_4 compound is formed at 977 K in a peritectic reaction L + ZnS → Ag_2ZnGeS_4 and shows the phase transition at 869 K. The $Ag_4ZnGe_2S_7$ intermediate compound is formed by the peritectic reaction L + Ag_2ZnGeS_4 → $Ag_4ZnGe_2S_7$ at 853 K and decomposes below 695 K to Ag_2ZnGeS_4 and Ag_2GeS_3.

The results of DSC–TG measurement of the Ag_2ZnGeS_4 compound is presented in Fig. 4.

From the analysis results shown in Figs. 3 and 4 it follows that temperature of phase transition for Ag_2ZnGeS_4 obtained from the DTA measurement ($T_{tra} = 869$ K) is in agreement with the DSC data ($T_{tra} = 866.9$ K). A slight difference between the DTA and DSC results for the phase transition of Ag_2ZnGeS_4 is due to the unequal

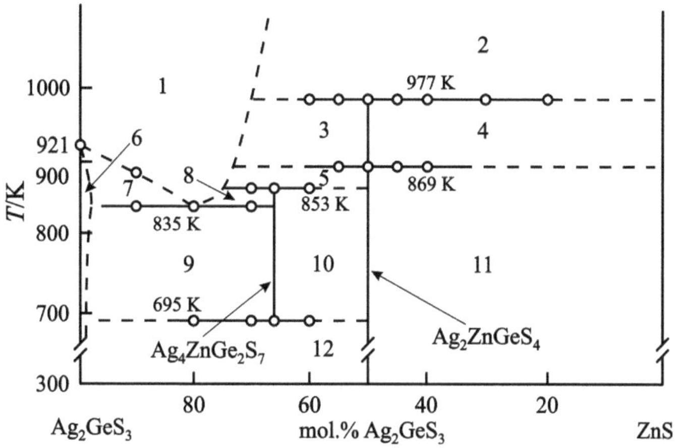

Fig. 3 Phase diagram of the Ag_2GeS_3–ZnS system. (1) L, (2) L + ZnS, (3) L + α-Ag_2ZnGeS_4, (4) α-Ag_2ZnGeS_4 + ZnS, (5) L + β-Ag_2ZnGeS_4, (6) α-Ag_2GeS_3, (7) L + α-Ag_2GeS_3, (8) L + $Ag_4ZnGe_2S_7$, (9) α-Ag_2GeS_3 + $Ag_4ZnGe_2S_7$, (10) $Ag_4ZnGe_2S_7$ + β-Ag_2ZnGeS_4, (11) β-Ag_2ZnGeS_4 + ZnS, (12) α-Ag_2GeS_3 + β-Ag_2ZnGeS_4

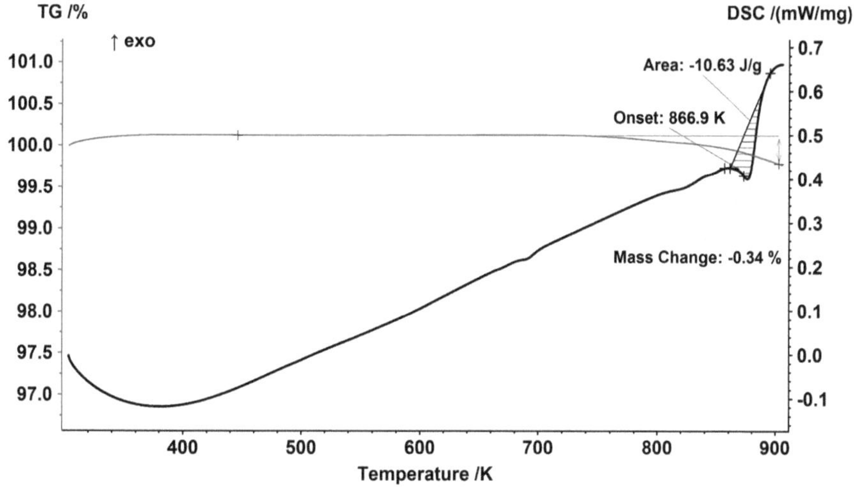

Fig. 4 DSC–TG curve as a function of temperature of the Ag_2ZnGeS_4 compound

change in the sample composition in the evacuated quartz glass ampoule versus under argon pressure in a furnace. Such a difference in the temperature for the phase transition during DTA and DSC measurements was observed earlier for the $Ag_2FeSn_3S_8$ compound [29]. The small deviations on the DSC curve at T ~ 690 K and above 800 K are also correlated with DTA thermal effects. These effects are due to slight equilibrium deviation of the quaternary compound from the stoichiometric composition. From DSC versus T relation, enthalpy of the Ag_2ZnGeS_4 phase transformation has been determined to be $\Delta_{tra}H = (5.12 \pm 0.05)$ kJ mol^{-1} at $T_{tra} = (866.9 \pm 1)$ K. Errors are average accuracies calculated during the temperature and sensitivity calibrations as described in the experimental section. Decimals were rounded to next whole number. As the TG curve in Fig. 4 shows, the weight loss observed at $P(\mathrm{Ar}) = 1 \times 10^5$ Pa and above $T \sim 750$ K is possibly due to evaporation of sulfur from the Ag_2ZnGeS_4 compound.

Thermodynamic Functions

According to the phase equilibria presented in Fig. 1, the virtual reactions in the ECCs (i) and (ii), which contain different compositions of phases from regions (V) and (VI), can be used to calculate the thermodynamic properties of the Ag_2ZnGeS_4. The electrochemical process of the formation of [Ag_2ZnGeS_4 and GeS] from [Ag, GeS_2 and ZnS] in the positive electrode *D* of the phase region (V) can be written as follows:

$$2\mathrm{Ag} = 2\mathrm{Ag}^{+} + 2e^{-} \quad \text{left side electrode(reference system)}$$
$$2\mathrm{Ag}^{+} + 2e^{-} + 2\mathrm{GeS_2} + \mathrm{ZnS} = \mathrm{Ag_2ZnGeS_4} + \mathrm{GeS} \quad \text{right side electrode(sample system)}$$
$$2\mathrm{Ag} + 2\mathrm{GeS_2} + \mathrm{ZnS} = \mathrm{Ag_2ZnGeS_4} + \mathrm{GeS} \quad \text{overall cell reaction.} \tag{1}$$

The electrochemical process of the formation of [Ag_2ZnGeS_4 and Ge] from [Ag, GeS and ZnS] in the positive electrode *D* of the phase region (VI) can be written as follows:

$$2\mathrm{Ag} = 2\mathrm{Ag}^{+} + 2e^{-} \quad \text{left side electrode(reference system)}$$
$$2\mathrm{Ag}^{+} + 2e^{-} + 2\mathrm{GeS} + \mathrm{ZnS} = \mathrm{Ag_2ZnGeS_4} + \mathrm{Ge} \quad \text{right side electrode(sample system)}$$
$$2\mathrm{Ag} + 3\mathrm{GeS} + \mathrm{ZnS} = \mathrm{Ag_2ZnGeS_4} + 2\mathrm{Ge} \quad \text{overall cell reaction.} \tag{2}$$

The relationship of EMF versus temperature measured with cells (i) and (ii) was approximated by Eqs. (3) and (4), respectively.

$$E_1/\mathrm{mV} = (32.95 \pm 2.20) + (437.23 \pm 4.39) \times 10^{-3}T/K \quad 489 \le T \le 514 \tag{3}$$
$$E_2/\mathrm{mV} = (78.89 \pm 1.94) + (333.1 \pm 3.87) \times 10^{-3}T/K \quad 489 \le T \le 514 \tag{4}$$

The Gibbs energies, entropies and enthalpies of the reactions (1) and (2) can be calculated using EMF measurement results [23, 24] and the thermodynamic Eqs. (5)–(7)

$$\Delta_r G = -n \cdot F \cdot E \tag{5}$$
$$\Delta_r S = n \cdot F \cdot (dE/dT) \tag{6}$$
$$\Delta_r H = -n \cdot F \cdot [E - (dE/dT) \cdot T] \tag{7}$$

Table 1 Standard thermodynamic values of reactions (1) and (2) at $T = 298.15$ K and $P = 1$ atm

Reaction	$-\Delta_r G^\circ$	$-\Delta_r H^\circ$	$\Delta_r S^\circ$
	kJ mol^{-1}		J mol^{-1} K^{-1}
(1)	31.51 ± 0.68	6.36 ± 0.72	84.37 ± 0.85
(2)	34.39 ± 0.60	15.22 ± 0.64	64.28 ± 0.75

where $n = 2$ is the number of electrons involved in the reactions (1) and (2), $F =$ 96485.33289 C mol $^{-1}$ is Faraday constant, and E is the EMF of the ECCs.

By assuming $\left(\frac{\partial \Delta_r H}{\partial T}\right)_p = 0$ and $\left(\frac{\partial \Delta_r S}{\partial T}\right)_p = 0$ [24], the thermodynamic function of reactions (1) and (2) were calculated through extrapolation of the linear temperature dependences of the EMF of ECCs to $T = 298.15$ K and using Eqs. (5)–(7). The results of the calculations are listed in Table 1.

Standard Gibbs energy and entropy of reactions (1) and (2) are related to the Gibbs energy of formation and entropy of compounds and pure elements according to the following equations:

$$\Delta_{r(1)} G^\circ = \Delta_f G^\circ_{Ag_2ZnGeS_4} + \Delta_f G^\circ_{GeS} - 2\Delta_f G^\circ_{GeS_2} - \Delta_f G^\circ_{ZnS} \quad (8)$$

$$\Delta_{r(1)} S^\circ = S^\circ_{Ag_2ZnGeS_4} + S^\circ_{GeS} - 2S^\circ_{Ag} - 2S^\circ_{GeS_2} - S^\circ_{ZnS} \quad (9)$$

$$\Delta_{r(2)} G^\circ = \Delta_f G^\circ_{Ag_2ZnGeS_4} - 3\Delta_f G^\circ_{GeS} - \Delta_f G^\circ_{ZnS} \quad (10)$$

$$\Delta_{r(2)} S^\circ = S^\circ_{Ag_2ZnGeS_4} + 2S^\circ_{Ge} - 2S^\circ_{Ag} - 3S^\circ_{GeS} - S^\circ_{ZnS} \quad (11)$$

Equations (12)–(15) were obtained from Eqs. (8)–(11).

$$\Delta_f G^\circ_{Ag_2ZnGeS_4} = \Delta_{r(1)} G^\circ - \Delta_f G_{GeS} + 2\Delta_f G^\circ_{GeS_2} + \Delta_f G^\circ_{ZnS} \quad (12)$$

$$S^\circ_{Ag_2ZnGeS_4} = \Delta_{r(1)} S^\circ - S^\circ_{GeS} + 2S^\circ_{Ag} + 2S^\circ_{GeS_2} + S^\circ_{ZnS} \quad (13)$$

$$\Delta_f G^\circ_{Ag_2ZnGeS_4} = \Delta_{r(2)} G^\circ + 3\Delta_f G^\circ_{GeS} + \Delta_f G^\circ_{ZnS} \quad (14)$$

$$S^\circ_{Ag_2ZnGeS_4} = \Delta_{r(2)} S^\circ - 2S^\circ_{Ge} + 2S^\circ_{Ag} + 3S^\circ_{GeS} + S^\circ_{ZnS} \quad (15)$$

The reaction of formation of the Ag_2ZnGeS_4 compound from pure elements can be written as

$$2Ag + Zn + Ge + 4S = Ag_2ZnGeS_4 \quad (16)$$

Based on reaction (16), the entropy of formation of the Ag_2ZnGeS_4 compound can be calculated as

$$\Delta_f S^\circ_{Ag_2ZnGeS_4} = S^\circ_{Ag_2ZnGeS_4} - 2S^\circ_{Ag} - S^\circ_{Zn} - 3S^\circ_{Ge} - 4S^\circ_{S} \quad (17)$$

By combining Eqs. (8), (10), (12) and (14) and data of the pure components in reactions (1), (2) and (16) reported in [30], the standard Gibbs energy of formations

Table 2 Standard thermodynamic properties of the selected phases in the Ag–Zn–Ge–S system at $T = 298.15$ K

Phase	$-\Delta_f G^\circ$	$-\Delta_f H^\circ$	S°	[References] Note
	kJ mol^{-1}		J mol^{-1} K^{-1}	
Ag	0	0	42.677	[30]
Zn	0	0	41.631	[30]
Ge	0	0	31.087	[30]
S	0	0	32.056	[30]
GeS	76.995	76.149	65.982	[30]
GeS_2	154.588	156.900	87.446	[30]
ZnS	200.403	205.183	57.656	[30]
$Ag_2ZnGeS_4^*$	464.1 ± 2.5	449.2 ± 2.6	336.3 ± 1.3	This work
$Ag_2ZnGeS_4^{**}$	465.8 ± 2.3	448.9 ± 2.4	343.1 ± 0.9	This work

*Phase region (V)
**Phase region (VI)

of Ag_2ZnGeS_4 compound were calculated as a function of temperature in the phase regions (V) and (VI), respectively

$$\Delta_f G^\circ_{Ag_2ZnGeS_4}/\left(\text{kJ mol}^{-1}\right) = -(449.2 \pm 2.6) - (50.0 \pm 1.3) \cdot 10^{-3} T/\text{K} \quad (18)$$

$$\Delta_f G^\circ_{Ag_2ZnGeS_4}/\left(\text{kJ mol}^{-1}\right) = -(448.9 \pm 2.4) - (56.8 \pm 0.9) \cdot 10^{-3} T/\text{K} \quad (19)$$

The standard entropy of formation of the Ag_2ZnGeS_4 compound was also calculated by combining Eqs. (9), (11), (13) and (15) and entropy data of the pure components in reactions (1), (2) and (16) reported in [30]. The uncertainties in Eqs. (18) and (19) are standard uncertainties. A comparative summary of the calculated Gibbs energies and entropies of formations is presented in Table 2 together with the available literature values.

The calculated values of the Gibbs energy and enthalpy of the Ag_2ZnGeS_4 compound in both phase regions are consistent. The difference in the entropies is higher, however, it does not exceed ~2%. Based on these results, it was concluded that Ag_2ZnGeS_4 has stoichiometric composition.

Conclusions

Phase equilibria in the $ZnS–Ag_2GeS_3–Ge–GeS_2$ part of the Ag–Zn–Ge–S system were established by applying DTA, XRD, and EMF methods. The phase diagram along the Ag_2GeS_3–ZnS cross section was constructed and thermal stability of the Ag_2ZnGeS_4 compound was established. The measured EMF versus temperature values were used to calculate the Gibbs energies of formation of the Ag_2ZnGeS_4 compound in the Ag_2ZnGeS_4–GeS–GeS_2–ZnS and Ag_2ZnGeS_4–GeS–Ge–ZnS phase regions. The thermodynamic functions of the quaternary compound in both phase regions are consistent. This consistency indicates that Ag_2ZnGeS_4 has stoichiometric composition. The experimental thermodynamic functions of the quaternary phase determined in this work supplement data for complete thermodynamic modeling of the T–x phase diagrams of the Ag–Zn–Ge–S system.

Acknowledgements The authors are grateful to the Academy of Finland for financial support. This work was made under the project "Thermodynamic investigation of complex inorganic material systems for improved renewable energy and metals production processes" (Decision number 311537) as part of the activities of the Johan Gadolin Process Chemistry Centre at Åbo Akademi University. In addition, funding from the Academy of Finland project "Behavior and properties of molten ash in biomass and waste combustion" (Decision number 266384) for M. Moroz and D. Lindberg is greatly appreciated.

Conflict of Interest The authors declare that they have no conflict of interest.

References

1. Tsuji I, Shimodaira Y, Kato H, Kobayashi H, Kudo A (2010) Novel Stannite-type complex sulfide photocatalysts A^I_2-Zn-A^{IV}-S_4 (A^I = Cu and Ag; A^{IV} = Sn and Ge) for hydrogen evolution under visible-light irradiation. Chem Mater 22:1402–1409
2. Parasyuk OV, Fedorchuk AO, Kogut YM, Piskach LV, Olekseyuk ID (2010) The Ag_2S–ZnS–GeS_2 system: phase diagram, glass-formation region and crystal structure of Ag_2ZnGeS_4. J Alloy Compd 500:26–29
3. Himmrich M, Haeuseler H (1991) Far infrared studies on stannite and wurtzstannite type compounds. Spectrochim Acta Part A 47:933–942
4. Chen S, Gong XG, Walsh A, Wei SH (2009) Electronic structure and stability of quaternary chalcogenide semiconductors derived from cation cross-substitution of II-VI and I-III-VI_2 compounds. Phys Rev B 79:165211-10
5. Guo Q, Hillhouse HW, Agrawal R (2009) Synthesis of Cu_2ZnSnS_4 nanocrystal ink and its use for solar cells. J Am Chem Soc 131:11672–11673
6. Fontané X, Izquierdo-Roca V, Saucedo E, Schorr S, Yukhymchuk VO, Valakh MY, Pérez-Rodríguez A, Morante JR (2012) Vibrational properties of stannite and kesterite type compounds: Raman scattering analysis of Cu2(Fe, Zn)SnS4. J Alloy Compd 539:190–194
7. Fan F-J, Wu L, Yu S-H (2014) Energetic I-III-VI_2 and I_2-II-IV-VI_4 nanocrystals: Synthesis, photovoltaic and thermoelectric applications. Energy Environ Sci 7:190–208
8. Zhang K, Guo L (2013) Metal sulphide semiconductors for photocatalytic hydrogen production. Catal Sci Technol 3:1672–1690
9. Davydyuk GE, Myronchuk GL, Kityk IV, Danyl'chuk SP, Bozhko VV, Parasyuk OV (2011) Ag_2CdSnS_4 single crystals as promising materials for optoelectronic. Opt Mater 33:1302–1306
10. Parasyuk OV, Piskach LV, Olekseyuk ID, Pekhnyo VI (2005) The quasi-ternary system Ag_2S-CdS-GeS_2 and the crystal structure of Ag_2CdGeS_4. J Alloy Compd 397:95–98
11. Parasyuk OV, Gulay LD, Piskach LV, Gagalovska OP (2002) The Ag_2S–HgS–GeS_2 system at 670 K and the crystal structure of the Ag_2HgGeS_4 compound. J Alloy Compd 336:213–217
12. Piskach LV, Parasyuk OV, Olekseyuk ID, Romanyuk YE, Volkov SV, Pekhnyo VI (2006) Interaction of argyrodite family compounds with the chalcogenides of II-b elements. J Alloy Compd 421:98–104

13. Olekseyuk ID, Kogut YuM, Yurchenko OM, Parasyuk OV, Volkov SV, Pekhnyo VI (2009) Glass formation and optical properties of the glasses in the Ag_2S–HgS–GeS_2 system. Chem Met Alloy 2:49–54
14. Ipser H, Mikula A, Katayama I (2010) Overview: the emf method as a source of experimental thermodynamic data. Calphad 34:271–278
15. Kroupa A (2013) Modeling of phase diagrams and thermodynamic properties using Calphad method—development of thermodynamic databases. Comput Mater Sci 66:3–13
16. Robinel E, Carette B, Ribes M (1983) Silver sulfide based glasses (I): glass forming regions, structure and ionic conduction of glasses in GeS_2–Ag_2S and GeS_2–Ag_2S–AgI systems. J Non-Cryst Solids 57:49–58
17. Moroz M, Tesfaye F, Demchenko P, Prokhorenko M, Lindberg D, Reshetnyak O, Hupa L (2018) Determination of the thermodynamic properties of the $Ag_2CdSn_3S_8$ and Ag_2CdSnS_4 phases in the Ag–Cd–Sn–S system by the solid-state electrochemical cell method. J Chem Thermodyn 118:255–262
18. Mikolaichuk AG, Moroz NV, Demchenko PY, Akselrud LG, Gladyshevskii RE (2010) Phase relations in the Ag8SnS6-Ag2SnS3-AgBr system and crystal structure of Ag6SnS4Br 2. Inorg Mater 46:590–597
19. Preston-Thomas H (1990) The international temperature scale of 1990 (ITS-90). Metrologia 27:3–10
20. Diffractom. Stoe WinXPOW, Version 3.03 (2010) Stoe Cie GmbH Darmstadt
21. Kraus W, Nolze G (1996) POWDER CELL—a program for the representation and manipulation of crystal structures and calculation of the resulting X-ray powder patterns. J Appl Crystallogr 29:301–303
22. Villars P, Cenzual K (2014) Pearson's crystal data: crystal structure database for inorganic compounds. Release 2014/15. ASM International, Materials Park
23. Babanly MB, Yusibov YA, Babanly NB (2011) The EMF method with solid-state electrolyte in the thermodynamic investigation of ternary copper and silver chalcogenides. InTech 57–78
24. Osadchii EG, Echmaeva EA (2007) The system Ag-Au-Se: phase relations below 405 K and determination of standard thermodynamic properties of selenides by solid-state galvanic cell technique. Am Mineral 92:640–647
25. Moroz MV, Prokhorenko MV, Prokhorenko SV, Determination of thermodynamic properties of Ag_3SBr superionic phase using EMF technique. Russ J Electrochem 51:886–889
26. Moroz MV, Prokhorenko MV, Reshetnyak OV, Demchenko PYu (2017) Electrochemical determination of thermodynamic properties of saturated solid solutions of Hg_2GeSe_3, Hg_2GeSe_4, $Ag_2Hg_3GeSe_6$, and $Ag_{1.4}Hg_{1.3}GeSe_6$ compounds in the Ag–Hg–Ge–Se system. J Solid State Electrochem 21:833–837
27. Babanly MB, Mashadieva LF, Aliev ZS, Shevelkov AV, Yusibov YA (2012) Phase diagram and thermodynamic properties of compounds of the AgI-TlI-I system. J Alloy Compd 524:38–45
28. Tesfaye F, Taskinen P (2014) Electrochemical study of the thermodynamic properties of matildite (β-$AgBiS_2$) in different temperature and compositional ranges. J Solid State Electrochem 18:1683–1694
29. Moroz M, Tesfaye F, Demchenko P, Prokhorenko M, Lindberg D, Reshetnyak O, Hupa L (2018) Phase equilibria and thermodynamics of selected compounds in the Ag–Fe–Sn–S system. J Electron Mater 47:5433–5442
30. Barin I (1995) Thermochemical data of pure substance. VCH, Weinheim

Thermochemical Data of Selected Phases in the FeO_x–$FeSO_4$–$Fe_2(SO_4)_3$ System

Fiseha Tesfaye, In-Ho Jung, Min-Kyu Paek, Mykola Moroz, Daniel Lindberg and Leena Hupa

Abstract Several recent studies have shown the potential of oxy-fuel combustion to reduce NO_x(gas) and SO_2(gas) emissions. However, the mechanisms through which SO_2(gas) reduction takes place has yet to be fully understood. Therefore, the development of oxy-sulfate thermodynamic database for a better understanding and control of SO_2(gas) emission during oxy-fuel combustion processes is essential. The focus of this research is on the thermodynamic modelling of the iron oxide–sulfate system with the FactSage 7.2 software package. Thermodynamic properties of selected phases in the FeO_x–$FeSO_4$–$Fe_2(SO_4)_3$ system were critically reviewed, compiled and assessed over a wide temperature range (298–2000 K) to obtain accurate thermodynamic description of the system at different temperatures. New C_p functions, which include the recent experimental data, were optimized. The obtained results are presented and discussed.

Keywords Sulfate · Sulfur dioxide · Decomposition reaction
Thermochemical data

F. Tesfaye (✉) · M. Moroz · L. Hupa
Laboratory of Inorganic Chemistry, Åbo Akademi University, Johan Gadolin Process Chemistry Centre, Piispankatu 8, 20500 Turku, Finland
e-mail: fiseha.tesfaye@abo.fi

F. Tesfaye · I.-H. Jung · M.-K. Paek
Department of Materials Science and Engineering, Seoul National University, Gwanwak-ro 1, Seoul 08826, Republic of Korea

D. Lindberg
Department of Chemical and Metallurgical Engineering (CMET), School of Chemical Engineering, Aalto University, Kemistintie 1, Aalto 00076, Finland

G. Lambotte et al. (eds.), *Materials Processing Fundamentals 2019*, The Minerals, Metals & Materials Series, https://doi.org/10.1007/978-3-030-05728-2_21

Introduction

Today, oxy-fuel combustion is considered as one of the major technologies for CO_2(gas) capture in power plants. A number of recent studies have also shown its potential to reduce NO_x(gas) and SO_2(gas) emissions. To control and optimize the emission reduction mechanisms, the development of oxy-sulfate thermodynamic database for the oxy-fuel combustion processes is essential. Therefore, the focus of this research is on the thermodynamic modelling of the FeO_x–$FeSO_4$–$Fe_2(SO_4)_3$ system with the FactSage 7.2 software package as a part of the larger database of the CaO–MgO–Al_2O_3–SiO_2–FeO–Fe_2O_3–sulfate system.

In the present work, thermodynamic data of $FeSO_4$ and $Fe_2(SO_4)_3$ were compiled and critically assessed below 2000 K to obtain accurate thermodynamic parameters. All calculations were carried out using the FactSage 7.2 thermochemical software [1], including the pure substances and self-developed databases.

Phase Relations in the Ternary Fe–S–O System

Recently, the relatively well known Fe–O and Fe–S systems were reevaluated by Hidayat et al. [2] and Walder and Pelton [3], respectively. The ternary Fe–S–O system was investigated by Shishin et al. [4]. In this system, only two stable ternary compounds, $FeSO_4$ and $Fe_2(SO_4)_3$ were reported. At ambient pressure conditions, both ternary phases decompose below 1500 K to form the gas and oxide phases before melting [4]. Chemical structures of the anhydrous ferrous sulfate ($FeSO_4$) and ferric sulfate ($Fe_2(SO_4)_3$) are illustrated in Fig. 1. Barany and Adami [5], Pankratz and Weller [6] and Majzlan et al. [7] measured the thermodynamic properties of $Fe_2(SO_4)_3$, and Moore and Kelly [8] measured the low temperature thermodynamic properties of $FeSO_4$. Using the susceptibility measurement technique, Frazer and Brow [9] and Kirfel et al. [10] measured magnetic data (μ_B) for α-$FeSO_4$ and β-$FeSO_4$ to be 4.1 ± 0.4 and 5.44 ± 0.27, respectively.

The anhydrous iron (III) sulfate ($Fe_2(SO_4)_3$) crystallizes in two modifications, monoclinic and trigonal. The stability relationship between the two phases is not clear. Only the trigonal polymorph has been reported in nature, under the name mikasaite. Mikasaite is formed by precipitation from hot gases escaping fractures in coal beds [7]. $Fe_2(SO_4)_3$ is rare as a mineral, but it is an important compound for the development of a thermodynamic database for a number of natural hydrated iron sulfates [11] and oxy-fuel combustion applications. More details on the thermodynamic properties of the phases in the FeO_x–$FeSO_4$–$Fe_2(SO_4)_3$ system are presented and discussed in the subsequent sections.

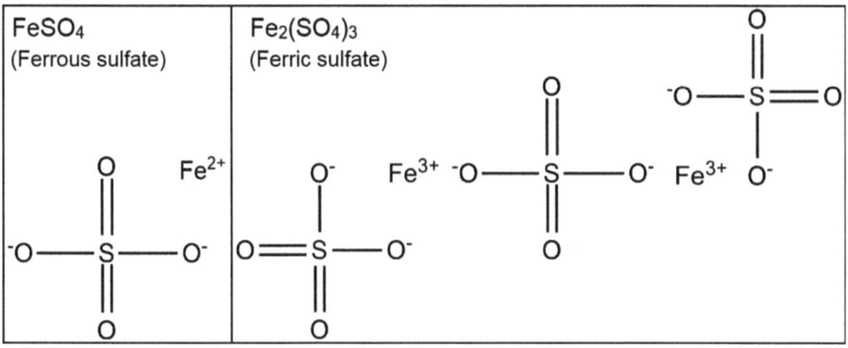

Fig. 1 Schematic diagram of the chemical structures of the ferrous and ferric sulfates, adopted from Chase [13]

Thermochemical Data of $FeSO_4$ and $Fe_2(SO_4)_3$

Enthalpy of Formation (ΔH_f°) of $FeSO_4$

Thomsen [12] measured the enthalpy of Reaction (1). In the NIST-JANAF thermochemical table [13], this enthalpy of Reaction (1) was used to estimate $\Delta_f H°(298.15\ K, FeSO_4) = -924.66\ kJ\ mol^{-1}$.

$$FeCl_2(200H_2O) + H_2SO_4(200H_2O) = FeSO_4(200H_2O) + 2HCl(100H_2O). \quad (1)$$

In the NIST-JANAF thermochemical data [13], $\Delta H_f°(298.15\ K, FeSO_4)$ values from the decomposition pressure of $FeSO_4$ reported by D'Ans [14], Greulich [15], and Neumann and Heintke [16] were also estimated by the second and third laws with an average value of $-925.1\ kJ\ mol^{-1}$. $\Delta H_f°(298.15\ K, FeSO_4) = -928.85 \pm 8.4\ kJ\ mol^{-1}$ and $\Delta H_f°(0\ K) = -919.33 \pm 8.4\ kJ\ mol^{-1}$ are the values recommended by Chase [13].

$\Delta H_f°(Fe_2(SO_4)_3)$

Based on the data reported for the chemical Reaction (2) by Barany and Adami [5] and using enthalpies of formations of the other components involved in the reaction, $\Delta H_f°(298.15\ K, Fe_2(SO_4)_3) = -2583 \pm 1.7\ kJ\ mol^{-1}$ was estimated and presented in [13].

$$Fe_2O_3 + 3(H_2SO_4 \cdot 14.855H_2O)(sol.) = Fe_2(SO_4)_3(cr) + 47.565H_2O(l). \quad (2)$$

The equilibrium pressures for decomposition of $Fe_2(SO_4)_3$ have been determined by several investigators [16–23] at different temperatures via chemical Reactions (3) and (4). The data were critically reviewed by Kellogg [24].

$$Fe_2(SO_4)_3 = Fe_2O_3 + 3SO_3, \quad (3)$$

$$SO_3(g) = SO_2(g) + 0.5O_2. \quad (4)$$

The measured pressure of the chemical equilibrium is the total pressures of $P(SO_3)$, $P(SO_2)$, and $P(O_2)$. In order to calculate the enthalpy change of Reaction (3), the partial pressures of $SO_3(g)$ from the total vapor pressure data at each temperature was evaluated. Based on the derived values for $P(SO_3)$, $\Delta H_r°(298.15\ K, Fe_2(SO_4)_3)$ values were derived using the third law and presented in the NIST-JANAF thermochemical data [13]. These average values of the calculations give -2584.164 kJ mol^{-1}. This value is higher only by ~1 kJ mol^{-1} than the value they obtained through the data of Barany and Adami [5].

After experimental studies in the temperature range of 273–395 K, Majzlan et al. [7] determined $\Delta H_f°(298.15\ K, Fe_2(SO_4)_3) = -2585.2 \pm 4.9$ kJ mol^{-1}, which is higher by ~2 kJ mol^{-1} than the value of Barany and Adami [5].

Heat Capacity (C_P) and Entropy ($S°$)

$FeSO_4$

The low-temperature C_p of the anhydrous $FeSO_4$ in the temperature range of 53–294.6 K was determined by Moore and Kelly [8] experimentally. Based on standard entropy at 50.12 K, they calculated, $S°(FeSO_4, 298.15\ K) = 107.57 \pm 0.84$ J K^{-1} mol^{-1}. Since the report by Moore and Kelley [8] did not mention the magnetic entropy contribution, an attempt was made by Knacke et al. [25] to add magnetic entropy (S_{mag}) to $S°(FeSO_4, 298.15\ K)$ and reevaluate the decomposition pressure data. They used the theoretical value of magnetic entropy (S_{mag}) for $FeSO_4$ $R\ln(2S_{spin})$, where S_{spin} is the magnetic spin. Since the electrons in the iron ions in $FeSO_4$ are in a high-spin configuration, they may have used $S_{spin} = 5/2$ which implies $S_{mag} = R\ln5 = 13.4$ J K^{-1} mol^{-1}. Therefore, the value $S°(298.15\ K) = (107.57 + 13.4)$ J K^{-1} mol^{-1} $= 120.96$ J K^{-1} mol^{-1} was adopted for $FeSO_4$. The C_p above 294.9 K was estimated by comparison with those for $MnSO_4$. In the calculations, the high-temperature $C_p(MnSO_4)$, 870.3–1082.3 K, were taken from the experiments of Southard and Shomate [26].

$Fe_2(SO_4)_3$

The heat capacities were established by comparison with those for $FeSO_4$, assuming their average specific heats, J K^{-1} g^{-1}, to be the same. The value of $S°(298.15\ K)$ was estimated so that the second and third law $\Delta H_r°(298.15\ K)$ values, derived from decomposition pressure data, are in reasonable agreement. The value $S°(298.15\ K) = 120.957 \pm 1.3\ J\ K^{-1}\ mol^{-1}$ was recommended in the NIST-JANAF thermochemical table [13].

Majzlan et al. [7] studied thermodynamic properties of the monoclinic $Fe_2(SO_4)_3$ by acid solution, adiabatic, and semi-adiabatic calorimetry methods. They reported $\Delta H_f°(298.15\ K,\ Fe_2(SO_4)_3) = -2585.2 \pm 4.9\ kJ\ mol^{-1}$ obtained by an appropriate thermochemical cycle with enthalpies of solution of monoclinic $Fe_2(SO_4)_3$, α-$MgSO_4$, γ-FeOOH, H_2O, and MgO in 5NHCl and at 298 K. They also compiled C_p values from 0.5 to 400 K and represented the data in the temperature range of 273–395 K by Eq. (5).

$$C_p/J \cdot K^{-1} \cdot mol^{-1} = 213 + 0.312 \cdot (T) - 2.959 \cdot 10^6 \cdot (T)^{-2}.\ (273-395K) \quad (5)$$

Melting and Decomposition Temperatures

The decomposition temperature 1451 K for $Fe_2(SO_4)_3$ at which the total pressure of the gaseous products equals 1 atm, was calculated by [13] through the graphical extrapolation of the decomposition pressure data measured by Warner and Ingraham [27]. By a similar method, they also calculated $T_{decomp.}(FeSO_4) = 944$ K from the data of Greulich [15]. Reactions (3) and (6) are the decomposition reactions for which the vapor pressure experimental data are available.

$$FeSO_4 \rightleftarrows FeO + SO_3. \quad (6)$$

The thermal decomposition of $FeSO_4{\cdot}6H_2O$ was studied by Masset et al. [28] applying the mass spectroscopy coupled with DTA/TG thermal analysis technique under inert atmosphere. After the dehydration, they reported to have observed two decomposition steps for $FeSO_4$ between 798 and 983 K according to the following Reactions (7) and (8):

$$6FeSO_4 \rightarrow Fe_2(SO_4)_3 + 2Fe_2O_3 + 3SO_2(798-923\ K), \quad (7)$$

$$Fe_2(SO_4)_3 \rightarrow Fe_2O_3 + 3SO_2 + 1.5O_2(898-983K). \quad (8)$$

$$Fe_2(SO_4)_3 \rightarrow 2FeSO_4 + SO_2 + O_2, \quad (9)$$

$$FeSO_4 \rightarrow 0.5Fe_2O_3 + SO_2 + 0.25O_2. \quad (10)$$

Gallagher et al. [29] studied the thermal decomposition of $FeSO_4{\cdot}xH_2O$ using the conventional thermogravimetry, differential thermal analysis, and evolved gas analysis techniques. In an oxidizing atmosphere, they observed that $FeSO_4$ oxidizes as

$$2FeSO_4 + 0.5O_2 \rightarrow Fe_2O(SO_4)_2, \quad (11)$$

$$Fe_2O(SO_4)_2 \rightarrow Fe_2O_3 + 2SO_2. \quad (873-948K) \tag{12}$$

In the inert atmosphere

$$2FeSO_4 \rightarrow Fe_2O_2SO_4 + SO_2, \ (748-848\ K) \tag{13}$$

$$Fe_2O_2SO_4 \rightarrow Fe_2O_3 + SO_2 + 0.5O_2. \quad (823-948K) \tag{14}$$

Other proposed decomposition reactions under N_2(gas) protective atmosphere are for Reaction (7), 891 K [30] (heating rate 20 K/min) and 960 K [31], and for Reaction (3), 951 K [30] (heating rate 20 K/min) and 1028 K [31].

A comparative summary of the decomposition temperatures of $FeSO_4$ and $Fe_2(SO_4)_3$ determined by different researchers and methods are compiled in Table 1.

Table 1 Summary of the phase transition and decomposition temperatures of $FeSO_4$ and $Fe_2(SO_4)_3$

Reaction	$T_{trans.}$/K	$T_{decomp.}$/K	Ref.	Remark
α-$FeSO_4 \rightarrow \beta$-$FeSO_4$	623	–	[10]	High-pressure phase transition (at $P_{tot} \approx 15$ kbar)
Reaction (3): $Fe_2(SO_4)_3 \rightarrow Fe_2O_3 + 3SO_3$	–	1461	[13]	Determined by the gas pressure measurements of [27]
	–	1017	This work	
	–	951	[30]	Heating rate 20 K min^{-1}
	–	1028	[31]	–
	–	1024	[7]	–
Reaction (6): $FeSO_4 \rightarrow FeO + SO_3$	–	944	[13]	–
Reaction (7): $6FeSO_4 \rightarrow Fe_2(SO_4)_3 + 2Fe_2O_3 + 3SO_2$	– – –	798–923 891 960	[28] [30] [31]	Inert atmosphere Heating rate 20 K min^{-1} –
Reaction (8): $Fe_2(SO_4)_3 \rightarrow Fe_2O_3 + 3SO_2 + 1.5O_2$		898–983	[28]	Inert atmosphere
Reaction (11): $2FeSO_4 + 0.5O_2 \rightarrow Fe_2O(SO_4)_2$ Reaction (12): $Fe_2O(SO_4)_2 \rightarrow Fe_2O_3 + 2SO_2$	–	870–948 873–948	[29]	Oxidizing atmosphere
Reaction (13): $2FeSO_4 \rightarrow Fe_2O_2SO_4 + SO_2$ Reaction (14): $Fe_2O_2SO_4 \rightarrow Fe_2O_3 + SO_2$	–	748–848 823–948	[29]	Inert atmosphere

Gibbs Energy Data

The solid gas reactions below 1100 K were studied using the EMF technique by Skeaff and Espelund [32], Hsieh and Chang [33], Musbah and Chan [34], Kobe and Couch [35], Rosenqvist and Hofseth [36], Schaefer [37], in a continuously flowing gas atmosphere, using the vapor pressure measurements technique by Warner and Ingraham [27] and using the DTA technique by Alcock et al. [40]. Gibbs energies of reactions determined by the different methods are summarized in Table 2. These EMF experiments may not correspond to equilibrium with the gas phase, which can contain other gaseous species such as SO_3, SO_2 and O_2, but rather to a fixed potential of SO_2, i.e. the other gaseous species do not form fast enough to significantly change the composition of the flowing SO_2. If the equilibrium with the gas phase were attained, it would be almost pure SO_2 over the range of P_{O2} from 10^{-4} to 10^{-14} atm. At higher oxygen potentials, amounts of SO_3 and O_2 become substantial, while at lower P_{O2} the partial pressure of S_2 starts to increase. The composition of the gas flowing out of the EMF furnaces was estimated by Shishin et al. [4] which indicated that large amounts of condensed phases in EMF cells would have to react in order to produce the gas phase of equilibrium composition, which is not what happened in the experiments.

Jacob and Iyengar [39] studied the Fe_2O_3 + $Fe_2(SO_4)_3$ phase equilibria by the EMF method. They measured EMF values between 800 and 1000 K. Based on their results, they have concluded that oxysulfates do not form below 1100 K in the Fe–S–O

Table 2 A comparative summary of Gibbs energies of reactions determined by different authors

Equilibrium reaction	ΔG_f°(kJ mol^{-1})	T/K	Ref.
$Fe_2(SO_4)_3 \rightleftarrows Fe_2O_3 + 3SO_2 + 1.5O_2$	970.05–0.724 · T	673–1073	[44]
	772.32–0.724 · T	920–1020	[32]
	870.235–0.8245 · T (±1.4)	800–1000	[39]
	704.61–0.697 · T	878–955	[40]
	802.86–0.762 · T	906–995	[27]
	753.098–0.744 · T	900–1000	[27]*
	730.5–0.688 · T	800–900	[43]
$Fe_2(SO_4)_3 \rightleftarrows Fe_2O_3 + 3SO_3$	576.895–0.546.1 · T (±0.5)	800–1000	[39]
	557.42–0.558 · T	900–1000	[27]*
$Fe_2(SO_4)_3 \rightleftarrows 2FeSO_4 + SO_2 + O_2$	395.64–0.352 · T	703–904	[32]
$FeSO_4 \rightleftarrows 0.5Fe_2O_3 + SO_2 + 1/4O_2$	258.74–0.202 · T	773–903	[44]
$FeSO_4 \rightleftarrows 0.5Fe_2O_3 + SO_2 + 1/4O_2$	203.47–0.202 · T	779–900	[32]

*Re-calculated from the experimental data of Warner and Ingraham [27]

system. The determined Gibbs energies of reactions involving the gases SO_3, SO_2, O_2 are compiled in Table 2.

Combining the high-temperature change in enthalpy of reaction $Fe_2(SO_4)_3 = Fe_2O_3 + 3SO_2(g) + 1.5O_2(g)$ reported by Jacob and Iyengar [39] with the pure substances enthalpy data of FactSage 7.2 for $SO_2(g)$ and Fe_2O_3, we have calculated $\Delta H_f°(Fe_2(SO_4)_3) = -2586.55$ kJ mol^{-1}. This value is higher only by -1.35 kJ mol^{-1} than that of Majzlan et al. [7]. Warner and Ingraham [27] have also determined the enthalpy of decomposition of $Fe_2(SO_4)_3$ from the vapor pressure measurements to be $\Delta H°_{decomp.} = 566.51$ kJ mol^{-1}.

Results and Discussion

The $C_p(SO_3)$ functions obtained through the reactions $MeO + SO_3 = MeSO_4$ (Me = Ca, Mg, Mn) were calculated by applying the Neumann–Kopp's rule described in Alcock et al. [41]. With the obtained C_p values, the corresponding $C_p(FeSO_4)$ values were calculated according to Reaction (6). The results are illustrated in Fig. 2.

The heat capacities of $FeSO_4$ were calculated between 273 and 400 K by comparison with those experimental data reported for $Fe_2(SO_4)_3$, assuming their average

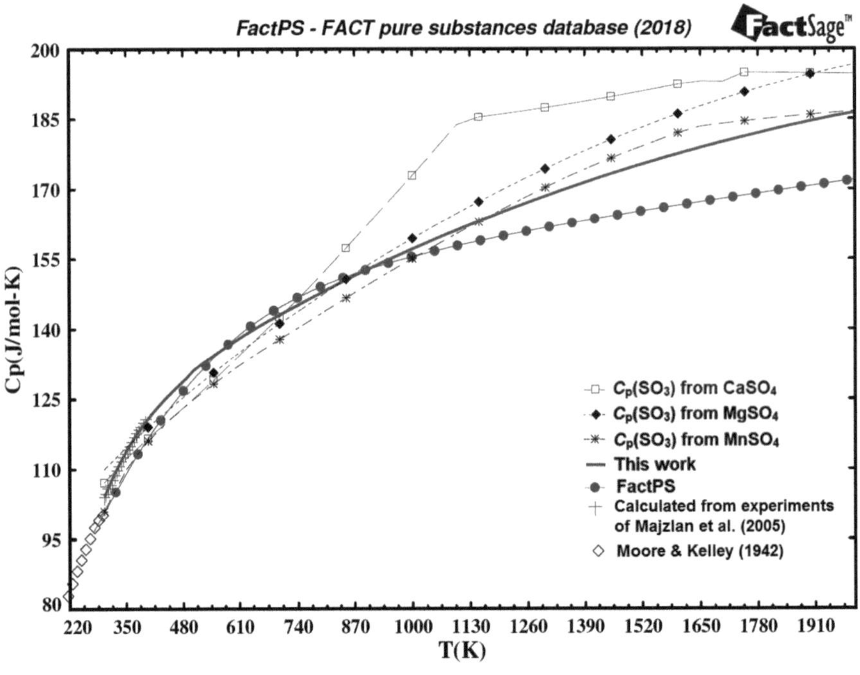

Fig. 2 C_p versus T diagram of $FeSO_4$

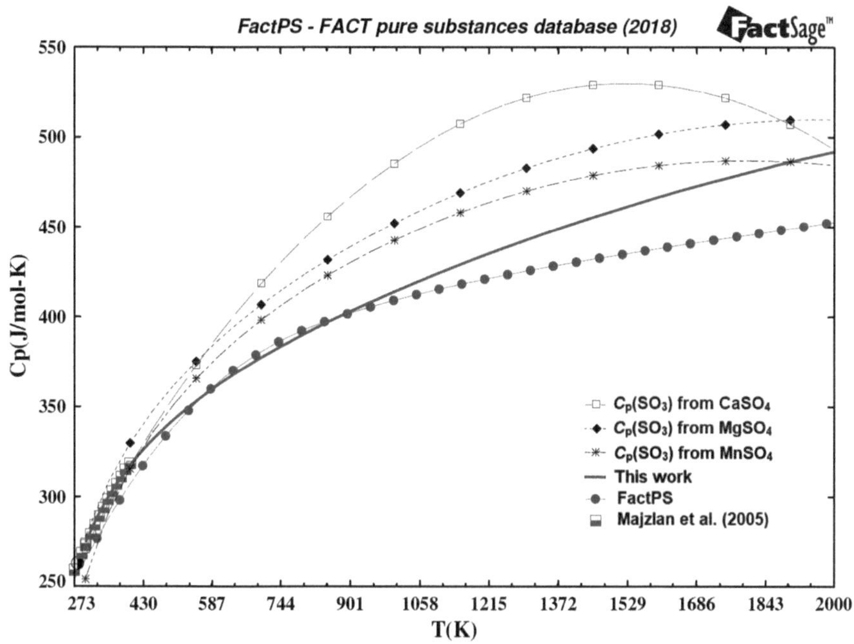

Fig. 3 C_p versus T diagram of $Fe_2(SO_4)_3$

specific heats, c_p(J K^{-1} g^{-1}), to be the same. We have also calculated the experimental $C_p(FeSO_4)$ values reported by Moore and Kelly [8] for the temperature range 273–295 K. By combining these all C_p values with the $C_p(FeSO_4)$ obtained through the derived $C_p(SO_3)$ from the reaction $MnSO_4 = MnO + SO_3$ using the data of Barin Sadakane et al. [42], optimal $C_p(FeSO_4)$ was calculated at each temperature, in the temperature range of 273–2000 K. By using the normal format for C_p polynomials, we fitted all the data together in two temperature ranges, 273–500 K (Eq. 15) and 500–2000 K (Eq. 16). Likewise, we have calculated $C_p(Fe(SO_4)_3)$ and the results are illustrated in Fig. 3. The optimized C_p polynomials for $Fe(SO_4)_3$ in the temperature range 273–2000 K is expressed by Eq. (17).

$$C_p(FeSO_4, 273{-}500\,K) = 118.67 + (0.047) \cdot T - (24.903) \cdot 10^5 \cdot T^{-2} - (5.943) \cdot 10^{-6} \cdot T^2, \quad (15)$$

$$C_p(FeSO_4, 500{-}2000\,1muK) = 108.213 + (0.0612) \cdot T - (12.842) \cdot 10^5 \cdot T^{-2} - (10.980) \cdot 10^{-6} \cdot T^2, \quad (16)$$

$$C_p(Fe_2(SO_4)_3, 273{-}2000K) = 305.363 + (0.136) \cdot T - (62.0743) \cdot 10^5 \cdot T^{-2} - (20.7454) \cdot 10^{-6} \cdot T^2. \quad (17)$$

As shown in Fig. 2, at high-temperatures, the C_p values are remarkably higher than the values currently used in the FactSage 7.2 database, which is based on the NIST-JANAF thermochemical data [13].

Recently, Majzlan et al. [7] determined the change in enthalpies of formations of $Fe_2(SO_4)_3$) in the temperature range of 273–395 K. This value is higher by about -2.2 kJ·mol^{-1} than the value recommended by NIST-JANAF [13]. In this work, we adopted the value $\Delta H_f°(298.15\ K, Fe_2(SO_4)_3) = -(2585.2 \pm 4.9)$ kJ mol^{-1} reported by Majzlan et al. [7], which was obtained through rigorous experimental studies.

By applying our new database for FeO_x–$FeSO_4$–$Fe_2(SO_4)_3$, $\Delta G_r°(T)$ for an equilibrium reaction $Fe_2O_3 + 3SO_2 + 1.5O_2 \rightleftarrows Fe_2(SO_4)_3$ was calculated in the temperature range of 760–1000 K, in which all the available experimental literature data fall. The obtained result is illustrated in Fig. 4 together with the selected experimental literature data presented in Table 2. Except for the values reported by Sadakane et al. [43] and Alcock et al. [40], the literature values are in agreement with our calculation.

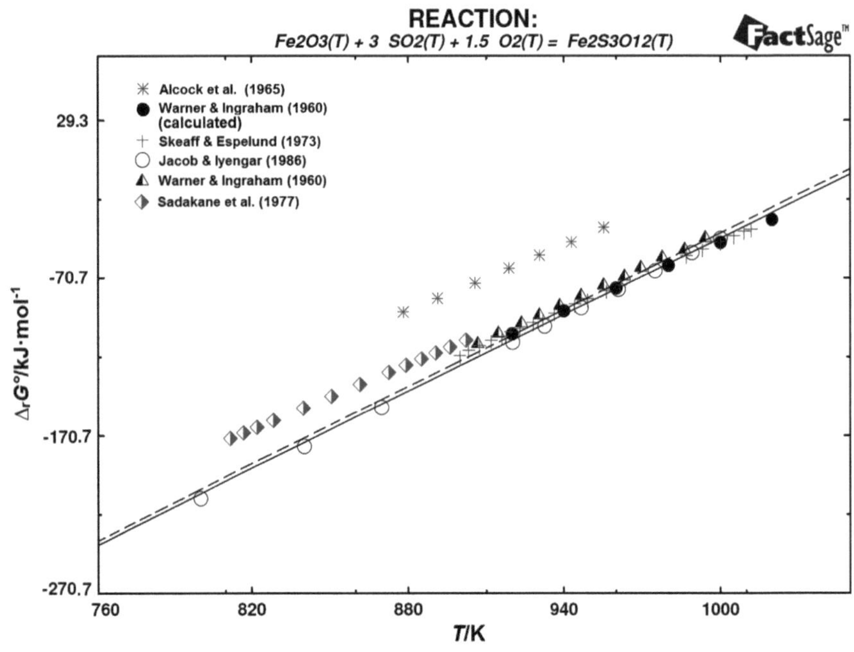

Fig. 4 $\Delta G_r°$ versus T diagram for the ferric sulfate formation reaction together with the literature experimental data. The solid line is calculated with the database developed in this work and the dashed line is calculated using the pure substances database in the commercial version of FactSage 7.2 [1]

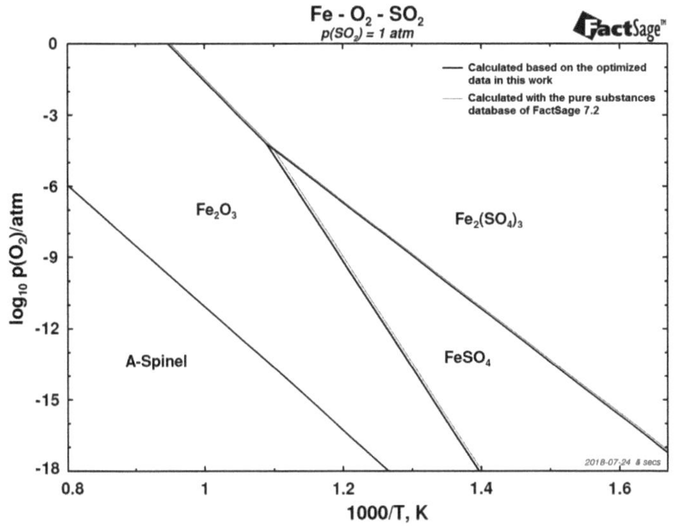

Fig. 5 Calculated potential phase diagram of Fe–O_2–SO_2 at $P(SO_2) = 1$ atm

Potential Phase Diagrams

Based on the results obtained for $FeSO_4$ and $Fe_2(SO_4)_3$, in this work, and the literature data for the other phases, chemical potential phase diagram has been calculated for the Fe–O_2–SO_2 system between 580 and 1250 K.

The potential phase diagram of $\log_{10}P(O_2)$ versus T shown in Fig. 5 was calculated at $P(SO_2) = 1$ atm and $P_{Total} = 2$ atm. The results obtained are in good agreement with those of Shishin et al.'s [4], hence they may have also used a total pressure close to 2 atm in their calculations.

By graphical extrapolation of the decomposition pressure data measured by Warner and Ingraham [27], we calculated $T_{decomp.}(Fe_2(SO_4)_3) = 1017$ K at which the total pressure of the gaseous products equals one atmosphere. This value is 434 K lower than the value reported in the NIST-JANAF thermochemical data [13] using the same source of experimental data. This large deviation indicates that there were most likely calculation errors in the NIST-JANAF thermochemical data [13]. Our calculations are in agreement with the results reported by [7, 31] and presented in Table 2.

Summary and Conclusions

The development of an oxy-sulfate thermodynamic database for a better understanding and control of SO_2 (gas) emission during oxy-fuel combustion processes is essential. Therefore, the focus of this research was on the thermodynamic modelling of the iron oxide–sulfate system with the FactSage 7.2 software package. Thermodynamic properties of selected phases in the FeOx–$FeSO_4$–$Fe_2(SO_4)_3$ system were critically reviewed, compiled and assessed over a wide temperature range of 298–2000 K to obtain accurate thermodynamic parameters. The obtained results were generally in agreement with the literature values in the low-temperature conditions and deviate in the high-temperature conditions. New high-temperature C_p functions, which include the recent experimental data, were optimized. Regarding the decomposition temperatures of $FeSO_4$ and $Fe_2(SO_4)_3$, large deviation were observed among the literature values. This warrants that there is a need for new experiments to accurately determine the decomposition temperatures. According to our assessment of the available literature values, we recommend 944 K for the decomposition of $FeSO_4$ and 1017 K for the decomposition of $Fe_2(SO_4)_3$ according to the reactions $FeSO_4 \rightarrow FeO + SO_3(g)$ and $Fe_2(SO_4)_3 \rightarrow Fe_2O_3 + 3SO_3(g)$, respectively, at ambient pressure condition.

Acknowledgements The authors are grateful to the Academy of Finland for financial support. This work was made under the project "Thermodynamic investigation of complex inorganic material systems for improved renewable energy and metals production processes" (Decision number 311537) as part of the activities of the Johan Gadolin Process Chemistry Center at Åbo Akademi University. This work is also a part of the project clean and efficient utilization of demanding fuels (CLUE), with support from the industrial partners: ANDRITZ, Fortum, International Paper, UPM-Kymmene Corporation, and Valmet Technologies Oy.

References

1. Bale CW et al (2009) FactSage 7.2 thermochemical software and databases—recent developments. Calphad 33:295–311
2. Hidayat T, Shishin D, Jak E, Decterov S (2015) Thermodynamic reevaluation of the Fe-O system. Calphad 48:131–144
3. Walder P, Pelton AD (2005) Thermodynamic modeling of the Fe-S system. J Phase Equilib Diffus 26:23–38
4. Shishin D, Jak E, Decterov SA (2015) Critical assessment and thermodynamic modeling of the Fe-O-S System. J Phase Equilib Diffus 36:224–240
5. Barany R, Adami LH (1965) Heats of formation of anhydrous ferric sulfate and indium sulfate. US Bur Mines Rep Invest 6687:8
6. Pankratz LB, Weller WW (1969) Thermodynamic data for ferric sulfate and indium sulfate. US Bur Mines Rep Invest 7280:9
7. Majzlan J et al (2005) Thermodynamics of monoclinic $Fe_2(SO_4)_3$. J Chem Thermodyn 37:802–809
8. Moore GE, Kelley KK (1942) The specific heats at low temperatures of anhydrous sulfates of iron, magnesium, manganese, and potassium. J Am Chem Soc 64:2949–2951
9. Frazer BC, Brow PJ (1962) Antiferromagnetic structure of $CrVO_4$ and the anhydrous sulfates of divalent Fe, Ni, and Co. Phys Rev 125(4):1283–1291
10. Kirfel A, Schafer W, Will G, Buschow KHJ (1977) The high-temperature polyrnorph β-FeSO4. Phys Stat Sol (a) 40:447
11. Hemingway BS, Seal RR II, Chou I.-M (2002) Thermodynamic data for modeling acid mine drainage problems: compilation and estimation of data for selected soluble iron-sulfate minerals. US Geological Survey Open-File Report, 02–161

12. Thomsen J (1886) Thermochemische untersuchunger. Verlag von Johann Ambrosius Barth, Leipzig
13. Chase, MW (ed) (1998) NIST-JANAF thermochemical tables. J Phys Chem Ref Data, Monograph No. 9 Part II:959–1951
14. D'Ans J (1905) PhD. dissertation, Darmstadt, Data quoted by B. Neumann and G. Heintke, loc. Cit
15. Greulich E (1927) Z Anorg Chem 168:197–202
16. Neumann B, Heintke G (1937) Z Elektrochem 43:246
17. Keppeler G, D'Ans J (1980) Z Phys Chem 62:89
18. Wöhler L, Plüddemann W, Wöhler P (1908) Eine neue Methode zur Tensionsbestimmung von Sulfaten. Ber Deut Chem Ges 41:703–717
19. Bodenstein M, Suzuki T (1910) Z Elektrochem 16:912
20. Wöhler L, Grünzweig M (1913) Die Sulfat-Tensionen und die Affinität der seltenen Erden. Ber Deut Chem Ges 46:1726–1732
21. Grunzweig M (1913) PhD dissertation, Darmstadt, Germany
22. Blanks RF (1961) PhD dissertation, University of Melbourne, Australia
23. Ingraham TR (1963) Internal Rept. EMT-63-17, Department of Mines and Technical Surveys, Ottawa, Canada
24. Kellogg HH (1964) Critical review of sulfation equilibria. Trans TMS-AIME 230:1622–1644
25. Knacke O, Kubaschewski O, Hesselman K (1991) Thermochemical properties of inorganic substances, 2nd edn. Springer, Berlin, (KKH 91) pp. 1–1113
26. Southard JC, Shomate CH (1942) Heat of formation and high-temperature heat content of manganous oxide and manganous sulfate. High-temperature heat content of manganese. J Am Chem Soc 64(8):1770–1774
27. Warner A, Ingraham TR (1960) Decomposition pressures of ferric sulphate and aluminum sulphate. Can J Chem 38:2196–2202
28. Masset P, Poinso J-Y, Poignet J-C (2006) TG/DTA/MS Study of the thermal decomposition of FeSO4·6H_2O. J Therm Anal Calorim 83:457–462
29. Gallagher PK, Johnson DW, Schrey F (1970) Thermal decomposition of Iron(II) sulfates. J Am Ceram Soc 666–670. https://doi.org/10.1111/j.1151-2916.1970.tb12038.x. Accessed 26 July 2018
30. Pannetier G, Bregeault JM, Djega-Maria-Dassou Gerald (1964) Thermal dissociation of ferrous sulfate heptahydrate. C R Acad Sci 258:2832–2835
31. Saflullin NS, Gitis EB, Panesenko NM (1968) Thermochemical transformations of $FeSO_4 \cdot 7H_2O$ during heating in oxidizing and inert media. Russ J Inorg Chem 13:1493
32. Skeaff JM, Espelund AW (1973) An E.M.F. method for the determination of sulphate-oxide equilibria results for the Mg, Mn, Fe, Ni, Cu and Zn systems. Can Metall Q 12:445–454
33. Hsieh KC, Chang YA (1986) A Solid-State Emf Study of Ternary Ni-S-O, Fe-S-O, and Quaternary Fe-Ni-S-O. Metall Trans B 17:133–146
34. Musbah OA, Chang YA (1988) A solid-state EMF study of the Fe-S-O and Co-S-O ternary systems. Oxid Met 30:329–343
35. Kobe KA, JrEJ Couch (1954) Enthalpy-concentration diagram for system ferrous sulfate–water. Ind Eng Chem 46:377–381
36. Rosenqvist T, Hofseth A (1980) Phase relations and thermodynamics of the copper-iron-sulfur-oxygen system at 700–1000 °C. Scand J Metall 9:129–138
37. Schaefer SC (1980) Electrochemical determination of gibbs energies of formation of manganese sulfide and iron sulfide ($Fe_{0.9}S$), RI, 8486, Albany Researgh Center Bureau of Mines, Albany, OR
38. Espelund AW, Jynge H (1977) The zinc-iron-sulfur-oxygen system equilibriums between sphalerite or wurtzite, pyrrhotite, spinel, zinc oxide and a gas phase. Scand J Metall 6:256–262
39. Jacob KT, Iyengar GNK (1986) Thermodynamic study of Fe_2O_3-$Fe_2(SO_4)_3$ equilibrium using an oxyanionic electrolyte (Na_2SO_4-I). Metall Trans B 17:323–329
40. Alcock CB, Sudo K, Zador S (1965) The free energies of formation of the sulfates of cobalt, copper, nickel, and iron. Trans TMS-AIME 233:655–661

41. Tesfaye F et al (2014) Thermal stabilities and properties of equilibrium phases in the Pt-Te-O system. J Chem Thermodyn 106:47–58
42. Barin I (1989) Thermochemical data of pure substances, Part II. VCH Verlagsgesellschaft, Weinheim/VCH Publishers, New York
43. Sadakane K, Kawakami M, Goto KS (1977) Measurement of the standard free energy of formation of sulfides and sulfates using oxygen concentration cells with $ZrO_2 \cdot CaO$. Tetsu-to-Hagané 63(3):432–440
44. Turkdogan ET (ed) (1980) Physical chemistry of high temperature technology. Academic Press, New York

The Effect of Heat Treatment to $FePt/Fe_2O_3$ and FePt/Cu Magnetic Performance

Naidu Seetala, Deidre Henderson, Jumel Jno-Baptiste, Hao Wen and Shengmin Guo

Abstract The effect of heat treatment to $FePt/Fe_2O_3$ and FePt/Cu magnetic performance was examined in this paper. To obtain desired phase transformations for high magnetic storage media, a self-assembled layer of FePt nanoparticles was placed between two layers of Fe_2O_3 nanoparticles using surfactants on Si and Cu substrates to minimize aggregation during heat treatment. To eliminate the surfactant, a sample made by simply mixing FePt and Cu nanoparticles in hexane was deposited on a Cu substrate. Vacuum furnace annealing at 600 °C (1 h) or laser heat treatment at 20, 40, 80 W at a speed of 1–1.5 m/s were carried out on the samples. The coercivity of $FePt/Fe_2O_3$-layered samples increased from 148 Oe to 366 Oe and 246 Oe for furnace annealed samples on Si and Cu substrates, respectively; while, it remained almost unchanged in laser heat-treated samples on Cu substrate and slightly higher magnetization at 40 W compared to 20 W laser heat treatment of samples on Si substrate. The FePt/Cu nanoparticles mixer layer on Cu substrate was subjected to laser heat treatment at 40 and 80 W. The coercivity at both laser powers did not show any significant change. During the 80 W laser heat treatment, most of the particles escaped from the surface indicating a high temperature process. The results indicate that the furnace annealing at 600 °C brings the desired magnetic phase transformation in all cases, while the laser heat treatment even at high power does not bring the phase transformation due to very short time period for processing.

Keywords Magnetic storage media · Magnetization · FePt nanoparticles Laser heating · Vacuum annealing

---

N. Seetala (✉) · D. Henderson · J. Jno-Baptiste
Department of Mathematics and Physics, Grambling State University, Grambling, LA 71245, USA
e-mail: naidusv@gram.edu

H. Wen · S. Guo
Department of Mechanical and Industrial Engineering, Louisiana State University, Baton Rouge, LA 70803, USA

G. Lambotte et al. (eds.), *Materials Processing Fundamentals 2019*, The Minerals, Metals & Materials Series, https://doi.org/10.1007/978-3-030-05728-2_22

Introduction

FePt nanoparticles have attracted considerable attention because of their potential application for ultra-high-density data storage medium [1]. Starting from nanoparticles, the nanostructured materials can be developed using advanced manufacturing technologies such as additive manufacturing [2] and Selective Laser Melting (SLM) [3]. In this study, we tried to determine the best heat treatment method to produce ordered face-centered tetragonal phase (FTP) of FePt nanostructures starting from disordered face-centered cubic (FCC) FePt nanoparticles. The ordered FTP phase has high magnetic coercivity that is required for magnetic storage to store information for longer periods while the FCC phase has negligible amount of coercivity [4].

Chemically synthesized 6.5 nm FePt particles have the low-anisotropy disordered FCC (A1) phase and must be annealed to temperatures above 500 °C to transform to the high-anisotropy ordered FTP ($L1_0$) phase [4]. The particles aggregate during annealing. This study is to determine the best heat treatment methods to minimize aggregation of particles during the required annealing stage. It has been observed that the magnetic coercivity of FePt thin films increased by coating the FePt film with a monolayer array of γ-Fe_2O_3 nanoparticles [5]. Here, Fe_2O_3 nanoparticles have been included to isolate FePt nanoparticles during furnace annealing. In addition, high power laser heat treatment is also used for a reduction in heating time to help minimize the aggregation of those particles. Addition of Cu to FePt reduced the required $L1_0$ ordering temperature and increased the coercivity [6, 7]. In the second set, nonmagnetic Cu nanoparticles have been mixed with FePt nanoparticles to minimize aggregation of FePt nanoparticles during laser heating or vacuum furnace annealing required to bring the desired phase transformation.

Experimental

The nanoparticle layers were self-assembled using surfactants (5 mL of hexane and 0.06 mL of 50/50 oleic acid and oleylamine). A layer of FePt nanoparticles with average size of 6.5 nm was placed in between two layers of Fe_2O_3 nanoparticles with average size of 20 nm on Si substrates. After each layer deposition, the samples were analyzed, and then some samples were annealed in a vacuum ($<1 \times 10^{-5}$ torr obtained using a turbo molecular pump) furnace set at 600 °C for an hour while others were laser heat-treated at 20 or 40 W power with 1.5 m/s scanning speed.

The second set of nanoparticle structure was made without surfactants and mixing FePt nanoparticles (6.5 nm) with nonmagnetic Cu nanoparticles (70 nm) in hexane using an ultrasonicator and a thick layer of the mixture was dispensed onto a Cu substrate. Three samples were heat treated using a laser. One with only FePt nanoparticles was exposed to laser power set at 80 W and the other two with FePt and Cu nanoparticles mixer deposited were exposed to laser power set at 40 and 80 W, respectively. The laser scanning speed was set to 1 m/s for these three samples, slightly lower than that for the first set to increase the sample temperature slightly at the laser contact as the first set results indicated a need to increase the sample temperature. Vacuum annealing with the same parameters as above was performed on FePt and FePt/Cu nanoparticles deposited onto Cu substrates.

Model 880A signal processor Vibrating Sample Magnetometer (VSM) was used with a maximum applied field of 10,000 Gauss to study the magnetic properties. The applied magnetic field is varied from +10,000 Gauss to −10,000 Gauss and the magnetization is measured at each applied field using Ni standard calibrated pick-up coils.

Results and Discussion

The magnetic hysteresis curves for Fe_2O_3 nanoparticles on Si substrate before and after furnace annealing are shown in Fig. 1. The magnetization at 10,000 Gauss applied field and the coercivity (reverse magnetic field required to remove magnetization, i.e. the average of the x-axis zero magnetization intercepts of the curves) are compared between the samples. The magnetization decreased and coercivity slightly increased upon furnace annealing, indicating a phase transformation due to heating. The phase transformation is assumed to be from γ-Fe_2O_3 phase to α-Fe_2O_3 phase that is observed to start at 350 °C and complete by above 500 °C [8, 9]. The measurements are done to a maximum applied field of 10,000 Gauss, but the figures are shown up to 3,000 Gauss for clarity. The magnetization at 10,000 Gauss applied field decreased from 47 to 28.6 emu/g and the coercivity slightly increased from 151 to 208 Gauss upon furnace annealing of Fe_2O_3 nanoparticles.

Figure 2 shows the magnetic hysteresis curves for FePt nanoparticles layer between two layers of Fe_2O_3 nanoparticles on Si substrate, before and after furnace

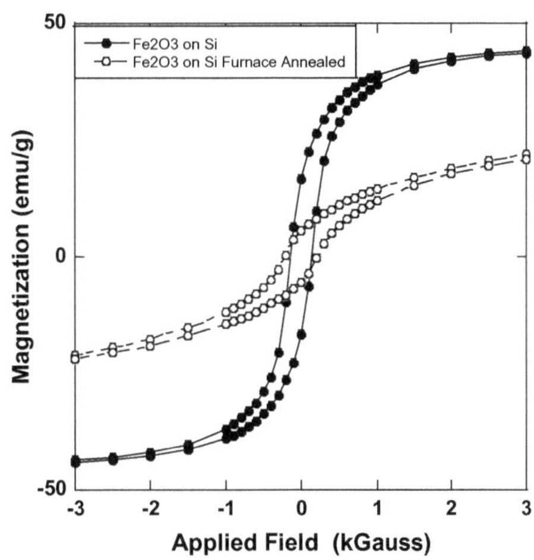

Fig. 1 Magnetization of Fe_2O_3 nanoparticle layer on Si substrate, before and after furnace annealing

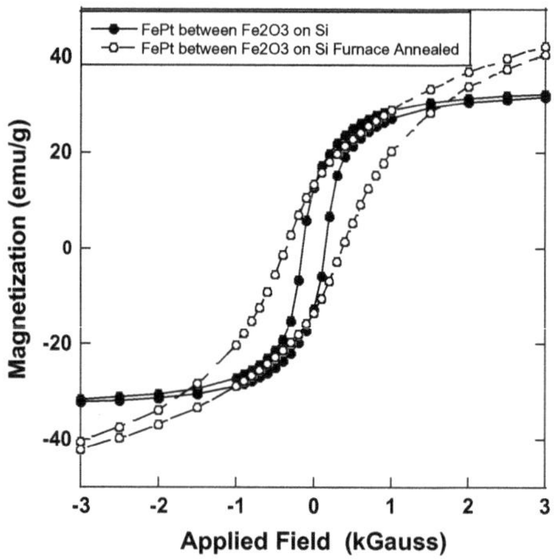

Fig. 2 Magnetization of FePt nanoparticle layer between two layers of Fe_2O_3 nanoparticles on Si substrate, before and after furnace annealing

annealing. Both magnetization and coercivity increased upon annealing at 600 °C. The magnetization at 10,000 Gauss applied field increased from 30 to 54 emu/g, while the coercivity increased 2.6 times from 142 to 366 Gauss upon furnace annealing to 600 °C.

Figure 3 shows the magnetic hysteresis curves for FePt nanoparticle layer between two layers of Fe_2O_3 nanoparticles on Si substrate, before and after laser heat treatment with two different laser powers (20 and 40 W) at a laser scan speed of 1.5 m/s. The magnetization decreased from 30 emu/g to 14 and 16 emu/g, respectively, for 20 and 40 W laser heat treatments. The coercivity remained almost unchanged at 142 and 143 Gauss, respectively, for 20 and 40 W laser heating, while the unheated sample coercivity was 142 Gauss. The difference between furnace annealing and laser heat treatment is evident where the magnetization increased upon furnace annealing and decreased upon laser heat treatment, while coercivity increased about threefold upon furnace annealing and remained almost unchanged upon laser heat treatment. This difference can be understood from the observations in Fig. 1 for Fe_2O_3 nanoparticles. When Fe_2O_3 nanoparticles were furnace annealed, it showed a decrease in magnetization with a slight increase in coercivity. If this change can occur at lower temperature than what is needed for FePt nanoparticles which shows both coercivity and magnetization increase upon furnace annealing, the laser heat treatment might be happening at much lower temperature that is just enough for Fe_2O_3 phase change but not high enough for FePt phase change.

We could not measure the temperature of the samples during laser heat treatment to confirm. The 40 W heating showed slightly higher magnetization compared to 20 W heating. This could be due to slightly higher temperature at higher wattage.

The magnetization curves for FePt nanoparticle layer between two layers of Fe_2O_3 nanoparticles on Cu substrate (figure not shown) are similar to that shown for Si substrate in Fig. 3 but almost no difference between 20 and 40 W laser heat treatments with slightly lower magnetization as shown in Table 1. This may be because of Cu

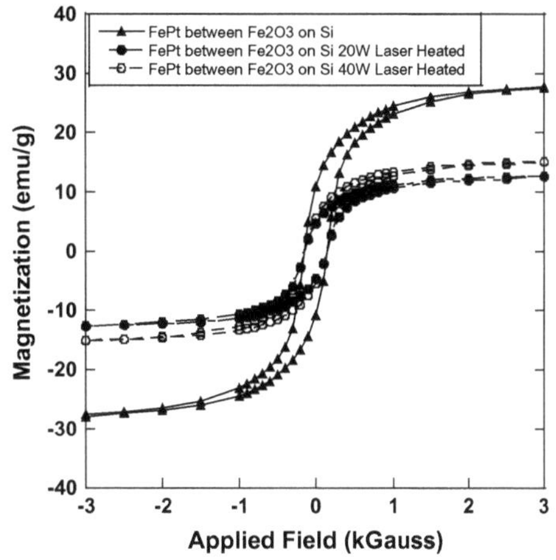

Fig. 3 Magnetization of FePt nanoparticle layer between two layers of Fe_2O_3 nanoparticles on Si substrate, before and after laser heat treatment

being a better thermal conductor compared to Si and the heat produced by laser is quickly dissipated to the substrate.

The surfactant started decomposing and producing a smoke during the laser heat treatment that may use up most of the laser energy. In addition, the surfactant is observed to reduce the magnetization [10]. So, in the second set, the surfactant is eliminated, and non-magnetic Cu nanoparticles were mixed with FePt nanoparticles using hexane. This made particles loose and some particles were lost during handing and laser heat treatment.

The magnetic hysteresis curves of FePt and FePt/Cu nanoparticle mixtures deposited on Cu substrates are shown in Figs. 4 and 5 for as-deposited samples and the laser heat-treated samples at 40 W or 80 W power. Figure 5 is a magnified view of Fig. 4 to show the coercivity comparison. Unfortunately, the magnetization values in Fig. 4 are not comparable between samples since the exact mass of FePt in the sample is not known for each sample as the deposited powder on the Cu plate came out loosely during handling and some powder escaped out during laser heating. There was no loss of particles from the surface during laser heat treatment at 40 W. However, most of the particles escaped from the surface during the 80 W laser heat treatment which may indicate that the temperature of the particles during 80 W laser heating reached close to vaporization temperature. Thus, the magnetization curves in Fig. 4 are just to show the nature of magnetic character qualitatively and should not be used to compare magnetization values. On the other hand, the coercivities seen in Fig. 5 are comparable as the coercivity is mostly independent of the mass of the sample.

The coercivity values for all samples are compared in Table 2. The coercivity values of laser heat-treated samples are almost the same as deposited samples or lower even for the 80 W laser power that is expected to have reached high temperatures for a very short time. The results indicate that even at very high temperatures the phase transformation requires some time span. The short time peri-

Table 1 Magnetization results of $FePt/Fe_2O_3$ nanoparticle structures upon furnace annealing and laser heating

Sample	H_c (Gauss)	M (emu/g) at 10 kGauss H
Nanoparticle layers on Si substrate		
Fe_2O_3 single layer	151.1	47.0
Furnace annealed single layer Fe_2O_3	208.1	28.6
FePt on Fe_2O_3	159.6	33.2
$Fe_2O_3/FePt/Fe_2O_3$	142.0	29.8
Furnace annealed $Fe_2O_3/FePt/Fe_2O_3$	365.5	54.1
20 W Laser heated $Fe_2O_3/FePt/Fe_2O_3$	142.3	13.7
40 W Laser heated $Fe_2O_3/FePt/Fe_2O_3$	143.0	16.2
Nanoparticle layers on Cu substrate		
Fe_2O_3 single layer	148.1	47.2
FePt on Fe_2O_3	152.4	32.3
$Fe_2O_3/FePt/Fe_2O_3$	148.4	25.1
Furnace annealed $Fe_2O_3/FePt/Fe_2O_3$	245.9	62.4
20 W Laser heated $Fe_2O_3/FePt/Fe_2O_3$	148.1	12.1
40 W Laser heated $Fe_2O_3/FePt/Fe_2O_3$	146.5	12.1

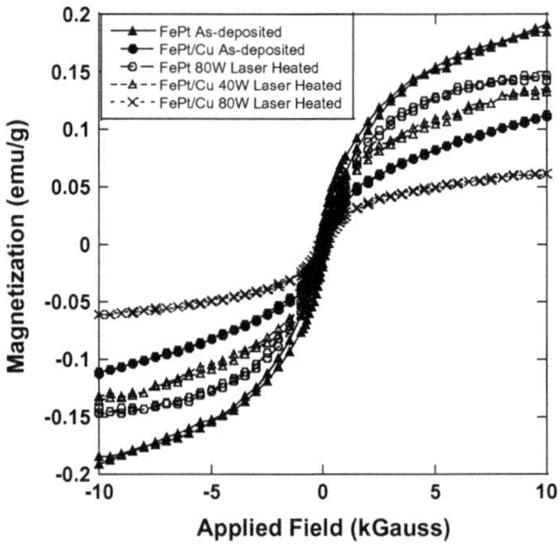

Fig. 4 Magnetization of FePt/Cu nanoparticles mixture, before and after laser heat treatment

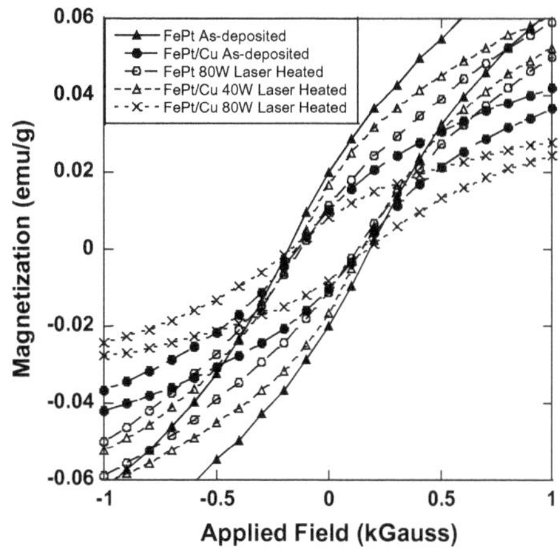

Fig. 5 Magnified view of Fig. 4

Table 2 Magnetic coercivity values of FePt and Cu (or Fe_2O_3) nanoparticles mixers before and after furnace annealing or laser heat treatment

Sample	H_c (Gauss)
FePt as-deposited	200
FePt furnace annealed	1350
FePt/Cu furnace annealed	300
FePt/Fe_2O_3 furnace annealed	365
FePt 80 W laser heat treated	130
FePt/Cu 40 W laser heat treated	150
FePt/Cu 80 W laser heat treated	170
FePt/Fe_2O_3 40 W laser heat treated	147

ods of laser heat treatment are not sufficient to bring phase transformation even at high temperatures of vaporization.

The magnetic hysteresis curves are shown in Figs. 6 and 7 for as-deposited and vacuum furnace annealed samples (also showing one laser heat-treated sample for comparison). Figure 7 is a magnified view of Fig. 6 to clearly show the smaller magnetic curves. The magnetization and coercivity values are much higher for furnace annealed samples compared to as-deposited and laser heat-treated samples, which indicate the phase transformation due to furnace annealing. Both magnetization and coercivity for furnace annealed FePt is higher than that of furnace annealed FePt/Cu. This agrees with the previous studies [11] which showed that the magnetization decreased by adding Cu to FePt. But, the coercivity increased by adding Cu to FePt in the temperature range up to 500 °C where coercivity reaches saturation; while for FePt without Cu coercivity value overtakes and continues increasing drastically. Another study [12] done at 600 °C showed that the FePt coercivity decreased with the increasing Cu content in FePt. Our results are in support of these findings especially for the laser heat-treated samples for which the expected temperatures are much higher (at 80 W) and the coercivity values are very low (Table 2).

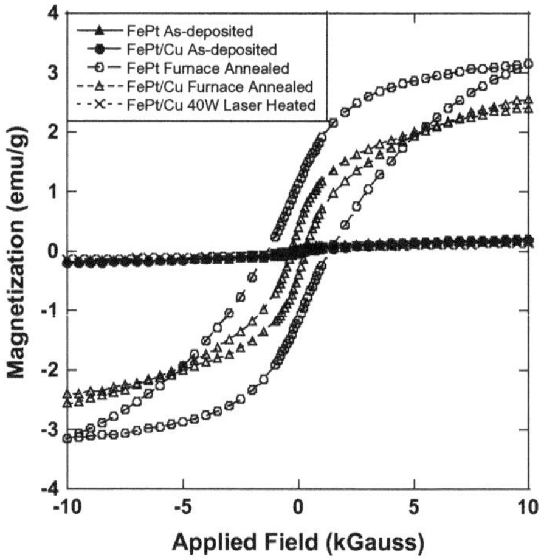

Fig. 6 Magnetization of FePt/Cu nanoparticles mixture, before and after furnace annealing or laser heat treatment

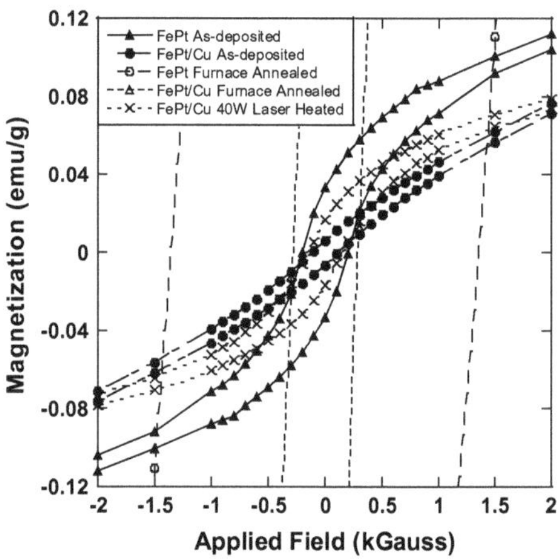

Fig. 7 Magnified view of Fig. 6

For the same reason of unknown mass of FePt nanoparticles particularly in the laser-heated samples, we do not compare magnetization values, but the coercivity values between the samples. The main distinction is that the coercivity increased about 7 times (see Table 2) by furnace annealing the FePt nanoparticles and 3 times by furnace annealing the FePt/Cu or FePt/Fe_2O_3 mixed nanoparticle samples. This clearly shows that the furnace annealing brings the phase transformation in pure FePt or mixed FePt with either nonmagnetic Cu nanoparticles or soft magnetic Fe_2O_3 nanoparticles, while almost no phase transformation is observed due to laser heat treatment.

Conclusions

Fe_2O_3 or Cu nanoparticles have been used to isolate FePt nanoparticles during furnace annealing or high-power laser heat treatment to minimize the aggregation of those FePt nanoparticles during heat treatment in order to get desired phase transformation for high magnetic storage media. The performance of laser-treated nanoparticles at different powers was investigated and compared to samples which were heated in a vacuum furnace. The coercivity increased about 7 times for the FePt nanoparticles and 3 times for the FePt/Cu or FePt/Fe_2O_3 mixed nanoparticle samples by furnace annealing. This clearly shows that the furnace annealing brought the phase transformation in pure FePt or mixed FePt with either nonmagnetic Cu nanoparticles or soft magnetic Fe_2O_3 nanoparticles while the laser heat treatment did not bring remarkable change to obtain ordered $L1_0$ phase transformation.

The results indicate that even at very high temperatures, the phase transformation requires some time span. The short time periods of laser heat treatment are not sufficient to bring about the phase transformation even at high temperatures of vaporization.

Acknowledgements This work is supported by the National Science Foundation through the NSF EPSCoR CIMM cooperative agreement award OIA-1541079 to the Louisiana Board of Regents.

References

1. Sun S, Murray CB, Weller D, Folk L, Moser A (2000) Monodisperse FePt nanoparticles and ferromagnetic FePt nanocrystal superlattices. Science 287(5460):1989–1992
2. Huang S, Liu P, Mokasdar A, Hou L (2013) Additive manufacturing and its societal impact. Int J Adv Manuf Technol 67(5–8):1191–1203
3. Schleifenbaum H, Meiners W, Wissenbach K, Hinke C (2010) Individualized production by means of high power selective laser melting. Cirp J Manuf Sci Technol 2:161–169
4. Harrell JW, Nikles DE, Kang SS, Sun XC, Jia Z, Shi S, Lawson J, Thompson GB, Srivastava C, Seetala NV (2005) Effect of metal additives on $L1_0$ ordering of chemically synthesized FePt nanoparticles. Scr Mater 53(4):411–416
5. Yi DK, Zafiropoulou I, Papaefthymiou GC (2012) Coercivity enhancement of FePt thin films coated with a monolayer array of γ-Fe_2O_3 nanoparticles. Thin Solid Films 522:376–380
6. Maeda T, Kai T, Kikitsu A, Nagase T, Akiyama J (2002) Reduction of ordering temperature of an FePt-ordered alloy by addition of Cu. Appl Phys Lett 80:2147–2149
7. Platt CL, Wierman KW, Svedberg EB, Veerdonk R, Howard JK, Roy AG, Laughlin DE (2002) $L1_0$ ordering and microstructure of FePt thin films with Cu, Ag, and Au additive. J Appl Phys 92:6104–6109
8. Cao D, Li H, Pan L, Li J, Wang X, Jing P, Cheng X, Wang W, Wang J, Liu Q (2016) High saturation magnetization of γ-Fe_2O_3 nano-particles by a facile one-step synthesis approach. Sci Rep 6:32360. https://doi.org/10.1038/srep32360
9. Xu XN, Wolfus Y, Shaulov A, Yeshurun Y, Felner I, Nowik I, Koltypin Y, Gedanken A (2002) Annealing study of Fe_2O_3 nanoparticles: magnetic size effects and phase transformations. J Appl Phys 91:4611–4616

10. Mahdavi M, Ahmad MB, Haron MJ, Namvar F, Nadi B, Rahman MZ, Amin J (2013) Synthesis, surface modification and characterisation of biocompatible magnetic iron oxide nanoparticles for biomedical applications. Molecules 18:7533–7548
11. Takahashi YK, Ohnuma M, Hono K (2002) Effect of Cu on the structure and magnetic properties of FePt sputtered film. J Magn Magn Mater 246:259–265
12. George TA, Li Z, Yan M, Xu Y, Skomski R, Sellmyer DJ (2008) Nanostructure and magnetic properties of $L1_0$ FePt: X films. J Appl Phys 103:07D502(1–3)

Part VI
Poster Session

High-Temperature Study of Perovskite Evaporation

Sergey Shornikov

Abstract Evaporation of perovskite was studied by high-temperature Knudsen effusion mass spectrometric method at 1700–2200 K. Vapor species typical for simple oxides and a small amount of $CaTiO_3$ complex gaseous oxide were identified in the gas phase over perovskite. For the first time, the partial pressure of vapor species in the gas phase over perovskite are compared with those corresponding to simple oxides, showing the predominant effect of the calcium component of perovskite on evaporation character.

Keywords Knudsen effusion mass spectrometry
Thermodynamics of evaporation · Perovskite

Introduction

Calcium titanate $CaTiO_3$ (perovskite) is a relatively rare mineral on the Earth's crust, but it is a promising material as a matrix for reliable and long-term storage of actinides and rare earth elements present in radioactive waste [1]. Because of its high melting temperature and a number of specific physicochemical properties, perovskite is also an important material in the manufacturing of new types of high-refractory structural pure oxide ceramic materials that can find application in solar energy [2]. Perovskite is of interest for petrology, as well as for cosmochemistry, as a mineral composing refractory Ca–Al-rich inclusion in chondrites which are the earliest object of the Solar System with unusual isotopic characteristics [3, 4]. Two calcium titanates—$Ca_3Ti_2O_7$ and $Ca_4Ti_3O_{10}$, melting incongruently, were also found in the $CaO–TiO_2$ system (Fig. 1).

Evaporation processes and thermodynamic properties of simple oxides—CaO and TiO_2 were discussed in the details in reference books [6, 7].

S. Shornikov (✉)
Vernadsky Institute of Geochemistry and Analytical Chemistry, Russian Academy of Sciences, Moscow, Russia
e-mail: sergey.shornikov@gmail.com

G. Lambotte et al. (eds.), *Materials Processing Fundamentals 2019*, The Minerals, Metals & Materials Series, https://doi.org/10.1007/978-3-030-05728-2_23

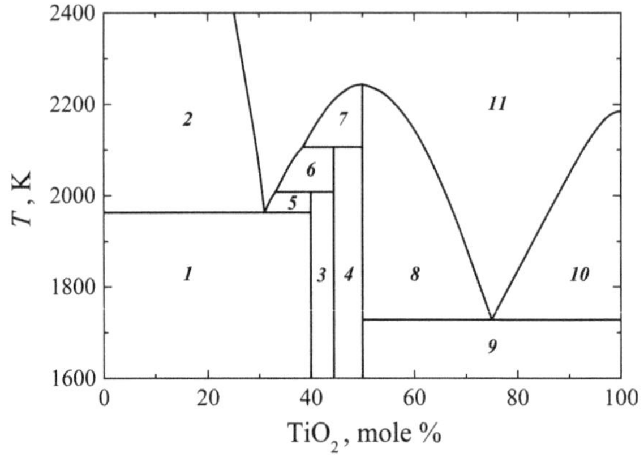

Fig. 1 The phase diagram of CaO–TiO_2 system [5]: ***1***—CaO + $Ca_3Ti_2O_7$; ***2***—CaO + liquid; ***3***—$Ca_3Ti_2O_7$ + $Ca_4Ti_3O_{10}$; ***4***—$Ca_4Ti_3O_{10}$ + $CaTiO_3$; ***5***—$Ca_3Ti_2O_7$ + liquid; ***6***—$Ca_4Ti_3O_{10}$ + liquid; ***7*** and ***8***—$CaTiO_3$ + liquid; ***9***—$CaTiO_3$ + TiO_2; ***10***—TiO_2 + liquid; ***11***—liquid

The gas phase over calcium oxide consists of vapor species: (O), (O_2), (O_3), (O_4), (Ca), (Ca_2) and (CaO), which can be formed by the following reactions:

$$[\mathrm{CaO}] = (\mathrm{CaO}) \tag{1}$$

$$(\mathrm{CaO}) = (\mathrm{Ca}) + (\mathrm{O}) \tag{2}$$

$$2(\mathrm{Ca}) = (\mathrm{Ca_2}) \tag{3}$$

$$2(\mathrm{O}) = (\mathrm{O_2}) \tag{4}$$

$$3(\mathrm{O}) = (\mathrm{O_3}) \tag{5}$$

$$4(\mathrm{O}) = (\mathrm{O_4}) \tag{6}$$

(here and further—the condensed phase is designated by square brackets, the gas phase is designated by round brackets).

In the gas phase over titanium oxide, similar vapor species are present: (O), (O_2), (O_3), (O_4), (Ti), (Ti_2), (Ti_3), (TiO) and (TiO_2), which are formed by similar reactions

$$[\mathrm{TiO_2}] = (\mathrm{TiO_2}) \tag{7}$$

$$(\mathrm{TiO_2}) = (\mathrm{TiO}) + (\mathrm{O}) \tag{8}$$

$$(\mathrm{TiO}) = (\mathrm{Ti}) + (\mathrm{O}) \tag{9}$$

$$2(\mathrm{Ti}) = (\mathrm{Ti_2}) \tag{10}$$

$$3(\mathrm{Ti}) = (\mathrm{Ti_3}). \tag{11}$$

Balducci et al. [8] using Knudsen effusion mass spectrometry found (Ti_2O_3) and (Ti_2O_4) vapor species in the gas phase over $CoTiO_3$ at 2210–2393 K. They can participate in the following equilibria:

$$2(TiO_2) = (Ti_2O_4) \tag{12}$$

$$(Ti_2O_4) = (Ti_2O_3) + (O). \tag{13}$$

Note that, predominant vapor species in the gas phase over these oxides are (Ca), (CaO), (TiO), (TiO_2), (O) and (O_2), the total amount of the remaining vapor species does not exceed 1% at 1700–2200 K.

The properties of the gas phase over perovskite are not studied in detail. The brief summary of the results of high-temperature study and the experiment conditions on perovskite evaporation is given below.

Using the differential mass spectrometry, Banon et al. [9] investigated the evaporation of 24 samples of $CaTiO_3$–Ti_2O_3–TiO_2 system from Knudsen molybdenum effusion cell at 1900–2200 K. Mass spectra were recorded at low ionizing voltage of 13 eV in order to avoid the possible fragmentation of TiO_2^+ molecular ion to Ti^+ and TiO^+ ions, which are also molecular. The predominant vapor species of the gas phase over the compositions studied was atomic calcium. The ($CaTiO_3$) complex gaseous oxide was not found. The partial pressure values of (Ca), (TiO), (TiO_2) and (O) over perovskite at 2150 K was calculated using thermochemical data [7]. The logarithm of partial pressures is shown as a function of the inverse temperature in Fig. 2 (for ease of understanding, the temperature was scaled accordingly).

The gaseous $CaTiO_3$ was not detected in mass spectrometric studies during high-temperature evaporation of CaO–TiO_2–SiO_2 melts using Knudsen effusion molybdenum and tungsten cells at 1700–2500 K [14–17]. This might be due to either low sensitivity of the equipment used to determine the $CaTiO_3^+$ ion [14, 17], or it was not the purpose of research [15, 16].

Lopatin and Semenov [18] studied the evaporation of a mixture of calcium carbonate and titanium dioxide from tungsten cells by Knudsen effusion mass spectrometric method at 2100–2500 K. The following ions were found in the mass spectra of vapor over the mixture: Ca^+, CaO^+, Ti^+, TiO^+, TiO_2^+ and $CaTiO_3^+$. The appearance energies of the ions in the mass spectra of vapor over the mixture allowed to establish the molecular nature of the origin of Ca^+, CaO^+, TiO^+, TiO_2^+ and $CaTiO_3^+$ ions. TiO^+ ion also contained a fragmentary component from TiO_2^+ ion. According to the authors, Ti^+ ion was completely fragmentary. The appearance energy of $CaTiO_3^+$ molecular ion was determined to be 9 ± 1 eV (the appearance energy of a gold ion in the mass spectrum of vapor was used as a standard). The values of partial pressures of vapor species (p_i) were calculated from a comparison with the accepted values of the partial pressure of gold vapor taken as standard (p_s) by the relation

$$p_i = \frac{I_i T_i}{I_s T_s} p_s \times \frac{\sigma_s \gamma_s \eta_s}{\sigma_i \gamma_i \eta_i}, \tag{14}$$

where I_i (I_s) is the intensity of the ion current of ith vapor species (standard substance) recorded at temperature of T_i (T_s). The calculation should also take into account the ratio of the ionization effective cross sections of the ith vapor species and the standard substance (σ_i/σ_s), isotope distributions (η_i/η_s) and individual ion efficiencies (γ_i/γ_s), depending on the various parameters of devices recording ion currents. The p_{CaO},

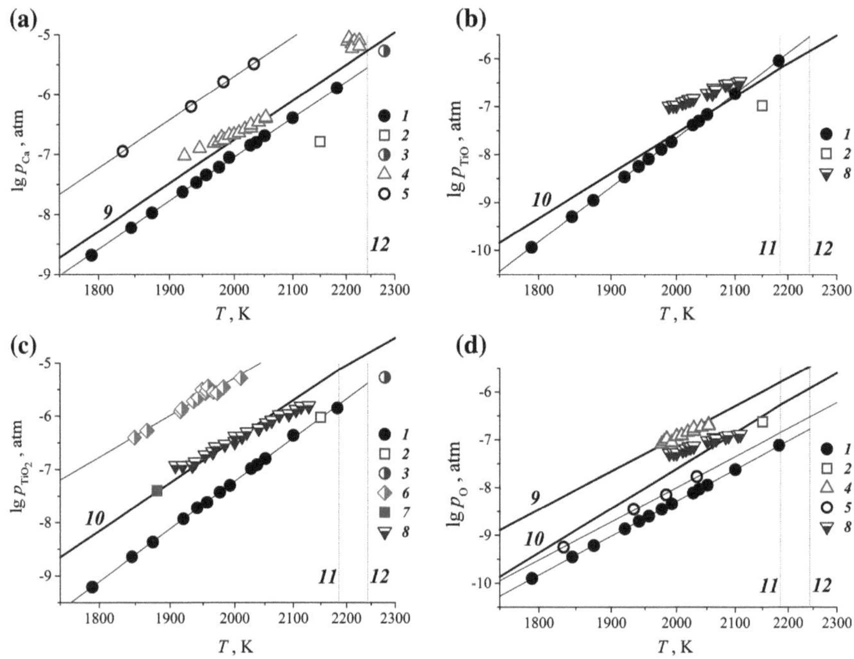

Fig. 2 Partial pressure of vapor species of Ca (**a**), TiO (**b**), TiO_2 (**c**) and O (**d**) over perovskite (***1–3***) and oxides of calcium (***4***, ***5***, ***9***) and titanium (***6–8***, ***10***) *versus* the inverse temperature, determined by Knudsen mass spectrometric effusion method: ***1***—at the present study; ***2***—Banon et al. [9]; ***4***—Samoilova and Kazenas [10]; ***5***—Shornikov [11]; ***6***—Groves et al. [12]; ***7***—Berkowitz et al. [13]; ***8***—Kazenas and Tsvetkov [6]; and using the vacuum furnace by Langmuir method: ***3***—Zhang et al. [4]; and calculated: ***9*** and ***10***—at the present study according to the thermochemical data [7]; the vertical dashed lines (***11***) and (***12***) indicate melting points of titanium oxide and perovskite, respectively

p_{TiO_2} and p_{CaTiO_3} values, calculated from the relation (14), were used to determine the temperature dependence of the gas phase equilibrium constant (Fig. 3)

$$(CaTiO_3) = (CaO) + (TiO_2) \tag{15}$$

and to calculate the values of the enthalpy of formation ($\Delta_f H_{298}$) and atomization ($\Delta_{at} H_{298}$) of $CaTiO_3$ vapor species.

Zhang et al. [4] studied the isotopic fractionation during evaporation of perovskite melt suspended on iridium wire in a vacuum furnace at 2278 K (according to Langmuir). The compositional changes of the residual melt during perovskite evaporation indicated the predominant evaporation of calcium components of the melt. It was possible to estimate the total vapor pressure over perovskite from the obtained data, which is also shown in Fig. 2.

Thus, the available information on the evaporation processes of perovskite is rather scarce and is limited to the established gas phase composition and the separate

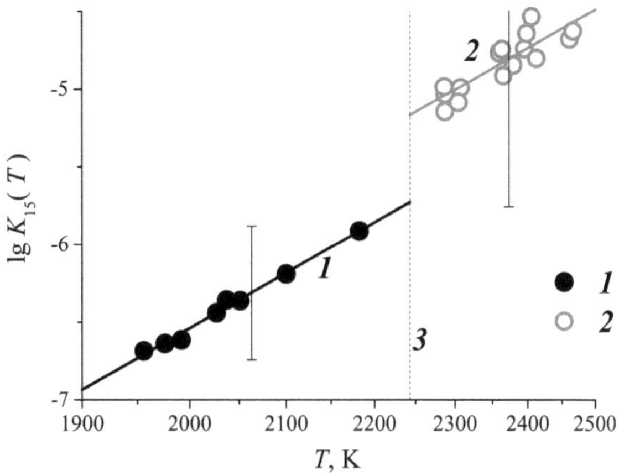

Fig. 3 The reaction equilibrium constant $(CaTiO_3) = (CaO) + (TiO_2)$ *versus* the inverse temperature determined by Knudsen mass spectrometric effusion method: ***1***—at the present study; ***2***—Lopatin and Semenov [18]; the vertical dashed line (***3***) denotes the melting point of perovskite

definitions of the partial pressure values of vapor species and the total vapor pressure over perovskite at different temperatures. The purpose of the present study was to obtain a reliable and the detailed thermodynamic data characterizing the high-temperature perovskite evaporation.

Experimental

The evaporation of perovskite from molybdenum effusion cell (according to Knudsen) was studied by the mass spectrometric method at 1700–2200 K. The used mass spectrometer MI–1201 was equipped with a modified ion source for high-temperature studies (up to 3000 K). The design of the developed ion source and the features of the mass spectrometric experiment are described in detail earlier [19].

Perovskite synthesis was performed according to ceramic technology followed by X-ray and chemical analysis of samples. $CaCO_3$ and TiO_2 were used as a starting material for the preparation of the required composition. They were mixed in a Jasper ball mill in acetone for 24 h. After drying, the resulting mixture was calcined in a silite furnace in air for 24 h at a temperature of up to 1300 K. After secondary grinding, high-temperature firing of the sample was carried out for 24 h at 1530 K.

In the mass spectra of vapor over perovskite at the ionizing electron energy of 20 eV, TiO_2^+, Ca^+, TiO^+ and O^+ ions prevailed, as well as other ions characteristic of the mass spectra over individual oxides [6, 7]. A small amount of $CaTiO_3^+$ ion was observed, which was fragmented by about a quarter into $CaTi^+$, $CaTiO^+$ and $CaTiO_2^+$ ions (I_{CaTi}: I_{CaTiO}: I_{CaTiO_2}: I_{CaTiO_3} = 6: 10: 13: 100). The ratio of the ion current intensities in the mass spectra of vapor over perovskite at 2182 K was the following: I_{Ca}: I_{CaO}: I_{Ti}: I_{TiO}: I_{TiO_2}: I_{CaTiO_3}: I_O: I_{O_2} = 80: 0.04: 0.1: 75: 100: 0.02: 40: 0.3. It corresponded to that observed by Samoilova and Kazenas [10] in the same temperature range at evaporation of CaO from alundum cell and by Semenov [20] at evaporation of TiO_2 from a tungsten cell.

The presence of MoO_i^+ ($i = 0$–3) ions in the mass spectra of vapor over perovskite were due to the evaporation of molybdenum cell at high temperature:

$$[Mo] = (Mo), \tag{16}$$

as well as the interaction of perovskite with the cell material ($I_{TiO_2}: I_{Mo}: I_{MoO}: I_{MoO_2}: I_{MoO_3} = 100: 0.8: 2.6: 7.3: 0.6$) that was detected according to the following equilibria:

$$(Mo) + (O) = (MoO) \tag{17}$$

$$(MoO) + (O) = (MoO_2) \tag{18}$$

$$(MoO_2) + (O) = (MoO_3). \tag{19}$$

Note that, Berkowitz et al. [13] found that during evaporation of titanium oxide from a molybdenum liner inserted into a tantalum crucible at 1881 K, p_{MoO_2} was initially 10–10^2 times higher than p_{TiO_2}. The value of p_{MoO_2} then gradually decreased and became comparable with p_{TiO_2} which is significantly different from the results of the present study. The high p_{MoO_2} observed in [13] probably was due to poor quality of the molybdenum liner material (or its alloy). Possibly it was made using powder technology from MoO_3 reduced to metallic molybdenum at temperatures of ~1300 K, which could lead to such an excess of partial pressure of (MoO_3) and its subsequent decrease as it evaporates from the surface layers of the liner material.

The appearance energies of ions in the mass spectra of vapor over perovskite were determined by the Warren method [21] and corresponded to the accepted values of the ionization energies of atoms and molecules typical for simple oxides [22]. The appearance energy of $CaTiO_3^+$ ion in the mass spectra of vapor over perovskite was equal to 8.5 ± 0.6 eV (the appearance energy of silver ion in the mass spectrum of vapor was used as a standard) and corresponded to that obtained by Lopatin and Semenov [18].

The established molecular composition of the gas phase over perovskite indicated on the predominant evaporation of perovskite according to the reactions (1), (2), (4), (7), (8) and (9), typical for evaporation of simple oxides [6, 7, 10, 20]. The presence of a small amount of ($CaTiO_3$) molecules in the gas phase over perovskite is probably due to the reaction:

$$[CaTiO_3] = (CaTiO_3). \tag{20}$$

The partial pressure values of vapor species in the gas phase over perovskite were determined by the Hertz–Knudsen equation, written in the following form [23]:

$$p_i = K_\alpha \frac{q_i}{s_{or} C_{or} t} \sqrt{\frac{2\pi RT}{M_i}}, \tag{21}$$

where q_i is the amount of ith component of the substance with the molecular weight (M_i), evaporated from the effusion cell during time (t) at temperature (T) through the effusion hole, characterized by the Clausing coefficient (C_{or}) and the hole area (s_{or}).

The K_α constant value was calculated, as in [20], taking into account the evaporation coefficient (α_i) of substance component associated with the molecule changing during its transition to the gas phase from the surface with an S_v area, using the Komlev equation [24]

$$K_\alpha = \frac{1}{C_{or}} + s_{or}\frac{1 - C_c\alpha_i}{S_v\alpha_i C_c}, \tag{22}$$

where C_c is the Clausing coefficient for the effusion cell.

The Clausing coefficient is associated with the collision of vapor species inside the effusion orifice channel of effusion cell and their reverse reflection from the channel walls. Its value does not exceed one and depends on the ratio of the diameter of the effusion hole to its thickness. Taking into account predominance of typical for CaO and TiO_2 vapor species in the gas phase over perovskite and small amounts of $CaTiO_3$, partial evaporation coefficients from [25] were used.

The partial pressures of vapor species over perovskite calculated using the relationships (22) and (23) with an error not exceeding 8% are shown in Table 1.

The partial pressure of atomic oxygen determined using these relationships agrees satisfactorily with those calculated using the thermochemical data [7] on $K_r(T)$ equilibrium constants of possible reactions (2), (4), (8), (9) and (20) in the gas phase over perovskite in the following relations:

$$p_O = \frac{p_{CaO}}{p_{Ca}} K_2(T) \tag{23}$$

$$p_O = \sqrt{\frac{p_{O_2}^2}{K_4(T)}} \tag{24}$$

$$p_O = \frac{p_{TiO_2}}{p_{TiO}} K_8(T) \tag{25}$$

$$p_O = \frac{p_{TiO}}{p_{Ti}} K_9(T) \tag{26}$$

$$p_O = \frac{p_{MoO_3}}{p_{MoO_2}} K_{19}(T). \tag{27}$$

It should be noted that the established correspondence of p_O values calculated according to the independent reactions in the gas phase over perovskite confirms the assumption about the molecular origin of the identified ions in the mass spectrum of vapor over perovskite.

Thus the defined partial pressures of vapor species over perovskite, as it follows from Fig. 2, can be represented as linear logarithmic dependence versus the inverse temperature as follows:

Table 1 Partial pressures of vapor species over perovskite (atm)

T, K	p_{Ca}	p_{CaO}	p_{Ti}	p_{TiO}	p_{TiO_2}	p_{O_2}	p_{CaTiO_3}
1791	2.08×10^{-9}	–	–	1.16×10^{-10}	6.19×10^{-10}	–	–
1845	5.96×10^{-9}	–	–	5.09×10^{-10}	2.28×10^{-9}	–	–
1875	1.07×10^{-8}	–	–	1.12×10^{-9}	4.31×10^{-9}	6.28×10^{-12}	–
1920	2.37×10^{-8}	3.93×10^{-12}	–	3.43×10^{-9}	1.17×10^{-8}	1.23×10^{-11}	–
1941	3.39×10^{-8}	7.06×10^{-12}	–	5.65×10^{-9}	1.89×10^{-8}	1.85×10^{-11}	–
1956	4.54×10^{-8}	9.71×10^{-12}	4.56×10^{-12}	8.09×10^{-9}	2.40×10^{-8}	2.45×10^{-11}	1.12×10^{-12}
1976	6.11×10^{-8}	1.41×10^{-11}	7.96×10^{-12}	1.29×10^{-8}	3.73×10^{-8}	3.56×10^{-11}	2.28×10^{-12}
1992	8.89×10^{-8}	1.83×10^{-11}	1.29×10^{-11}	1.85×10^{-8}	5.01×10^{-8}	4.78×10^{-11}	3.77×10^{-12}
2027	1.42×10^{-7}	3.37×10^{-11}	3.30×10^{-11}	4.14×10^{-8}	1.04×10^{-7}	8.93×10^{-11}	9.66×10^{-12}
2037	1.58×10^{-7}	4.84×10^{-11}	4.03×10^{-11}	5.02×10^{-8}	1.23×10^{-7}	1.06×10^{-10}	1.36×10^{-11}
2051	2.04×10^{-7}	6.24×10^{-11}	5.78×10^{-11}	6.90×10^{-8}	1.60×10^{-7}	1.35×10^{-10}	2.30×10^{-11}
2100	4.06×10^{-7}	1.58×10^{-10}	2.03×10^{-10}	1.88×10^{-7}	4.06×10^{-7}	3.08×10^{-10}	9.98×10^{-11}
2182	1.28×10^{-6}	6.55×10^{-10}	1.33×10^{-9}	9.16×10^{-7}	1.41×10^{-6}	1.12×10^{-9}	7.57×10^{-10}

$$\lg p_i = a_i/T + b_i. \tag{28}$$

Note that, the relationship (28) is the same as the expression for the reaction constant [7]:

$$R \ln K_r(T) = -\frac{\Delta_r G_T}{T} = -\frac{\Delta_r H_T}{T} + \Delta_r S_T, \tag{29}$$

which allows to determine the enthalpy ($\Delta_r H_T$) and entropy ($\Delta_r S_T$) of a reaction.

Results and Discussion

The partial pressures of the predominant vapor species of the gas phase over perovskite (Ca, TiO, TiO_2 and O) determined in this study are compared in Fig. 2 with those obtained in [4, 9], as well as with the results on evaporation of simple oxides

(CaO and TiO_2), under similar reducing conditions caused by the interaction of oxygen with molybdenum [11, 13], tungsten [6] and tantalum [12] effusion cells or in chemically neutral conditions (in the absence of this interaction) for alundum cell [10].

In the case of [9], the partial pressure values of vapor species over the perovskite at a temperature of 2150 K were calculated using TiO and TiO_2 activities and the Gibbs energy of perovskite (ΔG^{m}_{2150}) and thermochemical data [9] on equilibriums (1), (2), (7) and (8). Therefore, the obtained values characterize the evaporation of perovskite not in reducing conditions (from molybdenum cell), but, in contrary, in chemically neutral conditions (in the absence of interaction of perovskite with the cell material).

Figure 2 also shows the partial pressures of vapor species of the gas phase over calcium and titanium oxides calculated using thermochemical data of [7]. By comparing the experimental data obtained in [6, 10–13] and in this study, and the calculated results, one can see the effect of reducing properties of cell materials on gas phase speciation: tantalum [12], molybdenum [11, 13], tungsten [6] and alundum [10]. As we noted earlier [26], the greatest effect of cell materials on the vapor speciation is observed with the oxygen "deficient" species such as atomic calcium (Fig. 2a), titanium monoxide (Fig. 2b) and atomic oxygen itself (Fig. 2d). In the case of gaseous titanium dioxide (Fig. 2c), there are no differences in the values determined in evaporation experiments using molybdenum [13] and tungsten [6] cells.

The total vapor pressure over the perovskite melt at 2278 K obtained by Zhang et al. [4] is consistent with the extrapolated values of partial pressures of the predominant vapor species of the gas phase—atomic calcium and titanium dioxide (Fig. 2a, c) found in the present study. It is interesting to note slight changes in the slope of lg p_{TiO_2} versus the inverse temperature in Fig. 2c at the melting points of titanium oxide (2185 ± 10 K) and perovskite (2244 ± 10 K), obtained at present study and [4]. It suggests insignificant differences in the values of the enthalpy of melting of TiO_2 and perovskite.

Similar slopes of lg p_{Ca} and lg p_O versus the inverse temperature for calcium oxide and perovskite in Fig. 2a and 2d (lines **1**, **5** and **9**, respectively) indicate a predominant effect of calcium component on evaporation of calcium from perovskite. This also can explain the difference in the slope of lg p_{TiO} in Fig. 2b in the case of perovskite (line **1**) and rutile (line **10**) according to the equilibriums (2) and (8).

Figure 3 shows that the slope of the temperature dependences of the equilibrium constant of reaction (15) obtained in the present study is the same as determined by Lopatin and Semenov [18]. Some discrepancy between the absolute values of $K_{15}(T)$, which are within the calculation errors, is probably due to systematic deviations in calculations of partial pressures via relationship (14) used by the authors [18]. The possible cause of these deviations is approximate of estimation of the effective $CaTiO_3^+$ ionization cross section calculated on the additivity rule, and the uncertainty in the detecting efficiency of $CaTiO_3^+$ by the electron multiplier. The isotopic fractionation of perovskite [4] also accompanies the evaporation process and it will impact, although insignificantly, on the isotope distribution value used in the relation (14).

Table 2 Enthalpy and entropy of reactions involving $CaTiO_3$ vapor species

Reaction	T, K	$\Delta_r H_{298}$, kJ/mole	$\Delta_r H_T$, kJ/mole	$\Delta_r S_T$, J/(K × mole)	References
$(CaTiO_3) = (CaO) + (TiO_2)$	2287–2466	545 ± 8	284 ± 44	28 ± 19	[18]
	2000	–	298 ± 30	–	[26]
	1956–2182	–	287 ± 12	18 ± 6	The present study
$[Ca] + [Ti] + 3/2(O_2) = (CaTiO_3)$	2287–2466	–826 ± 26	–	–	[18]
	1956–2182	–	–760 ± 10	–242 ± 5	The present study
$(CaTiO_3) = (Ca) + (Ti) + 3(O)$	2287–2466	2225 ± 26	–	–	[18]
	2000	–	1983 ± 81	–	[26]
	1956–2182	–	1993 ± 15	396 ± 7	The present study
$[CaTiO_3] = (CaTiO_3)$	2000	–	1030 ± 22	–	[26]
	1956–2182	–	1027 ± 10	297 ± 5	The present study

The enthalpy and entropy of reactions involving $(CaTiO_3)$ gaseous complex oxide calculated by the relationship (29) are given in Table 2. They are in agreement with those found by Lopatin and Semenov [18] and our earlier estimates [26].

Conclusion

Evaporation of perovskite from a molybdenum cell was studied by Knudsen effusion mass spectrometry at 1700–2200 K. Vapor species typical for simple oxides were identified in the gas phase over perovskite along with small amount of $CaTiO_3$ complex gaseous oxide. For the first time, the partial pressures of vapor species in the gas phase over perovskite were determined. The comparison of these values with the available experimental literature data and those of simple oxides shows the predominant effect of the calcium component of perovskite on its evaporation behavior.

References

1. Stefanovsky SV, Yudintsev SV (2016) Titanates, zirconates, aluminates and ferrites as waste forms for actinide immobilization. Russ Chem Rev 85(9):962–994
2. Coontz R (2013) Newcomer juices up the race to harness sunlight. Science 342(6165):1438–1439

3. Wark D, Boynton WV (2001) The formation of rims on calcium-aluminum-rich inclusions: step I—flash heating. Met Planet Sci 36(8):1135–1166
4. Zhang J, Huang S, Davis AM, Dauphas N, Hashimoto A, Jacobsen SB (2014) Calcium and titanium isotopic fractionations during evaporation. Geochim Cosmochim Acta 140:365–380
5. Kirschen M, DeCapitani C (1999) Experimental determination and computation of the liquid miscibility gap in the system $CaO–MgO–SiO_2–TiO_2$. J Phase Equil 20(6):593–611
6. Kazenas EK, Tsvetkov YV (1997) Evaporation of the oxides. Nauka, Moscow
7. Chase MW (1998) NIST-JANAF themochemical tables
8. Balducci G, Gigli G, Guido M (1985) Identification and stability determinations for the gaseous titanium oxide molecules Ti_2O_3 and Ti_2O_4. J Chem Phys 83(4):1913–1916
9. Banon S, Chatillon C, Allibert M (1981) Free energy of mixing in $CaTiO_3–Ti_2O_3–TiO_2$ melts by mass spectrometry. Can Met Q 20(1):79–84
10. Samoilova IO, Kazenas EK (1995) Thermodynamics of dissociation and sublimation of calcium oxide. Russ Met 1:33–35
11. Shornikov SI (1993) Processes of vaporization and thermodynamic properties of the $CaO–Al_2O_3–SiO_2$ system and the materials based on this system. Thesis, Inst Silicate Chem RAS, Saint-Petersburg
12. Groves WO, Hoch M, Johnston HL (1955) Vapor–solid equilibria in the titanium–oxygen system. J Phys Chem 59(2):127–131
13. Berkowitz J, Chupka WA, Inghram MG (1957) Thermodynamics of the $Ti–Ti_2O_3$ system and the dissociation energy of TiO and TiO_2. J Phys Chem 61(11):1569–1572
14. Archakov IY, Shornikov SI, Tchemekova TY, Shultz MM (2000) The behavior of titanium dioxide in the slag melts. In: Proceedings of the 9th world conference on titanium, vol 3. pp 1464–1468
15. Shornikov SI, Archakov IY, Shultz MM (2000) Thermodynamic properties of the melts, containing titanium dioxide. In: Proceedings of the 9th world conference on titanium, vol 3. pp 1469–1473
16. Ostrovski O, Tranell G, Stolyarova VL, Shultz MM, Shornikov SI, Ishkildin AI (2000) High-–temperature mass spectrometric study of the $CaO–TiO_2–SiO_2$ system. High Temp Mater Proc 19(5):345–356
17. Stolyarova VL, Zhegalin DO, Stolyar SV (2004) Mass spectrometric study of the thermodynamic properties of melts in the $CaO–TiO_2–SiO_2$ system. Russ Glass Phys Chem 30(2):142–150
18. Lopatin SI, Semenov GA (2001) Thermochemical study of gaseous salts of oxygen-containing acids: XI. alkaline-earth metal titanates. Russ J General Chem 71(10):1522–1526
19. Shornikov SI, Archakov IY, Chemekova TY (2000) A mass spectrometric study of vaporization and phase equilibria in the $Al_2O_3–SiO_2$ system. Russ J Phys Chem A 74(5):677–683
20. Semenov GA (1969) The evaporation of titanium dioxide. Russ Inorg Mater 5(1):67–73
21. Warren JW (1950) Measurement of appearance potentials of ions produced by electron impact, using a mass spectrometer. Nature 165(4203):810–811
22. Gurvich LV, Karachevtsev GV, Kondratev VN, Lebedev YA, Medvedev VA, Potapov VK, Khodeev YS (1974) Bond dissociation energies, ionization potentials, and electron affinity. Nauka, Moscow
23. Shornikov SI (2002) Thermodynamic study of the mullite solid solution region in the $Al_2O_3–SiO_2$ system by mass spectrometric techniques. Geochem Int 40:S46–S60
24. Komlev GA (1964) On determination of saturated steam pressure by effusion method. Russ J Phys Chem 38(11):2747–2748
25. Shornikov SI (2015) Vaporization coefficients of oxides contained in the melts of Ca–Al–inclusions in chondrites. Geochem Int 53(12):1080–1089
26. Shornikov SI, Yakovlev OI (2015) Study of complex molecular species in the gas phase over the $CaO–MgO–Al_2O_3–TiO_2–SiO_2$ system. Geochem Int 53(8):690–699

Power Consumption Model for Electrolytic Preparation of Copper Powders Using Response Surface Methodology

Hongdan Wang, Wentang Xia and Bingzhi Ren

Abstract Electrolytic production of copper powders is a process with a high power consumption. This study establishes a power consumption model for copper powder electrolysis using the response surface methodology (RSM) under laboratory conditions, which provides the basis for reducing power consumption. First, seven process parameters were screened out using the Plackett–Burman design (PBD) experiments. The results show that the factors that have a significant effect on the power consumption of copper powder electrolysis are electrolyte temperature, Cu^{2+} concentration, H_2SO_4 concentration, inter-electrode spacing, and current density. A quadratic mathematical model of significant factors and power consumption was then developed using the Box–Behnken design (BBD) of RSM. Finally, the model was used to optimize the most energy-efficient process conditions. In addition, scanning electron microscopy (SEM) analysis indicates that the morphologies of electrolytic copper powders deposited under the optimized conditions generally have dendritic structure and the agglomerated copper particles are almost globular.

Keywords Electrolytic copper powders · Power consumption
Response surface methodology · Box–Behnken design · Plackett–Burman design

Introduction

Electrolytic copper powders are dendritic, that is, have a branched form, and possess such advantageous properties as high purity, specific surface area, and low apparent density. Electrolysis is one of the principal methods for producing copper powders, and copper powder electrolysis is similar to copper electrorefining [1]. In both processes, the anode and cathode are both made of copper plate, and a mixed solution of copper sulfate and sulfuric acid is used as the electrolyte. However, unlike in copper

H. Wang · W. Xia · B. Ren (✉)
School of Metallurgical and Materials Engineering, Chongqing University of Science and Technology, Chongqing 401331, P.R. China
e-mail: renbingzhi@163.com

G. Lambotte et al. (eds.), *Materials Processing Fundamentals 2019*, The Minerals, Metals & Materials Series, https://doi.org/10.1007/978-3-030-05728-2_24

electrorefining, copper powder electrolysis uses a larger current density and a lower cupric ion concentration, so that hydrogen evolution and copper electrowinning occur at the cathode simultaneously, thus producing fine and loose copper powders.

For a long time, researchers have focused on the influence of process parameters on copper powder morphology [2–6], particle size [6–8], apparent density [2, 4], etc., while research into electrolytic power consumption has been less extensive [9, 10]. Guo et al. [9] studied the effect of process parameters on the power consumption and grain size of electrolytic copper powders by the insoluble anodic electrolysis process using the single factor method and obtained the optimum process parameters. Akbarzadeh and Shakib [10] studied the influence of process parameters on the apparent density and power consumption of electrolytic copper powders by electrorefining and electrowinning, respectively. However, a satisfactory mathematical model describing a functional relationship between the power consumption of electrolytic copper powders and the process parameters has not yet been established.

The power consumption of electrolytic copper powders is calculated as follows:

$$W = \frac{1000U}{1.1852\eta} \tag{1}$$

where W is the direct current (DC) power consumption, kW h t^{-1}; U is the cell voltage, V; and η is the current efficiency, which refers to the ratio of the weight of cathode copper powder to that of the theoretical yield.

Equation (1) shows that the electrolytic power consumption is proportional to the cell voltage and inversely proportional to the current efficiency. However, electrolytic copper powders are produced at high current density and low cupric ion concentration. High current density will inevitably lead to high cell voltage. The cupric ion concentration of the electrolyte is low and hydrogen evolution occurs at the cathode, resulting in a decrease in current efficiency. These two effects combine to make the power consumption of copper powder electrolysis much higher than that of copper electrorefining. Therefore, it is necessary to establish a suitable power consumption model for copper powder electrolysis that will provide the basis for energy conservation.

Plackett–Burman design (PBD) is a two-level experimental design method based on the principle of non-complete equilibrium block, which can estimate the main effect of the factors with the minimum number of experiments and quickly and effectively select the most important factors from many factors for further study [11].

Response surface methodology (RSM) is based on a reasonable experimental design and gathers its data from experiments. A multiple quadratic regression equation is used to fit the functional relationship between the different factors and the response values. The optimal process parameters are obtained by analyzing the regression equation using a statistical method to solve the multivariable problem [12, 13]. RSM has been proven to be an economical and effective method for modeling and optimization of various processes [11, 14–21]. However, RSM approach has not been used to optimize copper powder eletrolysis process. Conventionally, RSM is applied after the vital controllable factors are determined and to find the factor settings that optimize the response [11, 21].

Based on the above-mentioned facts, in this paper Plackett–Burman design is used to screen out the significant factors affecting the power consumption of copper powder eletrolysis, and then the functional relationships between the process parameter and the average cell voltage, the current efficiency, and the power consumption were constructed through the Box–Behnken design (BBD) and response surface methodology under laboratory conditions. Furthermore, the influence of the process parameter on the power consumption was analyzed comprehensively and the optimal process parameter was determined. This provides a theoretical and experimental base for the energy saving of copper powder electrolysis process.

Experimental Details

Materials

The electrolyte used for copper powder electrolysis is composed of a $CuSO_4$–H_2SO_4 mixed solution system. The chemical reagents used to prepare the electrolyte include copper sulfate pentahydrate, concentrated sulfuric acid and deionized water. The chemical reagents used to check the concentrations of Cu^{2+} and H^+ in the electrolyte are $Na_2S_2O_3$ solution, NaOH solution, KI, etc. The antioxidant used to wash copper powders is benzotriazole (BTA) corrosion inhibitor.

The cathodes and anodes used for copper powder electrolysis are all made of pure copper plates. The effective area of the electrode is 100 mm $\times$ 80 mm, and the rest of the surface is covered with transparent tape. The anode and cathode plates are sanded to a smooth surface before electrolysis.

Apparatus and Experimental Procedure

Figure 1 shows a schematic diagram of the experimental apparatus for electrolytic preparation of copper powders. A 2 L Plexiglas electrolytic cell was connected to a 25 L plastic bucket by means of a plastic pipeline. The electrolyte was circulated via a pump between the cell and the bucket in the conventional manner (down-in and up-out). During the experiment, a heating stick was inserted in the bucket to keep a constant electrolyte temperature and an electro-rotator was used to maintain homogeneous composition and temperature of the electrolyte.

Three pairs of cathode and anode electrodes were set in the electrolytic cell to simulate the fluid flow in the actual electrolytic cell, but only the middle pair was connected to the power supply. No other additives were used in the electrolyte. The

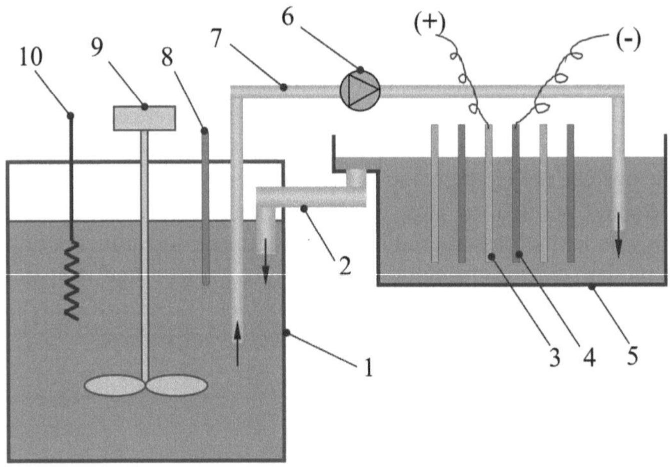

Fig. 1 Schematic diagram of apparatus for electrolytic preparation of copper powders: 1-plastic bucket, 2-electrolyte outlet, 3-anode, 4-cathode, 5-electrolytic cell, 6-electrolyte pump, 7-electrolyte inlet, 8-thermometer, 9-electro-rotator, 10-heating stick

current density was kept constant by using a DC power supply. The cell voltage was measured by a multimeter with recording software.

At the beginning of each experiment, the concentrations of cupric ion and sulfuric acid in the 25 L plastic bucket were measured. According to the difference between the measured and target values, measured quantities of concentrated sulfuric acid and distilled water were added, stirred, and the concentrations measured again to ensure that the cupric ion and sulfuric acid levels were constant.

At the end of each experiment, the electrolytic copper powders produced were washed with distilled water several times and subjected to an antioxidant treatment with 1% Benzotriazole ($C_6H_5N_3$) solution, dried at 80 °C in a vacuum, and weighed.

Experimental Design

There are many factors that affect the power consumption of copper powder electrolysis. These can be screened out through use of the PBD method [11]. In this study of the characteristics of the copper powder electrolysis process, as well as in previous studies [2–10], seven factors were investigated: electrolyte temperature (A), Cu^{2+} concentration (B), H_2SO_4 concentration (C), inter-electrode spacing (D), current density (E), electrolysis time (F), and circulation rate (G). In order to screen out the significant factors affecting the power consumption of copper powder eletrolysis, the PBD with seven factors and two levels was used and the experimental design and statistical analysis of the results were performed by using Design-Expert 8.0.6 software. The number of experiments was twelve and the measured responses were

Table 1 Coded and actual levels of experimental factors used in PBD experiments

Factor	Symbol	Coded factor level	
		Lower level/−1	Higher level/+1
Electrolyte temperature/°C	*A*	40	60
Cu^{2+} concentration/(g L^{-1})	*B*	8	12
H_2SO_4 concentration/(g L^{-1})	*C*	120	160
Inter-electrode spacing/mm	*D*	30	40
Current density/(A m^{-2})	*E*	1500	2000
Electrolysis time/min	*F*	45	60
Circulation rate/(L min^{-1})	*G*	0.3	0.4

average cell voltage (Y_{ACV}), current efficiency (Y_{CE}) and power consumption (Y_{PC}). The selected experimental factors and their lower and higher levels used in PBD experiments are given in Table 1.

Based on the experimental results of the PBD, 3–5 significant factors affecting the power consumption of copper powder electrolysis could be obtained. Furthermore, in accordance with the BBD method [11–14, 18], a multifactor, three-level response surface analysis experiment was designed by using Design-Expert 8.0.6 software. A multiple regression analysis was carried out on the experimental data and a quantitative relationship between the significant factors and the measured responses to them was established. The most energy-saving process conditions for copper powder electrolysis could be obtained.

SEM Analysis

After electrolytic copper powders were dried, they were dispersed so that they could be examined by using a scanning electron microscopy (SEM). A JSM–7800F Field Emission Scanning Electron Microscope (FESEM) (Japan) was employed for analyzing the morphology and shape of electrolytic copper powder particles.

Results and Discussion

Determining Significant Factors by Plackett–Burman Design

Based on the experimental factors and levels given in Table 1, the Plackett–Burman design (PBD) was carried out by using Design-Expert 8.0.6 software and twelve experiments were designed with it. Each experiment was performed three times and the results were averaged. The PBD experimental design matrix and results are given in Table 2.

Table 2 Experimental design matrix and results for copper powder electrolysis with Plackett–Burman design (PBD)

Run	Coded level of factor							Response		
	A	*B*	*C*	*D*	*E*	*F*	*G*	Y_{ACV}/V	Y_{CE}/%	Y_{PC}/(kW h t^{-1})
1	40	8	120	40	1500	60	0.4	1.308	84.0	1313
2	40	8	160	30	2000	60	0.3	1.226	79.5	1300
3	60	12	120	30	1500	60	0.3	0.957	96.1	840
4	60	12	120	40	2000	60	0.3	1.317	95.5	1163
5	40	12	160	40	1500	45	0.3	1.209	84.9	1201
6	60	8	160	40	1500	60	0.4	1.010	90.4	943
7	40	12	160	30	2000	60	0.4	1.222	86.2	1196
8	60	8	120	30	2000	45	0.4	1.181	90.5	1102
9	60	8	160	40	2000	45	0.3	1.216	84.3	1217
10	40	12	120	40	2000	45	0.4	1.520	89.2	1439
11	40	8	120	30	1500	45	0.3	1.192	81.7	1231
12	60	12	160	30	1500	45	0.4	0.922	93.4	833

According to the experimental results in Table 2, the main effects of each factor were evaluated on the experimental data (Table 3), and the factor with P-value < 0.05 was selected as the main effect factor [16]. As can be seen from Table 3, the factor with a P-value less than 0.05 had a significant effect on the response value. The factor with a P-value greater than 0.10 had no significant effect on the response value, and the smaller the P-value, the more significant the effect of the factor on the response value. Therefore, the factors that had the greatest influence on the average cell voltage were, successively, $E > A > D > C$, the factors that influenced the current efficiency were, successively, $A > B > C$, and the factors that influenced the power consumption were, successively, $A > E > D > B$. In order to systematically investigate the effects of the main factors on average cell voltage, current efficiency and power consumption, five significant factors, namely, electrolyte temperature (A), Cu^{2+} concentration (B), H_2SO_4 concentration (C), inter-electrode spacing (D) and current density (E), were selected for further RSM experiments.

Developing Mathematical Models by Box–Behnken Design

Based on the results and analyses of the PBD experiments, the above five significant factors were selected as the investigation factors and Y_{ACV}, Y_{CE} and Y_{PC} were taken as the measured responses in RSM experiments. Each factor takes three levels and is coded with (−1, 0, 1). The Box–Behnken design was implemented by using Design-Expert 8.0.6 software. There were 46 groups of experiments. The experimental design and results are given in Tables 4 and 5, respectively.

Table 3 Analysis of the main effects for the Plackett–Burman design experiments

Source	Y_{ACV}		Y_{CE}		Y_{PC}	
	F-value	P-value	F-value	P-value	F-value	P-value
Model	121.6	0.0002	42.3	0.0014	83.2	0.0004
A	272.8	<0.0001**	157.0	0.0002**	313.6	<0.0001**
B	0.047	0.838	95.7	0.0006**	23.6	0.008**
C	106.2	0.0005**	26.0	0.007**	19.8	0.011*
D	183.7	0.0002**	0.082	0.789	75.0	0.001**
E	278.2	<0.0001**	2.21	0.211	139.3	0.0003**
F	9.79	0.035*	4.77	0.094	9.0	0.040*
G	0.51	0.513	10.6	0.031*	2.05	0.226

Note Adj R^2 = 0.987 for average cell voltage, 0.963 for current efficiency, and 0.981 for power consumption

*Significant at the 5% level; **very significant at the 1% level

Table 4 Coded and actual levels of experimental factors used in RSM experiments (Box–Behnken Design)

Factor	Symbol	Coded factor level		
		−1	0	1
Electrolyte temperature/°C	A	35	50	65
Cu^{2+} concentration/(g L^{-1})	B	5	10	15
H_2SO_4 concentration/(g L^{-1})	C	100	150	200
Inter-electrode spacing/mm	D	20	40	60
Current density/(A m^{-2})	E	700	1600	2500

Based on the experimental results in Table 5, analysis of variance (ANOVA) and quadratic polynomial regression fitting were performed by using Design-Expert 8.0.6 software. The test for significance of polynomial functions is given in Table 6, and it was found that the P-values of some of the quadratic terms were much larger than 0.1, which were insignificant factors. To express the regression equation more concisely, these quadratic terms with P-values greater than 0.1 were ignored and the regression equation fitting was carried out once again. The final equations are shown as the following.

For average cell voltage,

$$\begin{aligned} Y_{ACV} = (&8930.3 + 24.983A + 78.229B - 72.841C + 202.98D + 3.2199E \\ &- 7.2967AB - 1.5463AD + 0.11542BE - 0.7343CD - 0.015289CE \\ &+ 0.068DE + 0.3255C^2 - 0.0003127E^2) \times 10^{-4} \end{aligned} \quad (2)$$

For current efficiency,

Table 5 Experimental design matrix and results for copper powder electrolysis with Box–Behnken design (BBD)

Run	Coded factor level					Response		
	A	*B*	*C*	*D*	*E*	Y_{ACV}/V	Y_{CE}/%	Y_{PC}/(kW h t^{-1})
1	0	0	0	0	0	1.135	89.9	1065
2	0	0	−1	0	−1	0.891	96.7	777
3	−1	0	0	1	0	1.585	86.1	1554
4	0	1	−1	0	0	1.351	95.6	1193
5	−1	1	0	0	0	1.262	92.1	1156
6	0	1	0	1	0	1.339	99.6	1135
7	−1	0	1	0	0	1.190	79.3	1266
8	0	−1	0	0	−1	0.864	78.8	925
9	0	0	0	0	0	1.119	91.0	1038
10	1	0	0	0	1	1.241	93.3	1122
11	0	0	0	0	0	1.090	92.1	999
12	0	−1	1	0	0	1.116	71.6	1315
13	−1	0	0	0	1	1.640	81.4	1700
14	0	0	0	−1	1	1.074	85.5	1060
15	0	0	−1	1	0	1.674	92.0	1536
16	1	0	0	0	−1	0.559	97.6	483
17	0	0	0	0	0	1.088	93.0	988
18	1	0	0	1	0	1.191	95.9	1048
19	1	0	1	0	0	0.898	96.1	788
20	0	0	0	0	0	1.138	93.7	1025
21	0	0	1	1	0	1.253	95.7	1104
22	0	0	0	0	0	1.146	92.2	1049
23	0	1	1	0	0	0.985	94.4	880
24	1	0	−1	0	0	1.162	98.7	993
25	1	0	0	−1	0	0.769	96.3	675
26	−1	−1	0	0	0	1.286	69.1	1570
27	0	−1	−1	0	0	1.399	76.1	1551
28	0	0	1	0	1	1.330	82.2	1365
29	0	1	0	0	−1	0.641	97.1	557
30	0	0	1	0	−1	0.764	94.4	683
31	0	1	0	−1	0	0.843	98.1	726
32	0	0	1	−1	0	0.876	84.4	876
33	1	−1	0	0	0	1.075	83.4	1087
34	1	1	0	0	0	0.832	98.6	712
35	0	−1	0	−1	0	0.921	75.3	1031

(continued)

Table 5 (continued)

Run	Coded factor level					Response		
	A	*B*	*C*	*D*	*E*	Y_{ACV}/V	Y_{CE}/%	Y_{PC}/(kW h t^{-1})
36	0	0	−1	0	1	1.732	88.6	1650
37	0	0	0	−1	−1	0.675	96.4	591
38	0	−1	0	0	1	1.440	65.9	1845
39	0	1	0	0	1	1.425	93.4	1287
40	−1	0	−1	0	0	1.519	87.4	1466
41	−1	0	0	−1	0	0.978	84.4	977
42	0	0	0	1	−1	0.897	96.1	787
43	−1	0	0	0	−1	0.904	90.2	845
44	0	0	−1	−1	0	1.004	93.3	908
45	0	−1	0	1	0	1.394	72.3	1627
46	0	0	0	1	1	1.786	88.0	1713

Table 6 Test for significance of polynomial functions based on Box–Behnken design in RSM experiments

Term	Y_{ACV}		Y_{CE}		Y_{PC}	
	F-value	*P*-value	*F*-value	*P*-value	*F*-value	*P*-value
A	650.5	<0.0001	154.0	<0.0001	619.7	<0.0001
B	62.3	<0.0001	592.6	<0.0001	514.9	<0.0001
C	504.3	<0.0001	17.4	0.0003	152.3	<0.0001
D	1482	<0.0001	2.73	0.1111	632.0	<0.0001
E	2804	<0.0001	91.1	<0.0001	1750	<0.0001
AB	17.9	0.0003	4.61	0.042	0.27	0.605*
AC	1.58	0.220*	2.30	0.142*	0.004	0.950*
AD	12.9	0.001	0.32	0.578*	7.74	0.010
AE	1.13	0.298*	1.49	0.233*	8.76	0.007
BC	2.61	0.119*	0.87	0.340*	1.13	0.299*
BD	0.18	0.673*	1.57	0.221*	6.58	0.017
BE	16.2	0.0005	6.57	0.017	6.81	0.015
CD	32.3	<0.0001	12.2	0.002	30.2	<0.0001
CE	28.4	<0.0001	1.26	0.273*	6.84	0.015
DE	89.8	<0.0001	0.54	0.470*	39.4	<0.0001
A^2	0.38	0.542*	0.93	0.344*	0.89	0.355*
B^2	0.02	0.878*	106.0	<0.0001	76.4	<0.0001
C^2	88.3	<0.0001	2.07	0.163*	50.1	<0.0001
D^2	0.98	0.331*	0.08	0.776*	0.09	0.764*
E^2	7.86	0.010	2.45	0.130*	0.34	0.564*

Note *not significant, *P*-value > 0.1

Table 7 ANOVA of the three developed models for copper powder electrolysis

Term	Y_{ACV}		Y_{CE}		Y_{PC}	
	F-value	P-value	F-value	P-value	F-value	P-value
Model	443.1	<0.0001	104.3	<0.0001	314.6	<0.0001
Lack of fit	1.08	0.520	2.04	0.219	1.39	0.385
Adj-R^2	0.992		0.954		0.990	

$$Y_{CE} = (5255.1 + 63.34A + 744.9B - 16.43C - 43.74D - 0.9955E - 2.59AB + 0.05155BE + 0.3165CD - 23.88B^2) \times 10^{-2} \quad (3)$$

For power consumption,

$$Y_{PC} = 1112.67 - 1.9687A - 88.709B - 6.5444C + 29.391D + 0.6328E - 0.1688AD - 0.00399AE - 0.46701BD - 0.01055BE - 0.10CD - 0.001058CE + 0.00634937DE + 4.14837B^2 + 0.033301C^2 \quad (4)$$

The results of ANOVA consisting of P-value and F-value for the above three models are given in Table 7. It can be found that the adjusted coefficients of determination of the three models are all greater than 0.95, the F-values of the three models are 443.1, 104.3 and 314.6, respectively, and the P-values are all less than 0.0001. This shows that the models fit well. In addition, the value of the model lack of fit is not significant (P-value > 0.1), which also shows that the models are highly accurate and can predict the response values.

Response Surface Analysis

In actual production, current density has a direct impact on the yield, while cupric ion concentration affects the particle size of electrolytic copper powders [6–8]. These two parameters are usually not adjusted to save energy. In this study, a current density of 1600 A m^{-2} and a Cu^{2+} concentration of 5 g L^{-1} were chosen to correspond to factory conditions. Therefore, the three factors of electrolyte temperature (A), H_2SO_4 concentration (C), and inter-electrode spacing (D) were the three that were adjusted.

Figure 2 is a response surface contour plots according to the power consumption regression model [Eq. (4)] of copper powder electrolysis, which can be used to evaluate the interaction of the three significant factors on power consumption and determine the optimal level of each factor. Figure 2a shows that the effect of the interaction of electrolyte temperature and sulfuric acid concentration on power consumption of copper powder electrolysis at an inter-electrode spacing of 20 mm. It can be clearly seen that with increasing electrolyte temperature, the power consumption is reduced. It is because an increase in electrolyte temperature leads to a decrease

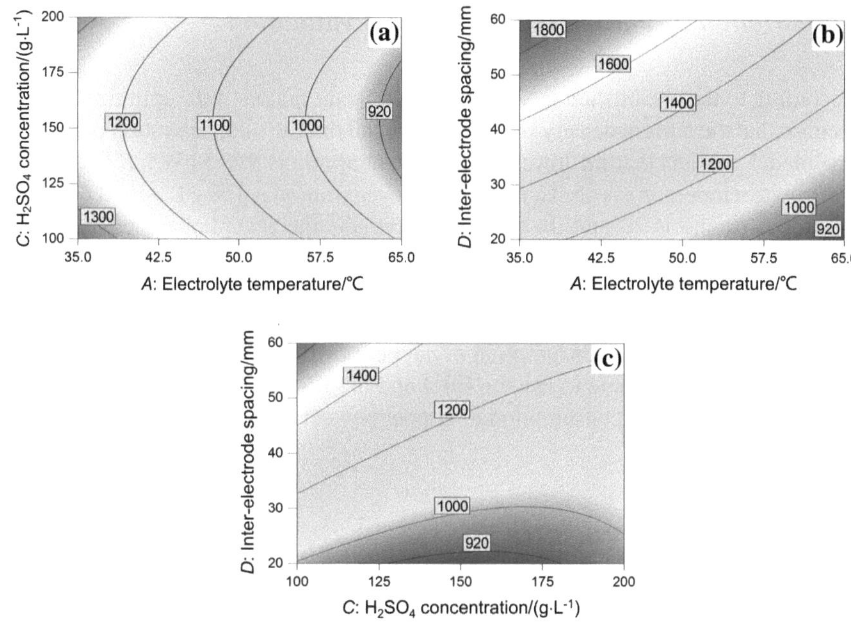

Fig. 2 Response surface contour plots showing the effect of two variables on power consumption of copper powder electrolysis: **a** electrolyte temperature and H_2SO_4 concentration, **b** electrolyte temperature and inter-electrode spacing, **c** H_2SO_4 concentration and inter-electrode spacing

in cell voltage [Eq. (2)] and an increase in current efficiency [Eq. (3)]. In addition, the interaction between electrolyte temperature and sulfuric acid concentration was not obvious, but the effect of electrolyte temperature was more significant than that of sulfuric acid concentration. Figure 2b shows that the effect of the interaction of electrolyte temperature and inter-electrode spacing on power consumption of copper powder electrolysis at an H_2SO_4 concentration of 150 g L^{-1}. It can be clearly seen that as electrolyte temperature increases and inter-electrode spacing decreases, power consumption decreases; otherwise, it increases. There is a certain interaction between electrolyte temperature and inter-electrode spacing. Figure 2c shows that the effect of the interaction of sulfuric acid concentration and inter-electrode spacing on power consumption of copper powder electrolysis at an electrolyte temperature of 65 °C. It can be clearly seen that when the inter-electrode spacing is at a range of 20–30 mm, power consumption decreases first and then increases with increasing sulfuric acid concentration. When the inter-electrode spacing is at a range of 30–60 mm, power consumption decreases with increasing sulfuric acid concentration. There is a certain interaction between the sulfuric acid concentration and the inter-electrode spacing, while the inter-electrode spacing has a greater effect than the sulfuric acid concentration.

Optimization of Parameters and Model Validation

According to the established model, three significant factors were optimized on the premise that the current density is 1600 A m^{-2} and the Cu^{2+} concentration is 5 g L^{-1}. The model predicts that the lowest power consumption is 896.5 kW h t^{-1} when the electrolyte temperature is 65 °C, the H_2SO_4 concentration is 150 g L^{-1}, and the inter-electrode spacing is 20 mm. In order to investigate the accuracy and effectiveness of the prediction model, a validated experiment was performed under the above-optimized conditions. The average power consumption of three parallel experiments was 905.7 kW h t^{-1} and the prediction accuracy was 98.98%. The model is able to better predict the power consumption of copper powder electrolysis. It can be seen therefore that it is feasible to use the PBD and the RSM-BBD method to study the reduction in the power consumption of copper powder electrolysis.

Analysis of Microstructures

The morphologies of electrolytic copper powders elctrodeposited under the above-optimized conditions are shown in Fig. 3. The copper powders generally have dendritic structure (Fig. 3a) and the agglomerated copper particles are almost globular (Fig. 3b). These results are in agreement with those found in literature [2, 22]. This shows that electrolytic copper powders obtained under the optimized conditions have good performance.

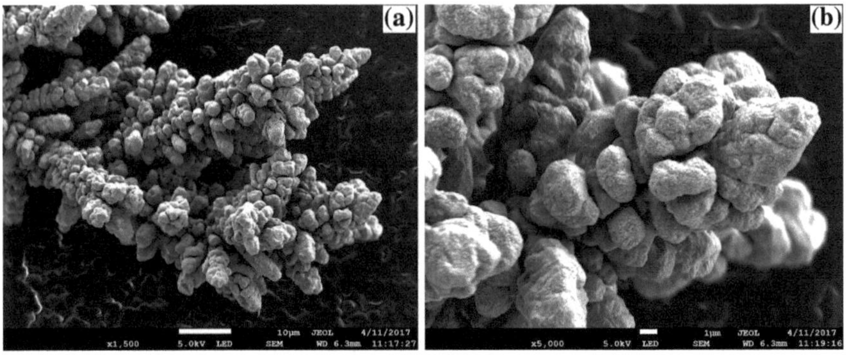

Fig. 3 SEM Photomicrographs of electrolytic copper powders obtained under the optimized conditions: **a** 1500X, **b** magnified part in (**a**), 5000X

Conclusions

(1) Five significant factors including electrolyte temperature, Cu^{2+} concentration, H_2SO_4 concentration, inter-electrode spacing and current density are screened out from the many process parameters that affect the power consumption of copper powder electrolysis by using Plackett–Burman design (PBD).
(2) The prediction models for the average cell voltage, the current efficiency and the power consumption of copper powder electrolysis [Eqs. (2)–(4)] are constructed by using response surface methodology and Box–Behnken design (RSM-BBD).
(3) Under the premise that a current density of 1600 A m^{-2} and a Cu^{2+} concentration of 5 g L^{-1} are applied, the optimized conditions are electrolyte temperature 65 °C, H_2SO_4 concentration 150 g L^{-1}, and inter-electrode spacing 20 mm. The lowest power consumption of copper powder electrolysis is 896.5 kW h t^{-1}, which is basically consistent with the results of the verification experiment under the optimized conditions. This indicates that it is feasible to use the PBD and the RSM-BBD method to study the reduction in the power consumption of copper powder electrolysis.
(4) The electrolytic copper powders obtained under the optimized conditions generally have dendritic structure and the agglomerated copper particles are almost globular.

Acknowledgements Support from the National Natural Science Foundation of China [grant number 51674057], the Chongqing Research Program of Basic Research and Frontier Technology [grant numbers cstc2016jcyjA0142, cstc2017jcyjAX0236], the Scientific and Technological Research Program of Chongqing Municipal Education Commission [grant numbers KJ1601326, KJ1713343], the Research Foundation of Chongqing University of Science & Technology [grant number CK2016B19] is gratefully acknowledged.

Disclosure Statement No potential conflict of interest was reported by the authors.

References

1. Jin S, Yang B (2004) Research on influence factors of preparing copper powders by electrolysis. China Powder Sci Technol 11(3):21–23
2. Orhan G, Gezgin GG (2012) Effect of electrolysis parameters on the morphologies of copper powder obtained at high current densities. J Serb Chem Soc 77(5):651–665
3. Owais A (2009) Effect of electrolyte characteristics on electrowinning of copper powder. J Appl Electrochem 39(9):1587–1595
4. Maksimović VM, LjJ Pavlović, Pavlović MG, Tomić MV (2009) Characterization of copper powder particles obtained by electrodeposition as function of different current densities. J Appl Electrochem 39(12):2545–2552
5. Nikolić ND, LjJ Pavlović, Pavlović MG, Popov KI (2008) Morphologies of electrochemically formed copper powder particles and their dependence on the quantity of evolved hydrogen. Powder Technol 185(3):195–201
6. He W, Duan X, Zhu L (2009) Characterization of ultrafine copper powder prepared by novel electrodeposition method. J Cent South Univ Technol 16:708–712

7. Xue J, Wu Q, Wang Z, Yi S (2006) Function of additives in electrolytic preparation of copper powder. Hydrometallurgy 82(3–4):154–156
8. Wang M, Wang Z, Guo Z (2010) Preparation of electrolytic copper powders with high current efficiency enhanced by super gravity field and its mechanism. Trans Nonferrous Met Soc China 20(6):1154–1160
9. Guo X, Chen S, Liang Y, Tian Q (2011) Effect of process conditions on the preparation of copper powders by electrodeposition. J Univ Sci Technol 33(2):182–187
10. Akbarzadeh E, Shakib SE (2011) Comparison of effective parameters for copper powder production via electrorefining and electrowinning cells and improvement using DOE methods. Int J Miner Metall Mater 18(6):731–740
11. Hao X, Yu X, Liu Z, Wang B (2006) The application of response surface methodology in optimization of microbial media. Food Res Dev 27(1):38–41
12. Myers RH, Montgomery DC, Anderson-Cook CM (2016) Response surface methodology. Wiley, New York
13. Montgomery DC (2012) Design and analysis of experiments. Wiley, New York
14. Shanavas S, Dhas JER (2017) Parametric optimization of friction stir welding parameters of marine grade aluminium alloy using response surface methodology. Trans Nonferrous Met Soc China 27(11):2334–2344
15. Singh J, Chauhan A (2017) Influence of reinforcement parameters and ageing time on mechanical behavior of novel Al2024/SiC/Red mud composites using response surface methodology. JOM 69(12):2471–2479
16. Xu T, Zhao L, Li S (2017) Gold self-leaching optimization of neutral roasted products of gold concentrate by response surface methodology. China J Nonferrous Met 27(3):629–636
17. Moghaddam MY, Shafaei SZ, Noaparast M, Ardejani FD, Abdollahi H, Ranjbar M, Schaffie M, Manafi Z (2015) Empirical model for bio-extraction of copper from low grade ore using response surface methodology. Trans Nonferrous Met Soc China 25(12):4126–4143
18. Kul M, Oskay KO, Şimşir M, Sübütay H, Kirgezen H (2015) Optimization of selective leaching of Zn from electric arc furnace steelmaking dust using response surface methodology. Trans Nonferrous Met Soc China 25(8):2753–2762
19. Yang K, Ye X, Su J, Su H, Long Y, Lü X, Wen Y (2013) Response surface optimization of process parameters for reduction roasting of low-grade pyrolusite by bagasse. Trans Nonferrous Met Soc China 23(2):548–555
20. Azizi D, Shafaei SZ, Noaparast M, Abdollahi H (2012) Modeling and optimization of low-grade Mn bearing ore leaching using response surface methodology and central composite rotatable design. Trans Nonferrous Met Soc China 22(9):2295–2305
21. Awe SA, Khoshkhoo M, Kruger P, Sandström Å (2012) Modelling and process optimisation of antimony removal from a complex copper concentrate. Trans Nonferrous Met Soc China 22(3):675–685
22. Boz M, Hasheminiasari M (2013) The effect of process parameters on copper powder particle size and shape produced by electrolysis method. Steel Compos Struct 15(2):151–162

Tensile Properties and Microstructure of Squeeze Cast Magnesium Matrix Composite Reinforced with 35 Vol. % of AL_2O_3 Fibers

Luyang Ren, Xuezhi Zhang and Henry Hu

Abstract Magnesium alloy AM60 matrix-based composite reinforced with 35 vol. % of Al_2O_3 fibers were fabricated by preform infiltration and squeeze casting technique under an applied pressure of 90 MPa. For the purpose of comparison, the matrix alloy without reinforcement was also squeeze cast under the same process conditions. Examination of microstructures by both optical microscopy (OM) and scanning electron microscopy (SEM) revealed that alumina fibers were distributed uniformly in the preform. The mechanical properties of the composite were evaluated in comparison with those of the matrix alloy AM60. The Rockwell hardness (HRB) hardness increased from 5.12 to 84.94 as the fiber volume fraction rose from 0 to 35%. The results of tensile testing indicated that the addition of Al_2O_3 fibres to magnesium alloy AM60 led to a significant improvement in mechanical properties. In particular, the modulus of the composite increased to116 GPA by almost three times over that (40 GPa) of the alloy, while the ultimate tensile strength (UTS) rose from 171.36 to 202.56 MPa by 18%. However, the notable increase in both the modulus and strength was at sacrifice in elongation.

Keywords Magnesium · Composite · Alumina fiber
Microstructure mechanical properties · Squeeze casting

Introduction

Magnesium is one of the lightest engineering materials with a density of 1.74 g/cm^3. Magnesium is usually used with aluminum as an alloying element to improve the machinability and the corrosion resistance. Recently, the need of lightweight materials for fuel saving in automotive industry has led to extensive research in the development of magnesium alloy [1]. To further enhance the mechanical properties of

L. Ren · X. Zhang · H. Hu (✉)
Department of Mechanical, Automotive and Materials Engineering, University of Windsor, Windsor, ON N9B 3P4, Canada
e-mail: huh@uwindsor.ca

G. Lambotte et al. (eds.), *Materials Processing Fundamentals 2019*, The Minerals, Metals & Materials Series, https://doi.org/10.1007/978-3-030-05728-2_25

magnesium alloys, metal matrix composites are introduced with the improvements in hardness, strength, toughness and wear resistance. In composites, magnesium alloy holds the reinforcement in position as a structural material to transfer of load to reinforcement. On the other hand, the reinforcement provides strengths to the matrix. The interface between the matrix and the reinforcement has significant influence on the final properties of the composites [2–4].

Preform squeeze casting process is considered as one of the cost-effective composite fabrication processes. Two steps are involved to fabricate magnesium-based composite by preform squeeze casting technique. First, preform is made and pre-treated prior to the infiltration of magnesium melt. The purpose and advantage of the preform are to uniformly and randomly distribute the reinforcement in order to achieve the desired mechanical properties. Second, pressure is applied to infiltrate the melt into the preform and the solidification process is under pressurized condition [5–7]. Recent studies have demonstrated that the preform and squeeze casting process was capable of producing porosity-free magnesium-based composites, which were reinforced with alumina fibers. However, in the open literature, there are limited reports on the effect of high volume fractions of alumina fibers as reinforcement on microstructure development and mechanical properties of magnesium matrix composites.

In this article, the progress of an ongoing research work on magnesium alloy AM60 matrix-based composite reinforced with 35 vol. % of Al_2O_3 fibers fabricated by preform infiltration and squeeze casting technique is presented. The microstructure, tensile behavior and fracture behavior of the composite are studied. The informative results are compared with those for the matrix.

Experimental Procedures

Material Preparation and Fabrication of Composite

Mg alloy AM60 was used as a matrix metal, of which chemical composition is given in Table 1. Alumina (Al_2O_3) short fibers were used as raw material for the preparation of reinforcement. Its chemical composition and properties are listed in Table 2.

The fabrication of preform involved mixing ceramic short fibers, introducing binding compounds, forming the shape of preforms under pressure, drying and sintering. The volume percentage of fibers in a preform was pre-determined based on desired

Table 1 Chemical composition of AM60 alloy (wt%)

Al	Zn	Mn	Si	Cu	Fe	Ni	Mg
6.0	0.22	0.4	0.1	0.01	0.004	0.002	Bal.

Table 2 Chemical composition and properties of alumina short fiber [5]

Material	Elasticity modulus (GPa)	Density (g/cm^3)	Diameter (μm)
Al_2O_3 fiber	200	2.2 ~ 3.5	4.0

porosity levels of cellular solids. The detailed procedure for preform preparation can be found in Ref. [5].

A 75-ton, vertical hydraulic press was used for direct squeeze casting of hybrid composite. Matrix alloy AM60 was melted in an electrical resistance furnace using a mild steel crucible. Melting magnesium alloy requires the use of protective gas to avoid oxidation and burning. The gas mixture used in this study was sulfur hexafluoride (SF_6) and carbon dioxide (CO_2) mixture.

During composite fabrication, preforms were first preheated to 300 °C. The molten matrix alloy AM60 at 720 °C was infiltrated into the preheated preform under an applied pressure of 90 MPa. Composite 35 vol. % of Al_2O_3 fibers (F/AM60) were cast in a cylindrical disk with a diameter of 0.10 m and a thickness of 0.02 m.

Microstructure Analysis

Specimens were sectioned, mounted, and polished from the center of the squeeze disk and prepared following the standard metallographic procedures. A Buehler optical image analyzer system was used to determine the primary characteristics of the specimens. The detailed features of the microstructure were characterized at high magnification using a JSM-5800LV scanning electron microscope (SEM) with a maximum resolution of 100 nm in a backscattered mode. Fractured surfaces of tensile specimens were analyzed with the SEM to ascertain the nature of fracture mechanisms.

Mechanical Testing

Macrohardness measurements were conducted on the polished AM60 matrix alloy and composites. The Rockwell F scale was used for macrohardness measurement in accordance with ASTME18-94 standard for both the matrix alloy and composites. Wilson M1CI mode of Rockwell hardness tester was employed. The other mechanical properties of the matrix alloy and composites were evaluated by tensile testing, which was performed at ambient temperature on an INSTRON (Grove City, PA) machine equipped with a computer data acquisition system. Following ASTM B557, subsize flat tensile specimens (25 mm in gage length, 6 mm in width, and 10 mm in thickness) were machined from the squeeze cast disks. The tensile properties, including elastic modulus (E), 0.2% yield strength (YS), ultimate tensile strength (UTS), and elongation to failure (e_f), were obtained based on the average of three tests.

Results and Discussion

Characterization of Magnesium Matrix Composites

The preform fabrication procedure as well as the distribution and orientation of the fibers are critical in determining the eventual properties of the magnesium composites. To ensure that the fibers were uniformly distributed in the composite, the composites samples were etched at the different time periods to allow the fibers to reveal their distribution at different depths of the composites, as shown in Fig. 1.

As shown in Fig. 2, the fibers were distributed in a random and isotropic orientation and no agglomeration observed. This uniformity of the fiber distribution provided great contribution to the mechanical properties of the magnesium matrix composite. Figure 2 shows the orientation of the fiber before and after the infiltration. As illustrated, the orientations of the fibers were unchanged and fibers were not deformed even after the application of the high pressure.

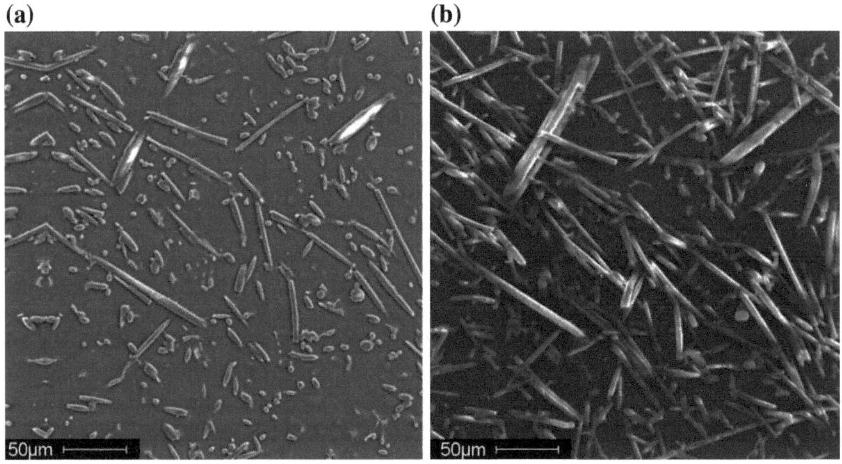

Fig. 1 Magnesium composite sample etched by **a** 10 s and **b** 110 s

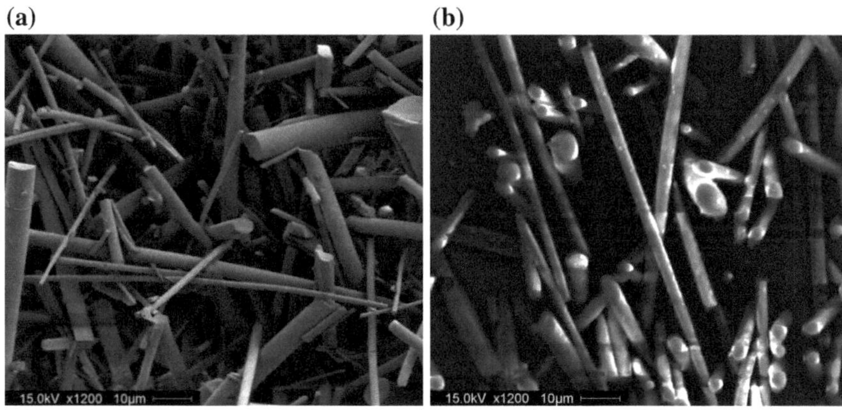

Fig. 2 SEM showing the orientation of the fiber (**a**) before and (**b**) after the infiltration

Fig. 3 Hardness and elastic modulus variation as a function of fiber volume fraction for the composite

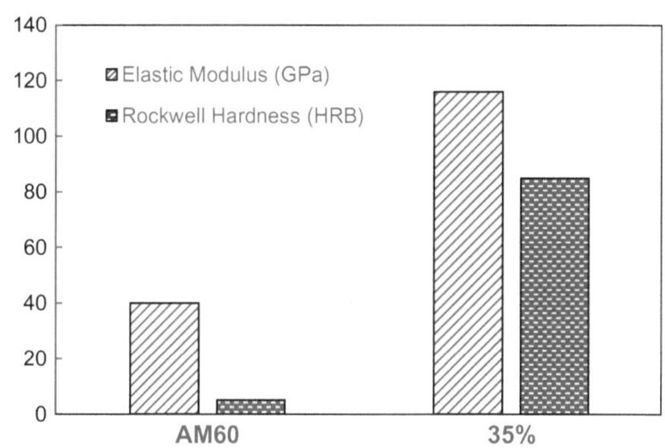

Hardness

Figure 3 compares the Rockwell hardness (HRB) as a function of fiber volume fraction for the unreinforced matrix alloy to that of the fiber reinforced composite. The preference of using Rockwell rather than Vickers hardness scale was due to the fact that the large indentation was capable of covering both the fiber and matrix alloy as one entire entity which ensures the consistency of the measurement. If the indent was too small, the areas were taken could be only either the reinforcements or the matrix alloy and consequently cause the large variation of the hardness values. From Fig. 3, it is noted that the fiber reinforcement significantly increased the hardness as expected. The HRB hardness increased from 5.12 to 84.94 as the fiber volume fraction rose from 0 to 35%. This observation should be attributed to the presence of the reinforcing fibers with their superior strength and stiffness.

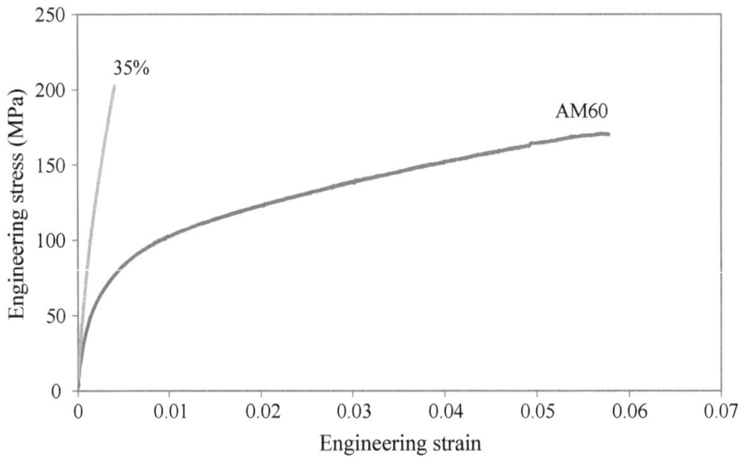

Fig. 4 Typical engineering stress versus strain curves for AM60 alloy and 35% fiber reinforced composite

Tensile Properties

Figure 4 shows the typical engineering stress and strain curves for AM60 and its composite reinforced with Al_2O_3 fiber of volume fraction of 35%. It is evidently illustrated that the addition of the alumina fiber led to an increase in the strengths and Elastic's modulus, as shown in Fig. 3, but there was a significant reduction in elongation. As the fiber volume fraction increased from 0 to 35%, the yield strength of the composite tended to disappear due to the depletion of plasticity in the material. The brittleness of the composite became evidently high with 35% of fiber reinforcement.

From Table 3, it can be seen that as the fiber volume fraction increased to 35%, there was a slight increment in UTS, only by 0.7%. Compared with the unreinforced AM60 alloy, the elongation dropped dramatically for the larger amount of fiber reinforced composite, i.e., 6.1% for the AM60 but 0.4% for 35% composite. It decreased to almost 93%. The elastic modulus was measured from the linear portion of the engineering stress versus strain curve. It is known that the higher the Elastic's modulus, the higher the stiffness of the material. The elastic modulus for the 35% composite was 116 GPa, which was 190% higher than that of the AM60 alloy. It appears that with the increase in fiber volume fraction, more loads were transferred to the reinforcement, which resulted in a higher tensile strength. The decreasing ductility of the composite could be attributed to the increased amount of fiber. The reasons might be that since the fiber was the final load barrier, the high stress concentration developed in cracked fibers prior to fracture embrittled the composite, and the probability of debonding between the reinforcement and matrix could be increased as the amount of fiber increased.

Table 3 Mechanical properties of AM60 and composite with 35% of Al_2O_3 fibers

Material	YS (MPa)	UTS (MPa)	e_f (%)	E (GPa)	Rockwell Hardness HRB
AM60	81.21	171.36	6.1	40	5.12
35%	–	202.56	0.4	116	84.94

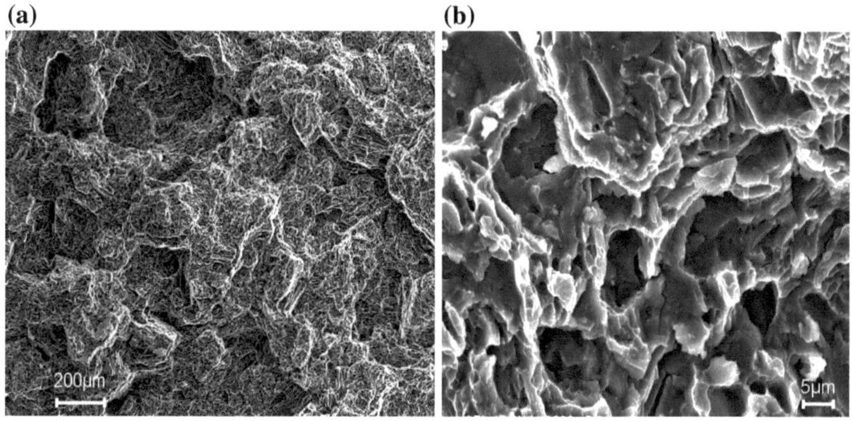

Fig. 5 SEM fractographs of the unreinforced magnesium alloy AM60, (**a**) low and (**b**) high magnification

Fracture Behavior

The fracture behavior of the unreinforced magnesium alloy AM60 and it composite with fiber volume fraction of 35% were investigated using SEM. Figure 5 illustrates the typical fracture surface of the unreinforced alloy AM60 with low and high magnifications. There were shallow dimples on the surface, which generally displays ductile behavior. It was well documented [6–8] that the fracture of unreinforced alloys was associated with the microscopic void nucleation and growth. The voids nucleate at the locations with coarse intermetallic particles and other hard phases in the microstructure, which then grow and coalesce resulting in final fracture. For composites with higher fiber volume fractions, such as 35%, it was found that the fracture was mainly caused by the debonding between the fiber and the matrix alloy. High volume of fibers led to relatively poor infiltration of the molten metal into the close-packed fiber network, while the fiber surfaces could not be fully covered by the matrix alloy. Figure 6 shows the fracture surface 35% fiber reinforced composite, and Arrow 1 in Fig. 6 indicated the locations of pullout fibers. These fibers pullout might be the main mechanism of the failure for the composites with higher fiber volume fractions.

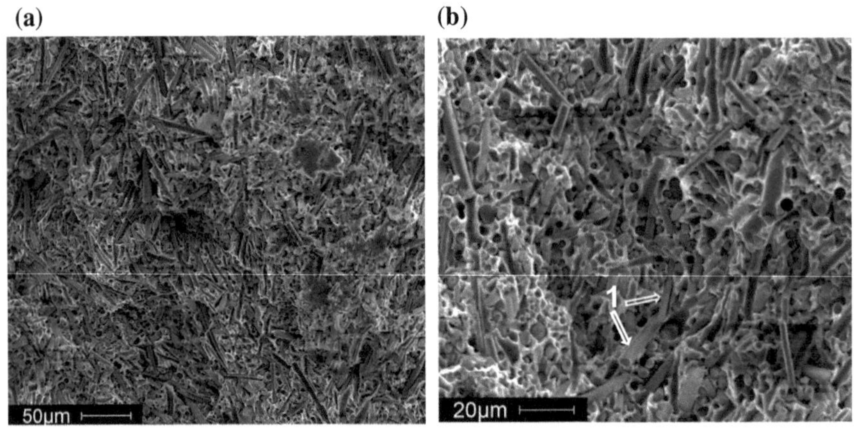

Fig. 6 Fractographs of 35% fiber reinforced composite, (**a**) low and (**b**) high magnification

Conclusions

Excellent strengths and modulus of the composites were achieved by adding 35% of the Al_2O_3 fiber into the matrix alloy AM60. These good results were accomplished by applying the combined preform and squeeze casting process, which was to infiltrate the liquid magnesium alloy AM60 into the preform under an applied pressure. The microstructure analysis revealed that the fibers were uniformly dispersed in the matrix alloy without any agglomeration. The property evaluation indicates that the fiber-reinforced composite exhibited a tensile strength of 202.56 MPa, an elastic modulus of 116 GPa and a HRB hardness of 84.94, which gave an improvement of 18, 190, and 1559% over those of the unreinforced alloy, respectively. For the composite with 35% volume fraction of fibers, there was a slight increment in the UTS, but there was no result of YS due to the high brittleness of the composite with high volume fraction of Al_2O_3 fiber. The elongation dropped dramatically for the fiber reinforced composite in comparison with the matrix alloy. Examination of the fracture surfaces of the composites and the matrix alloy via SEM showed that the composite fractured in brittle mode comparing with matrix alloy. The interface debonding and fiber cracking should be responsible for the final fracture of the composite.

Acknowledgements The authors would like to thank Natural Sciences and Engineering Research Council of Canada (NSERC) and University of Windsor for their financial support.

References

1. Hu H (1998) Squeeze casting of magnesium alloys and their composites. J Mater Sci 33:1579–1589
2. Banerji A, Hu H, Alpas AT (2013) Sliding wear mechanisms of magnesium composites AM60 reinforced with Al_2O_3 fibres under ultra-mild wear conditions. Wear 301:626–635
3. Ye HZ, Liu XY (2004) Review of recent studies in magnesium matrix composites. J Mater Sci 39:6153–6171
4. Jayalakshmi S, Satish VK, Seshan S (2002) Tensile behaviour of squeeze cast AM100 magnesium alloy and its Al_2O_3 fibre reinforced composites. Compos A: Appl Sci Manuf 33(8):1135–1140
5. Zhang X (2012) Tensile properties and microstructure of Mg (AM60)/Al_2O_3 metal matrix composite with varying volume fractions of reinforcement. M.A.Sc. thesis, University of Windsor
6. Yong MS, Clegg AJ (2004) Process optimization for a squeeze cast magnesium alloy. J Mater Process Technol 145(1):134–141
7. Zhang Q, Hu H (2009) Processing and characterization of Al-based hybrid composites. Paper presented at the 141st TMS annual meeting, San Francisco, 15–19 Feb 2009
8. Musson NJ, Yue TM (1991) The effect of matrix composition on the mechanical properties of squeeze cast aluminum alloy-Saffil metal matrix composites. Mater Sci Eng: A 135:237–242

Author Index

A
Abouelazayem, S., 49
Akinlabi, E. T., 143
Akinlabi, S. A., 143

B
Bentsman, Joseph, 23
Bracker, Gwendolyn, 171
Burggraf, Stefan, 171

C
Chen, Dengfu, 59
Chen, Fei, 131
Chen, Wei, 15
Chen, Yan Hui, 163
Chen, Zhelin, 23

D
Demchenko, Pavlo, 215
Ding, Hui Min, 163
Duan, Huamei, 59
Dugic, Izudin, 87
Dybalska, Agnieszka, 181

E
Eckert, S., 49
Eskin, Dmitry G., 181

F
Fang, Donggen, 205
Fatoba, O. S., 143

G
Glavinić, I., 49
Guo, Hanjie, 97
Guo, Shengmin, 241
Guo, Wei, 117, 131

H
He, Jilin, 193
Henderson, Deidre, 241
Herlach, Dieter, 171
Hlava, J., 49
Hu, Henry, 279
Hupa, Leena, 215, 227
Hyers, Robert, 171

J
Jayaganthan, R., 153
Jno-Baptiste, Jumel, 241
Joshi, Amit, 153
Jung, In-Ho, 227

L
Lee, Jonghyun, 171
Li, Chenyao, 193
Lindberg, Daniel, 215, 227
Li, Qin, 163
Li, Shaolong, 193
Li, Shaoying, 97
Liu, Shu, 163
Liu, Xuan, 205
Li, Xuepeng, 193

G. Lambotte et al. (eds.), *Materials Processing Fundamentals 2019*, The Minerals, Metals & Materials Series, https://doi.org/10.1007/978-3-030-05728-2

Li, Yanglong, 117, 131
Long, MuJun, 59
Luo, Ji Hui, 163

M
Magnusson, T., 3
Mao, Mingtao, 97
Matson, Douglas, 171
Moroz, Mykola, 215, 227

O
Olia, Hamed, 23
O'Malley, Ronald J., 73

P
Paek, Min-Kyu, 227
Patel, Jayesh B., 181
Petrus, Bryan, 23
Prokhorenko, Myroslava, 215

R
Raja, A., 153
Ratajczak, M., 49
Raviraj, 153
Reinartz, Marcus, 171
Rembold, Madeline, 23
Ren, Bingzhi, 265
Ren, Luyang, 279
Reshetnyak, Oleksandr, 215
Rettenmayr, Markus, 171

S
Saevarsdottir, G., 3
Scheller, Piotr Roman, 15
Schurmann, D., 49
Seetala, Naidu, 241
Shi, Xiao, 97
Shornikov, Sergey, 253
Shu, Yongchun, 193
Song, Jianxun, 193
Stefani, F., 49

T
Tangstad, M., 3
Tesfahunegn, Y. A., 3
Tesfaye, Fiseha, 215, 227
Thomas, Brian G., 23

W
Wang, Fengqin, 117, 131
Wang, Hongdan, 265
Wang, Hui, 117, 131
Wen, Hao, 241
Wen, Jie, 117, 131
Wen, Qiu Yue, 163
Wondrak, T., 49
Wu, Shixin, 59

X
Xiao, Xiao, 171
Xia, Wentang, 265
Xue, Jilai, 205
Xu, Pei, 59

Y
Yin, Yanbin, 107
Yogesha, K. K., 153
Yu, Hengsong, 59
Yu, Meng, 117, 131
Yu, Sheng, 59

Z
Zhang, Jiongming, 107
Zhang, Lifeng, 15, 37
Zhang, Pengju, 205
Zhang, Xubin, 15
Zhang, Xuezhi, 279
Zhou, Haichen, 37
Zhou, Qinghai, 107
Zhou, Tihe, 73
Zurob, Hatem S., 73

Subject Index

A
Adaptation model, 125
Alkali roasting, 205, 206
Alumina fiber, 284
Aluminum-Nickel, 172, 179
Annealing, 98, 99, 117, 121, 124, 127, 132, 138, 163–167, 217, 241–244, 246–249
Applications, 51, 55, 57, 74, 81, 84, 117, 128, 137, 144, 163, 193, 216, 228, 242, 282

B
Box–Behnken design, 265, 267, 270–273, 277

C
Carbide, 4, 7, 8, 97–102, 105
Carbide precipitation, 98
Casting simulation, 87, 92, 95
Characterization, 154, 206, 208, 212, 282
Composite, 98, 115, 116, 143, 144, 147, 149, 151, 279–286
Composition evolution, 110, 116
Computational model, 6
Contactless inductive flow tomography, 50
Containerless processing, 171
Continuous casting, 15, 23, 29, 35, 37, 38, 40, 44, 49–51, 54, 55, 59, 60, 164
Continuous unidirectional solidification, 163
Copper, 16, 17, 21, 74, 146, 154, 174, 265–270, 272, 274–277
Crankshaft steel, 107, 108, 110, 114–116
Cu–Ni–Sn alloy, 163
Current distribution, 4, 5
Current paths, 5, 13

D
Decomposition reaction, 231, 232
Defragmentation, 181, 183, 188–190
Dendrites, 146, 147, 151, 193, 200
Double cold reduction, 117, 128, 131, 132
Double cold reduction of tin plate, 131
Dynamic bulging model, 23–25, 27, 30, 33, 34

E
Electrolytic copper powders, 265, 266, 268, 269, 274, 276, 277
Electromagnetic brake, 49, 54, 55
Electromagnetic levitation, 171, 172, 174
Electromotive Force (EMF), 218
EMF method, 219, 233
Euler–Euler approach, 37

F
Fabrication of composite, 280
FePt nanoparticles, 241–245, 248, 249
Flatness control model, 123
Flow control, 49
Flow pattern, 16, 17, 37, 38, 41–45, 49–51, 54, 55, 176, 186, 187
  influence of casting speed, 41
  influence of gas volume fraction, 42
  influence of submergence depth, 44
Fluid and Turbulence model, 61
Fluid flow, 4, 15–17, 21, 54, 60, 61, 63, 66, 68, 171, 172, 174, 177–179, 182, 267
Fluid flow simulation, 62
Fluoride ions, 193, 196, 199, 200
Fractography, 153
Fracture behavior, 285

G. Lambotte et al. (eds.), *Materials Processing Fundamentals 2019*, The Minerals, Metals & Materials Series, https://doi.org/10.1007/978-3-030-05728-2

Fragmentation potency, 188
Friction model, 119, 134

G
Gibbs energy, 215, 223, 224, 233, 261
Grey cast iron, 87

H
Hardness, 90, 91, 95, 97–99, 143, 144, 146, 149, 151, 153, 155, 183, 279–281, 283, 285, 286
Heat Capacity (C_P), 230
Heat treatment, 98
High-cycle fatigue test, 155
High-shear, 181
High temperature, 6, 50, 73, 74, 79, 81, 82, 84, 85, 104, 147, 167, 176, 238, 241, 258
Hydraulic simulation, 69

I
Immersion depth of SEN, 59, 63–66, 68, 69
Inner diameter of SEN, 68
ISS-EML, 172, 180

K
Knudsen effusion mass spectrometry, 254, 262

L
L2 control model, 117, 118, 128, 132
Large inclusions, 107–109, 112, 114–116
Laser heating, 150, 242, 244–246
Laterite nickel ore, 205, 206
Leaching residue, 209
LiCl-KCl, 193, 195–197, 199, 200
Liquid metal, 49, 50, 52, 54, 56, 181–183, 189

M
Magnesium, 181, 205, 279–282, 285, 286
Magnesium matrix, 282
Magnetic storage media, 241, 249
Magnetization, 241, 243–248
Mathematical modeling, 6, 149
Mechanical testing, 281
Mesh generation, 8
Metal penetration, 87, 88, 90, 92–95
Micro-hardness property, 149
Microstructural characterization, 99
Microstructure, 74–77, 79, 82, 99, 143, 144, 146, 147, 149, 151, 153, 154, 163–166, 279–281, 285, 286
Microstructure analysis, 145, 276, 281
Microstructure mechanical properties, 143
Mixing uniformity, 183
Mold level fluctuation, 30, 32, 34
Multiphase flow, 37, 38

N
Nail board experiment, 50
Non-circular arc theory, 122, 131, 134, 137, 138
Numerical cases, 8
Numerical simulation, 37, 39, 60, 63–65, 68, 69
Numerical simulation verification, 39

P
Particle coarsening, 73, 74
Particle pinning, 79, 81, 84
Perovskite, 253, 255–262
Phase equilibria, 215–217, 219, 222, 224, 233
Physical model, 62
Plackett–Burman design, 269, 277
Potential phase diagrams, 237
Power consumption, 265–271, 274–277
Power distributions, 3–5, 8–13
Precipitation, 80, 98–100, 102–105, 110, 112, 146, 154, 209, 228
Pseudo-cavern, 181, 183, 186, 188

R
Reheating process, 78
Response surface methodology, 265–267, 277
Roasting, 206
Roll force, 134
Roll force model, 122, 132
Rolling, 73, 74, 77, 79, 81, 108, 109, 117–120, 122–128, 131–137, 139, 154
Rolling model, 109, 126, 128, 131, 132, 134
Rolling physics model, 119
Rotor-stator, 181, 183, 190

S
SEM analysis, 269
SEN parameters, 64, 67, 69
Shape, 7, 17, 20, 23, 25–27, 51, 52, 55, 116, 117, 123, 124, 126, 132, 137, 145, 157, 182, 186, 188, 190, 269, 280
Simulation, 8–11, 17, 18, 23, 30, 38–42, 52, 59, 62–64, 66, 67, 69, 83, 92–94, 131, 134, 138, 149, 188
Solidification, 15–17, 61, 73, 74, 77, 79–81, 85, 87, 88, 92, 94, 98, 101, 103, 104, 108, 110, 112, 115, 146, 147, 149, 151, 165, 166, 171, 172, 175, 179, 280
Solidification process, 80
Square wave voltammetry (SWV), 196, 197
Squeeze casting, 279–281, 286
Stator design, 181, 183, 185, 190

Steady-bulging model, 24
Steel–slag interface, 15–17, 19–21
Stopper rod flow model, 29, 31, 33
Submerged arc furnace, 3–5, 11
Sulfate, 227, 228, 236–238, 265, 267
Sulphur dioxide, 227, 232–234, 236, 237
Surface finish, 87

T
TBM cutter ring, 97
TEM analysis, 157
Temperature distribution, 24, 149, 150
Tempering, 97–101, 105
Tensile properties, 284
Ternary Fe–S–O system, 228
Thermal analysis method, 217
Thermochemical data, 229, 230, 236, 237, 255, 256, 259, 261
Thermochemical data of $FeSO_4$and $Fe_2(SO_4)_3$, 229
Thermodynamic data, 227
Thermodynamic functions, 222
Thermodynamic properties, 217, 222, 224, 227, 228, 231, 238, 253
Three-phase flow, 17, 21
Ti-6Al-4V alloy, 143, 145
Ti-Cu-Al coatings, 144

U
Ultrafine grains, 153, 154
Ultrahigh-speed continuous casting, 59, 60
Ultrasound, 50, 205–212
Unsteady bulging, 23–25, 27, 32–34
Unsteady state, 124

V
Vacuum annealing, 218, 243

W
Water leaching, 205–212
Wedge control model, 124

X
X-ray diffraction (XRD), 143, 145, 208, 210, 212, 217, 218, 224

Y
Yield stress model, 120

Z
Zeolite material, 207

MIX
Papier aus verantwortungsvollen Quellen
Paper from responsible sources
FSC® C105338

If you have any concerns about our products,
you can contact us on
ProductSafety@springernature.com

In case Publisher is established outside the EU,
the EU authorized representative is:
Springer Nature Customer Service Center GmbH
Europaplatz 3, 69115 Heidelberg, Germany

Printed by Libri Plureos GmbH
in Hamburg, Germany